The CRC Press
International Series on Computational Intelligence

Series Editor
L.C. Jain, Ph.D., M.E., B.E., (Hons), Fellow I.E. (Australia)

L.C. Jain, R.P. Johnson, Y. Takefuji, and L.A. Zadeh
Knowledge-Based Intelligent Techniques in Industry

L.C. Jain and C.W. de Silva
Intelligent Adaptive Control: Industrial Applications

L.C. Jain and N.M. Martin
Fusion of Neural Networks, Fuzzy Systems, and Genetic Algorithms: Industrial Applications

H.N. Teodorescu, A. Kandel, and L.C. Jain
Fuzzy and Neuro-Fuzzy Systems in Medicine

C.L. Karr and L.M. Freeman
Industrial Applications of Genetic Algorithms

L.C. Jain and Beatrice Lazzerini
Knowledge-Based Intelligent Techniques in Character Recognition

L.C. Jain and V. Vemuri
Industrial Applications of Neural Networks

H.N. Teodorescu, A. Kandel, and L.C. Jain
Soft Computing in Human-Related Sciences

B. Lazzerini, D. Dumitrescu, L.C. Jain, and A. Dumitrescu
Evolutionary Computing and Applications

B. Lazzerini, D. Dumitrescu, and L.C. Jain
Fuzzy Sets and Their Application to Clustering and Training

L.C. Jain, U. Halici, I. Hayashi, S.B. Lee, and S. Tsutsui
Intelligent Biometric Techniques in Fingerprint and Face Recognition: Practical Applications

Z. Chen
Computational Intelligence for Decision Support

L.C. Jain
Evolution of Engineering and Information System and Their Applications

H.N. Teodorescu and A. Kandel
Dynamic Fuzzy Systems and Chaos Applications

L. Medsker and L.C. Jain
Recurrent Neural Networks: Design and Applications

L.C. Jain and A.M. Fanelli
Recent Advances in Artifical Neural Networks: Design and Applications

M. Russo and L.C. Jain
Fuzzy Learning and Applications

J. Liu
Multiagent Robotic Systems

M. Kennedy, R. Rovatti, and G. Setti
Chatoic Electronics in Telecommunications

H.N. Teodorescu and L.C. Jain
Intelligent Systems and Techniques in Rehabilitation Engineering

I. Baturone, A. Barriga, C. Jimenez, D. Lopez, and S. Sanchez
Microelectronics Design of Fuzzy Logic-Based Systems

T. Nishida
Dynamic Knowledge Interaction

PREFACE

There is tremendous worldwide interest in the application of intelligent techniques to fingerprint and face recognition. This challenging field has many applications, limited only by the imagination of scientists and engineers. Some recognition systems can be used for identity verification for banking and security purposes.

These intelligent techniques involve expert systems, neural networks, fuzzy systems, and evolutionary computing paradigms. These techniques offer many advantages over conventional technology, such as abilities to learn and remember.

This book comprises thirteen chapters. Chapter 1 by Halici, Jain, and Erol introduces the reader to the field of fingerprint and face recognition. The literature on existing and possible applications are reviewed.

Chapter 2 by Roddy and Stosz explores methods of intelligent fingerprint processing for minutia and pore feature extraction and matching, as well as fingerprint image segmentation extraction and correlation and matching.

Chapter 3 by Drets and Liljenström offers different aspects of fingerprint sub-classification. A multi-resolution neural network approach to the singular point detection problem and a sub-classification procedure for loops are presented.

Chapter 4 by Hamamoto proposes a Gabor filter-based method for identifying fingerprints. The Gabor feature is obtained by convolving the filter with the image. The performance of the proposed method is demonstrated on the NIST database.

Chapter 5 by Maio and Maltoni presents a brief review of the principal minutiae extraction and filtering techniques. The authors propose two direct gray-scale methods.

Chapter 6 by Erol, Halici, and Ongun concentrates on feature selective filtering for ridge extraction. The basic idea is to obtain a fingerprint image in which the ridges and valleys are easily distinguishable. This chapter presents the enhancement algorithms for ridge or valley extraction and their experimental verification.

Chapter 7 by Howell presents the task of face recognition and draws up specific requirements to fulfill it. A literature review on face recognition is presented. This is followed with discussion on how task requirements affect the suitability of technique. A direct comparison of performance and generalization in several approaches using the same face database is presented.

Chapter 8 by Pandya and Szabo focuses on a neural network approach for face recognition. It presents an overview of different attempts to apply neural networks to a variety of face recognition tasks. The design of a neural network based transformation-invariant face recognition system is given as an example.

In Chapter 9, Howell discusses experimental work using a new variant of the RBF neural network, the 'Face Unit' network which learns to identify one particular individual only. This technique gives an alternative, parallel method of learning tasks which can then be used as additional evidence of identity.

Chapter 10 by Würtz presents several methods for the recognition of human faces. The coherence with the results of psycho-physical experiments on human face recognition and the applicability to general object recognition are discussed.

Chapter 11 by Wiskott, Fellous, Krüger, and Von der Malsburg presents a system for recognizing human faces from single images out of a large database containing one image per person. The face recognition is based on a new approach known as bunch-graph which is constructed from a small set of sample image graphs. A number of experiments are reported in this chapter.

Chapter 12 by King and Li presents a technique for synthesizing 2D gray-scale facial expressions using Radial Basis Function (RBF) neural networks. Three types of neural networks are constructed to synthesize facial expressions.

Chapter 13 by Onisawa and Kitazaki describes a model that recognizes emotions not only from facial expressions but also from human situations. The artificial neural network model and questionnaire data are used in this approach.

This book will be useful to researchers, scientists, practicing engineers and students who wish to develop successful fingerprint and face recognition systems.

We would like to express our sincere thanks to Berend Jan van der Zwaag and Ashlesha Jain, for their help in the preparation of the manuscript. We are grateful to the authors for their contributions, and thanks are due to Dawn Mesa, Helen Linna, Lourdes Franco and Suzanne Lassandro for their excellent editorial assistance.

L.C. Jain, Australia
U. Halici, Turkey
I. Hayashi, Japan
S.B. Lee, Korea
S. Tsutsui, Japan

CONTENTS

Chapter 1	*Introduction to Fingerprint Recognition* U. Halici, Turkey, L.C. Jain, Australia, and A. Erol, Turkey	1
Chapter 2	*Fingerprint Feature Processing Techniques and Poroscopy* A.R. Roddy and J.D. Stosz, U.S.A.	35
Chapter 3	*Fingerprint Sub-Classification: A Neural Network Approach* G.A. Drets, Uruguay, and H.G. Liljenström, Sweden	107
Chapter 4	*A Gabor Filter-Based Method for Fingerprint Identification* Y. Hamamoto, Japan	135
Chapter 5	*Minutiae Extraction and Filtering from Gray-Scale Images* D. Maio and D. Maltoni, Italy	153
Chapter 6	*Feature Selective Filtering for Ridge Extraction* A. Erol, U. Halici, and G. Ongun, Turkey	193
Chapter 7	*Introduction to Face Recognition* A.J. Howell, U.K.	217
Chapter 8	*Neural Networks for Face Recognition* A.S. Pandya and R.R. Szabo, U.S.A.	285

Chapter 9	*Face Unit Radial Basis Function Networks* A.J. Howell, U.K.	315
Chapter 10	*Face Recognition from Correspondence Maps* R.P. Würtz, Germany	335
Chapter 11	*Face Recognition by Elastic Bunch Graph Matching* L. Wiskott, Germany, J.-M. Fellous, U.S.A., N. Krüger and C. von der Malsburg, Germany	355
Chapter 12	*Facial Expression Synthesis Using Radial Basis Function Networks* I. King and X.Q. Li, Hong Kong	399
Chapter 13	*Recognition of Facial Expressions and Its Application to Human Computer Interaction* T. Onisawa and S. Kitazaki, Japan	425
Index		459

Chapter 1:

Introduction to Fingerprint Recognition

INTRODUCTION TO FINGERPRINT RECOGNITION

U. Halici
Computer Vision and Artificial Neural Networks Research Lab.
Dept. of Electrical and Electronics Eng.
Middle East Technical University, 06531, Ankara, Turkey
ugur-halici@metu.edu.tr

L.C. Jain
Knowledge-Based Intelligent Engineering Systems Centre
University of South Australia, Adelaide
Mawson Lakes S.A. 5095, Australia
Lakhmi.Jain@unisa.edu.au

A. Erol
Halici Software House, METU Technopark
Middle East Technical University, 06531, Ankara, Turkey
ali@halici.com.tr

One of the most interesting human abilities is the recognition of objects. Recognition is defined as a process involving perception and associating the resulting information with one or a combination of more than one of its memory contents. Visual perception means deriving information from a specific scene. From the psychological point of view, the actual perception process involves some information processing stages. The formation of the image on the retina is followed by the mental processing of the projected image. The actual process is not yet known, but several models exist.

Scientists commonly aim to design machines that emulate human abilities. However, current results are far from successful. Nevertheless, dividing human abilities to smaller tasks and implementing them reveals promising results.

Biometric systems have been an important area of research in recent years [10], [40]. There are two important utilizations of biometric systems: 1) Authentication or verification of a person's identity, i.e., a person proves that he is the person who he claims to be and 2) identification in which a person's identity is sought using the biometric sign available. Any physiological or behavioral characteristics can be used to make personal identification as long as it satisfies the following requirements [10], [32], [40].

1. Universality, which means that every person should have the characteristics;
2. Uniqueness, which indicates that no two persons should be the same in terms of the characteristics;
3. Permanence, which means that the characteristics should be invariant with time;
4. Collectability, which means that the characteristics can be measured quantitatively.

Examples of biometric signs are hand, fingerprint, iris [39], face, and speech [40]. In this book, we have included the most recent advances in fingerprint and face recognition. There are other systems used for identification purposes such as retinal image comparison, voice matching, and DNA matching, but these are currently not widely used.

1 The Use of Fingerprints in Personal Identification and Verification

Fingerprints, which have been used for about 100 years, are the oldest biometric signs of identity. Scientific studies on fingerprints were initiated in the late sixteenth century [35], but the foundations of modern fingerprint identification were established by the studies of Sir F. Galton [19] and E. Henry [25] at the end of nineteenth century. A fingerprint is formed of composite curve segments. The light areas of fingerprints are called *ridges* while the dark areas are called *valleys*. Galton's study introduced the *minutiae*, which are the local discontinuities in the ridge flow pattern, as discriminating features and showed the uniqueness and permanency of minutiae. Henry's study examined the global structure of fingerprints and established the famous "Henry System" of fingerprint classification which is an effective method of indexing fingerprints and is still in use in most identification systems. In the early twentieth century, fingerprints were formally accepted as valid signs of identity [35] by law-enforcement agencies. However manual fingerprint identification is tedious, time-consuming, and expensive as it needs to be performed by professional fingerprint experts. Therefore, in 1960, the FBI Home Office (UK) and the Paris Police Department initiated studies on automatic fingerprint identification systems [15].

The study by F. Galton [19] extensively examines the details that reside in fingerprints. Fingerprints were examined morphologically and experiments were carried out on different age groups within different races. Two fundamentally important conclusions were reached by Galton. The first was that a fingerprint of a person is permanent, i.e., it preserves its characteristics and shape from birth to death. The second result was that the fingerprints of individuals are unique. In the light of experimental evidence, it was proven that no two persons have the same fingerprints; even identical twins have different fingerprints despite signs of similarity. These fundamentally important results were the building blocks of research in this field over the last 90 years.

In his work, E.R. Henry examined the fingerprint's global structure and devised a classification method to partition the large fingerprint databases into five classes [25]. Unlike Galton, Henry did not extensively deal with the exact matching of fingerprints. However, his systematic way of partitioning fingerprint classes was so profound that it has traditionally been used by almost all of the government security forces and other users since then. The names given to these classes are Right Loop (R), Left Loop (L), Whorl (W), Arch (A), and Tented Arch (T) respectively. Samples taken from these classes can be seen in Figure 1.

By using the ideas presented above, fingerprints are partitioned by the *Henry Classification* and exact matching is carried out by comparing *Galton Features*.

(a) Right Loop

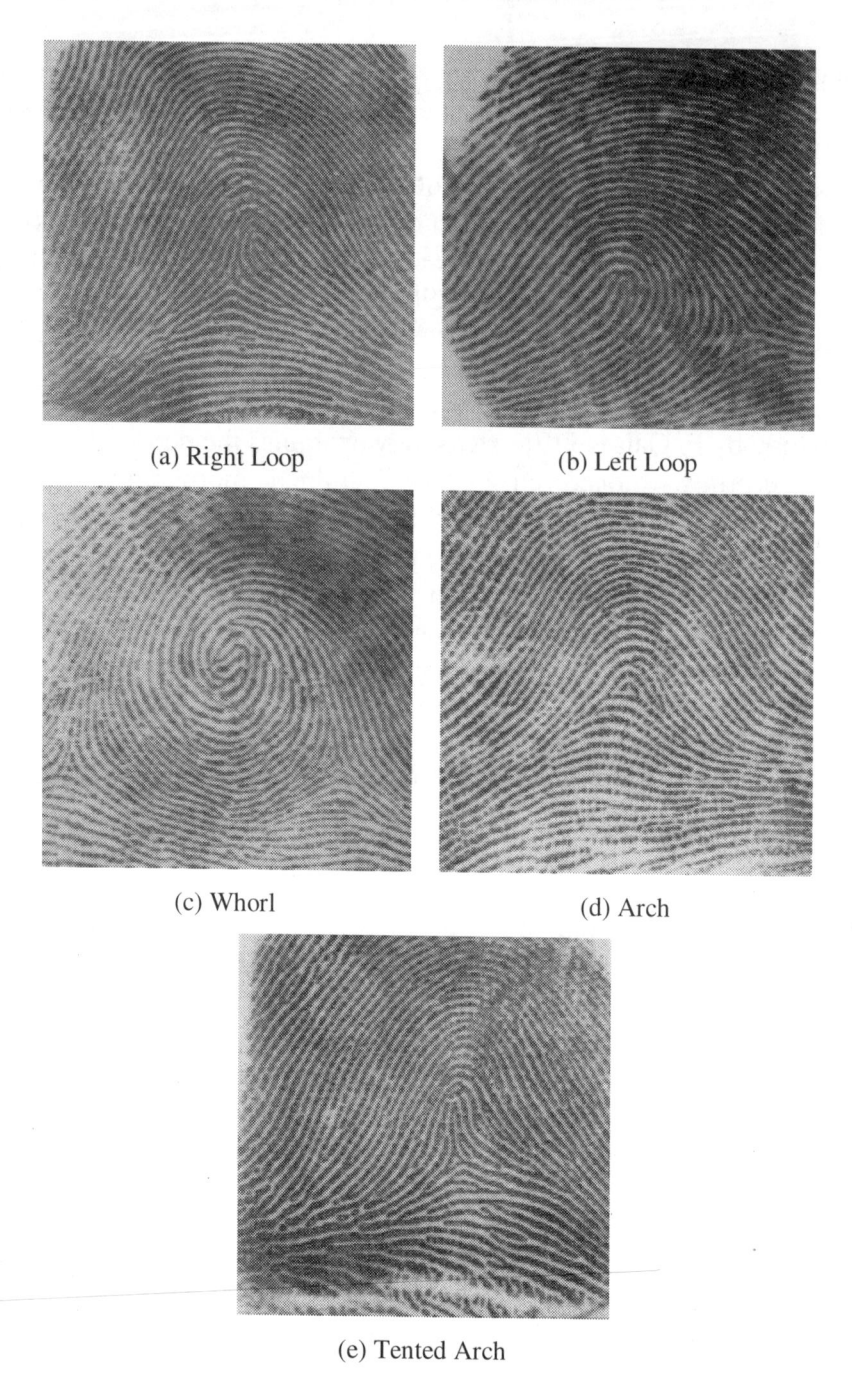

(b) Left Loop

(c) Whorl

(d) Arch

(e) Tented Arch

Figure 1. The five basic Henry classes of fingerprints.

Introduction to Fingerprint Recognition

The Galton Features are details formed on the ridges. A ridge can be defined as a single curve segment. The combination of several ridges forms a fingerprint pattern. The small features formed by the crossing and ending of ridges are called minutiae in the fingerprint literature. In his work, Galton defines four characteristics: the beginning and end of *ridges, forks, islands,* and *enclosures.* When searching for a new fingerprint in a database, a sufficient number of these minutiae should be located to decide the exact match.

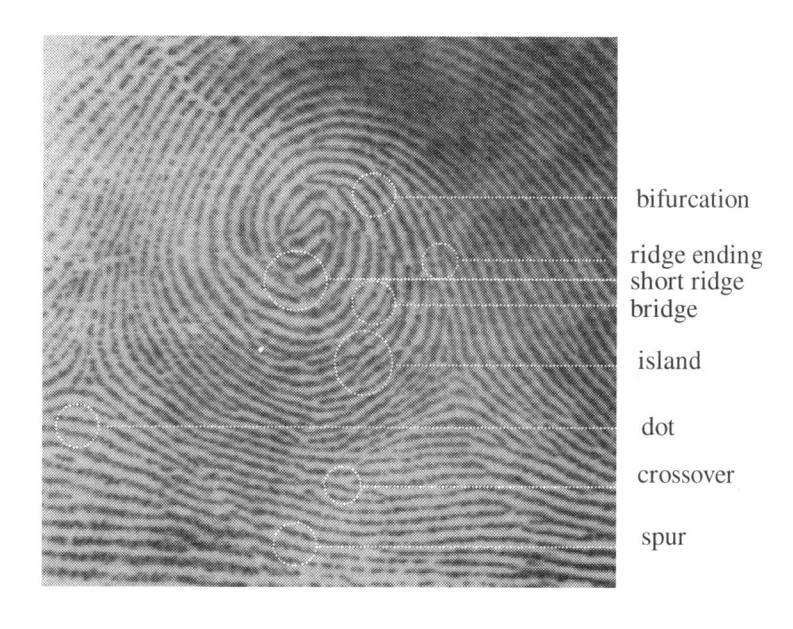

bifurcation

ridge ending
short ridge
bridge

island

dot

crossover

spur

Feature	Dot	Ridge End	Island	Bifurcation	Short Ridge	Crossover	Bridge	Spur
Sample								

Figure 2. Some basic fingerprint features.

After Henry and Galton, work on fingerprint identification and its specifications was extended and refined. The extended Galton Features [26] are shown in Figure 2. As can be seen, there are several extensions to the original Galton Features, but most are not used in automatic fingerprint identification systems. Instead, in accordance with the FBI representation of fingerprints [15], ridge endings and bifurcations are taken as the distinctive feature of fingerprints. In this method, the location and angle of the feature are taken to represent the fingerprint and used in the matching process.

Together with these, fingerprints contain two special types of features called *core* and *delta* points. These points are often referred to as singularity points of a fingerprint. The core point is generally used as a reference point for coding minutiae and defined as the topmost point on the innermost recurving ridge [54]. Example core and delta points are shown in Figure 3.

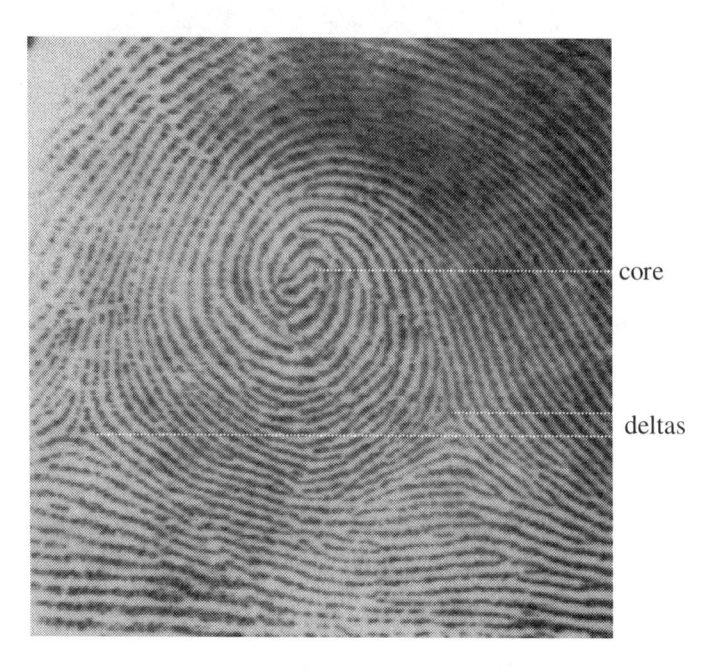

Figure 3. Core and delta points on a fingerprint.

With the increasing power of computers, automated systems have been developed to automate the tedious manual classification and matching methods of fingerprints. There are two types of fingerprint-based biometric systems in terms of their utilizations:

Automatic Fingerprint Authentication System (AFAS),
Automatic Fingerprint Identification System (AFIS).

Introduction to Fingerprint Recognition

The block diagram of a basic AFAS is shown in Figure 4. The input is an identity and a fingerprint image; the output is an answer of Yes or No indicating whether the input image belongs to the person whose identity is provided. The system compares the input image with the one addressed by the identity in the database [31].

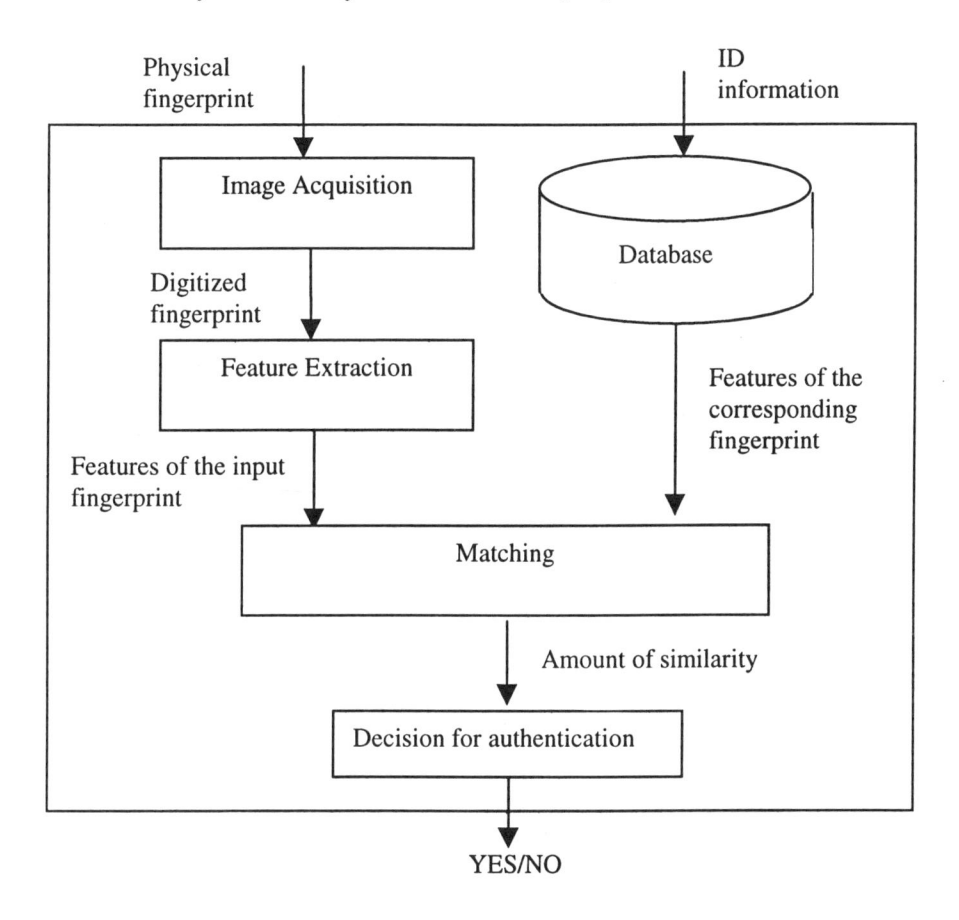

Figure 4. Block diagram of a basic fingerprint verification system.

In an AFIS (Figure 5), the input is just a fingerprint and the output is a list of identities of persons that can have the given fingerprint and a score for each identity indicating the similarity between the two fingerprints. It is possible to provide partial identity information to narrow the search space. The system compares the input image with many records in the database [12], [45], [46].

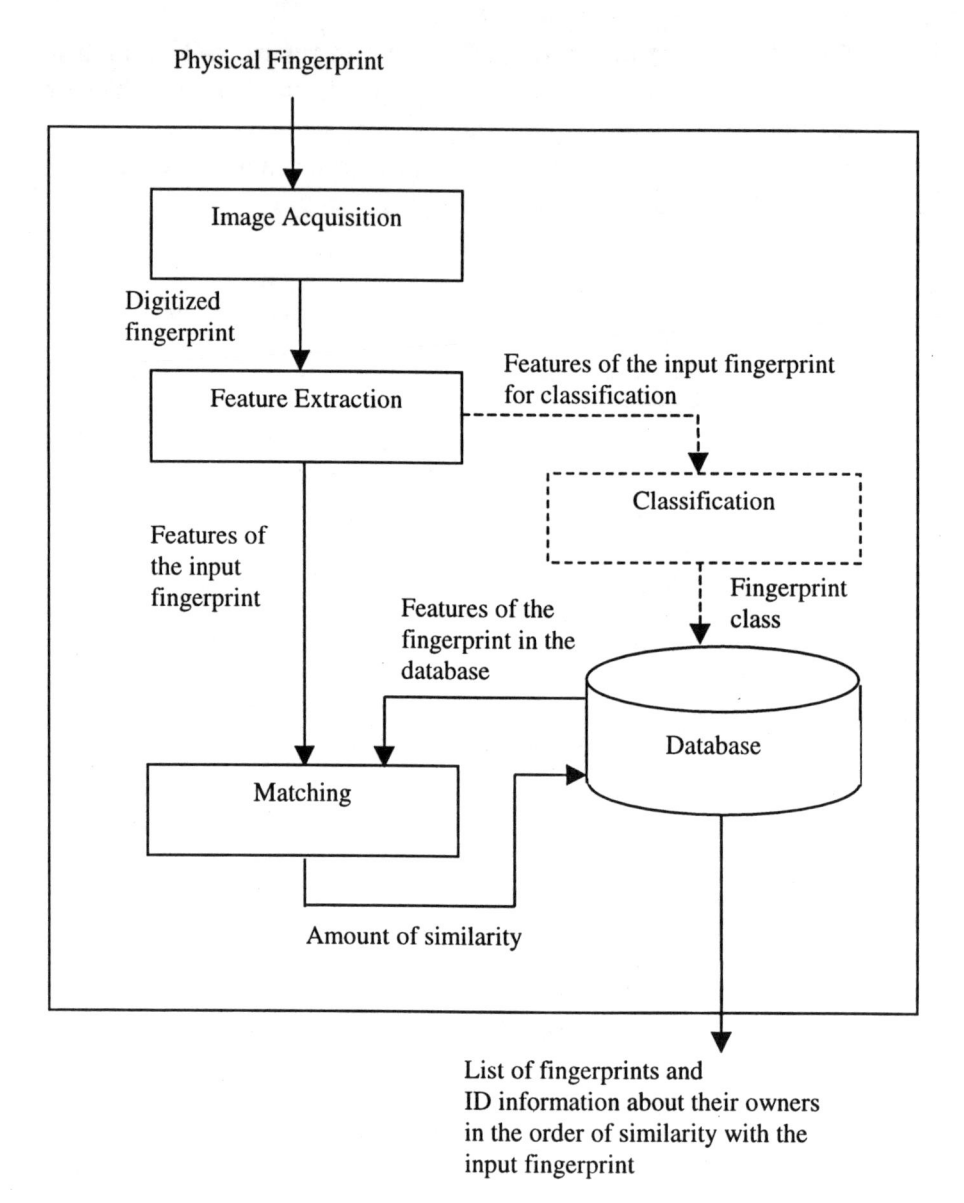

Figure 5. Block diagram of a basic fingerprint identification system.

Fingerprint based biometric systems are usually used for criminal identification and police work but now with the development in AFASs, they are highly utilized in civilian applications such as access control and financial security [35]. Therefore, AFASs are in great demand and a number of commercially available AFASs exist.

2 Fingerprint Identification and Authentication Systems

For each operation shown in Figures 4 and 5, several approaches are proposed in the literature [14]. These approaches are presented briefly below.

2.1 Image Acquisition

The oldest and most common method of capturing a fingerprint image is to obtain an impression by rolling an inked finger on paper and then scanning it using a flat-bed scanner. This method may result in highly distorted fingerprint images and thus it should be carried out by a trained professional. Another common way of obtaining fingerprint images is to scan the image directly using a CCD (Charge-Couple Device) camera [32]. The live scan method provides better images and does not need expertise, but highly distorted images are still possible because of dryness of skin, skin disease, sweat, dirt or humidity. In both of these methods, the image obtained is a high resolution (about 500 dpi) gray scale image of the fingerprint. In all of the available methods, the following variations between two copies of the same fingerprint are possible [14], [32]:

1) Translation because of different positioning of the fingerprint on the input device;
2) Rotation due to different positioning of the fingerprint on the input device;
3) Spatial scaling because of different downward pressure on the surface;
4) Contrast difference because of different downward pressure and ink density in ink-based methods;
5) Different regions of the same fingerprint as impression in the picture is usually a partial description of the whole fingerprint;
6) Shear transformation as the finger may exert a different shear force on the surface;
7) Local perturbations, i.e., local translation, rotation or scaling because of non-uniform pressure and shear force;

8) Breaks or smudges caused by non-uniform contact and non-uniform ink density in ink based methods;
9) Nonpermanent or semi-permanent distortions like skin disease, scars, sweat, etc.

2.2 Fingerprint Features

The most common representation used in fingerprint identification is the Galton features [19]. A ridge is defined as a single curved segment and a valley is the region between two adjacent ridges. The ridges and valleys in a fingerprint alternate, flowing in a local constant direction. Local discontinuities are called *minutiae* (see Figure 2). In his work, Galton defines four minutiae types. Later these features were refined and extended. Eighteen different kinds of fingerprint features are enumerated in [15]. In automatic systems, the set of minutiae types are restricted to two types: ridge endings and ridge bifurcations; as other types of minutiae can be expressed in terms of these two types [15]. A fingerprint is represented by the locations, types, and some attributes like orientation of minutiae [26], [27], [31], [32], [36], [43], [50], [58]. This representation reduces the fingerprint matching problem to a point or graph matching problem. One hundred years of study of fingerprints guarantees the uniqueness of minutiae based representation for a very large population of humans. There exist 50 to 150 minutiae on a complete fingerprint image, and in automated systems 10 matching minutiae are assumed to be sufficient to establish identity [5]. A problem with this representation is the lack of reliable minutiae extraction algorithms and the difficulty in quantitatively defining reliable match between two minutiae sets [32]. In Section 3, minutiae based feature extraction is explained in detail.

A less commonly used fingerprint feature is pores on the ridge surfaces which are also claimed to satisfy uniqueness condition [51]. They are also used as auxiliary features in some minutiae based systems. The fingerprint is represented by the locations of pores and local orientation of the ridge they reside on and the problem is again reduced to a point matching problem.

Coding the entire ridge pattern [26] and using the gray-scale image itself [1], [11] are other possible representations that satisfy the uniqueness condition but the computational load is higher in these representations and they are also more sensitive to the variations listed in Section 2.1. Especially in on-line verification, computational load is an important parameter. For this reason, in some systems, the above representations are derived from a small but consistent part of the fingerprint [1], [48]. Such simplifications result in loss of the degree of uniqueness but is reasonable for classification purposes and for authentication systems which are usually used by a few hundred people.

Features that satisfy the uniqueness condition can be used in both matching and classification but there are other features that are more suitable to be used in classification. The most common representation is the direction map (see Figure 6) which is a matrix of directions representing the ridge or valley orientation at each location on the fingerprint image [20], [56], [57].

Two special types of features related to the direction map are the *core* and *delta* points which are also called the singularity points [54]. The core point is defined as the topmost point on the innermost recurving ridge and a delta point is the center of a triangular region where three different direction flows meet (see Figure 3). The number and locations of these points are used in the definition of Henry Classes (see Figure 1). In [52] and [56] it has been shown that by just using the locations and the types of these points a model in the form of a direction field that approximates the direction map can be built.

In [9], B-spline functions are used to represent ridge curves, which is useful for compression of finger print images.

2.3 Fingerprint Classification

Fingerprint classification is used in AFISs. The purpose of classification is clustering a database of fingerprints into as many classes as possible. In identification, just the class corresponding to the input fingerprint is searched and thus the search time is reduced. It may also be used as a recognition system for a few people where each class corresponds to only one person.

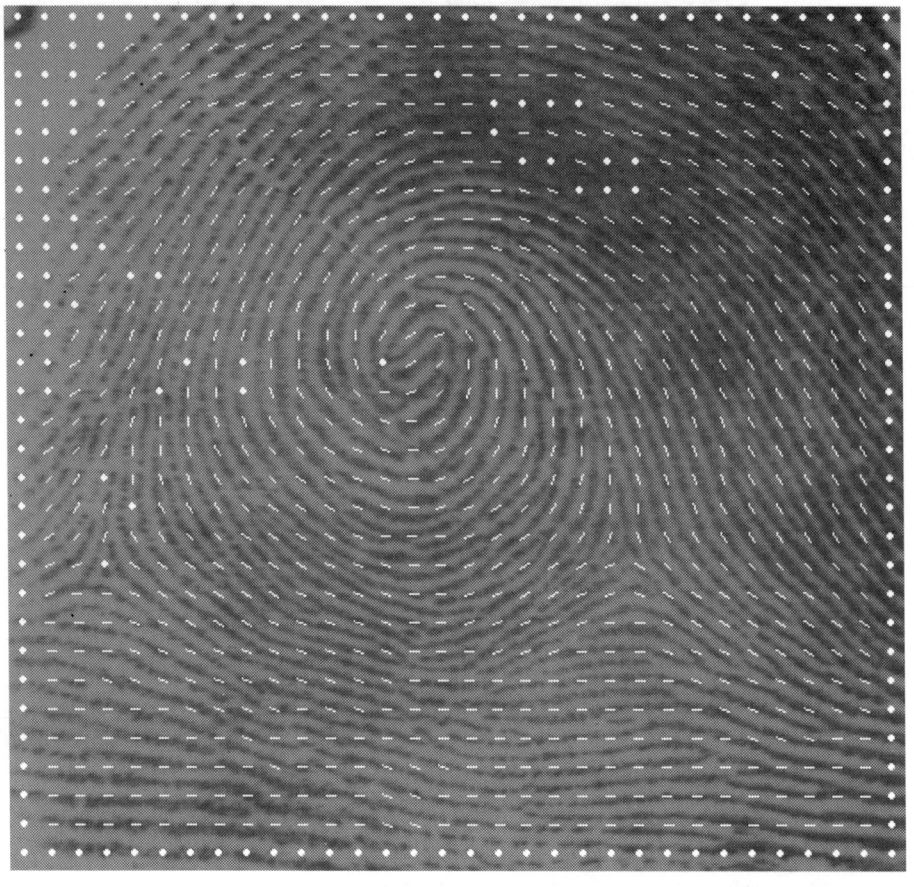

Figure 6. Block directional image overlaid on a sample fingerprint

The oldest classification scheme, which is used in manual fingerprint identification, is the Henry Classification scheme [25]. In this scheme there are five classes namely *arch*, *tented arch*, *right loop*, *left loop*, and *whorl* (see Figure 1). These classes are determined by the ridge flow on the core area and the number and relative locations of core and delta points (see Figure 3). Rarely, fingerprints cannot be assigned to any of the classes and are thus assigned to a class called *accidental*. The Henry classification is efficient for manual classification as humans can easily identify each class but not much is gained with clustering a database into just six classes. Also, these classes have uneven distributions [6].

Automatic fingerprint classification systems mostly try to implement the Henry Classification scheme but other approaches are possible. There are four major approaches that have been taken for automatic

There are four major approaches that have been taken for automatic fingerprint classification: syntactic, structural, neural network, and statistical approaches.

In the syntactic approach [6], [42], [43], the ridge patterns and minutiae are approximated as a string of primitives. Then pre-defined classes (e.g., Henry Classes) are modeled as production rules or a set of grammars are inferred from the training samples. When a new pattern arrives the string of primitives are formed and passed to a parser(s) whose output yields the class of the input pattern.

In the structural approach [26], [48], features based on minutiae are extracted and then represented using a graph data structure. Structural matching is done by exploiting the topology of features. Another structural approach to use the topology of singularities. In [33] and [50], the types and locations of core and delta points and ridge flow between paired core and delta points are used for classification of fingerprints into Henry Classes.

In the neural network [17], [22] approach a feature vector is constructed and classified by a neural network classifier. In [20], [21], and [45], K-L transform of normalized direction vectors are used as features. Then the classes are formed in an unsupervised manner using a hierarchical neural network structure [13], [21], [23] consisting of a modified version of self-organizing feature maps [34]. In [57], the K-L transformed direction vectors are used as input to a multilayer perceptron network which classifies the input pattern into one of the Henry Classes. In [11], wedge-ring features, which are extracted from the Fourier transform of the fingerprint image, are used as input to a multilayer perceptron network. Wedge-ring features are integral of the magnitude of Fourier coefficients over regularly spaced semi-circles and wedges in the frequency domain. In [41], textural features together with some directional features are used as an input to a fuzzy multilayer perceptron network which is trained to recognize Henry Classes.

Another approach is using statistical classifiers instead of neural network classifiers [11]. In [16], wedge-ring features obtained from the hexagonal Fourier transform of enhanced and thinned fingerprint image are used as an input to a nearest neighbor classifier.

2.4 Fingerprint Matching

Matching is the process of measuring the similarity between two fingerprint images [2], [32]. The most commonly used matching method is minutiae based matching. But there are also other approaches. An example is given in [5] which proposes the use of neural networks in fingerprint matching. A multilayer network whose input layer is constrained to implement feature selective filters is used for matching. The input to the network is low-pass filtered and averaged central region of two aligned fingerprints. The output of the network is the probability that two fingerprints belong to the same finger. The network is firstly trained by pairs of fingerprint images and tested on different pairs which have not been used in training. The network learns to match fingerprint images.

There are two approaches in minutiae based matching: point matching and structural matching. In point matching, two sets of minutiae code using their locations are aligned and the sum of similarity between the overlapping minutiae is calculated. The similarity between two minutiae is measured using the attributes of the minutiae. Alignment is an important problem in point matching and corresponds to the registration operation in most of the systems. In structural matching, the locations are mostly discarded and a graph which codes the relative locations of minutiae are constructed. The subgraphs around each minutiae are used to build feature vectors. As the locations are discarded, alignment is not necessary.

In [50], alignment is performed using a Hough transform method. All minutiae pairs are used to find out possible translations. Then possible rotations and translations are scored using the number of overlapping minutiae. The rotation angle and translation vector with maximum score is used in alignment. Each minutiae is represented by a bounding box centered at minutiae coordinate. The size of the bounding box is adjusted using the distance to the core point. The minutiae are paired if their bounding boxes overlap, and their angles and types match.

In [49], rotation is not considered in the alignment phase. The translation is handled using the upper core point in the fingerprint image, and then the overlapping minutiae are determined.

In [31] and [32], the alignment is performed using ridge matching. Together with the minutiae, the corresponding ridge is stored and they are matched one by one. Their similarity is measured using the correlation between the functions giving the distance of each ridge point to the line extending from the x-axis. When correlation exceeds a threshold, a translation vector and a rotation angle are calculated using the paired ridges. Polar coordinates are used for matching. The minutiae are ordered with respect to radial angle and a string is formed. The matching is handled using string matching.

In [26], structural matching is performed. In detection, not only bifurcations and ends but also other types of minutiae such as islands and crossovers are detected. For each minutia, a feature vector representing the frequency of each type of minutia in a circular neighborhood around the corresponding feature is constructed. Then by statistical analysis, discriminating feature vectors are chosen in a database of fingerprints. The feature is invariant to most of the possible distortions. Then the matching phase searches the new fingerprint features in the database. The system is especially designed for partially available fingerprint data.

In [28], a complex graph structure representing the ridges, minutiae, and their relative locations is constructed, and a graph matching algorithm suitable for the constructed graph is proposed. The matching algorithm is again robust to most of the possible distortions. It does not make use of minutiae locations.

3 Minutiae-Based Fingerprint Feature Extraction

As stated before, most common features used in matching are the minutiae on the fingerprint. A fingerprint image is preprocessed and minutiae on the fingerprint are extracted and coded using their location and attributes like orientation of the ridge they are located on. Then a list or a graph of minutiae are formed. The disadvantage with this approach is the lack of reliable minutiae extraction algorithms. All the available minutiae detection algorithms result in spurious minutiae. Spurious minutiae are inevitable because of distortions such as scars, overinking, and sweat, but mostly the algorithm itself creates the spurious minutiae. For that reason enhancement of the fingerprint image and spurious minutiae elimination forms an important part of the system.

Figure 7 is a block diagram of a very general minutiae based feature extraction. From the image a direction map (see Figure 6) which is used in enhancement and singularity point detection is constructed. Then the image is segmented to separate the fingerprint area and the background. After some enhancement the fingerprint image is binarized. Binarization corresponds to segmentation of ridges and valleys. The binarized image is enhanced further to get rid of the distortions like holes and breaks. Enhancement may be a simple low-pass filtering operation but using the direction map is a better idea as it provides good clues for enhancement operations like connecting ridge breaks which are inevitable. Binarization is followed by thinning operation which results in an image where the ridges are represented by one pixel wide curves. The results of enhancement, binarization, and thinning on a sample fingerprint image is given in Figure 8 [3], [4], [7] [8], [18], [24], [29], [30], [37].

The thinned image usually needs further enhancement to eliminate spurs and breaks. From the thinned image, minutiae are extracted and the false minutiae are eliminated. Minutiae reduction is performed using some structural or statistical heuristics. The reduced minutiae list provides one of the inputs to a matching module. The matching module outputs a score indicating the degree of similarity between the input minutiae list and the one in the database as explained in Section 2.4. In the matching stage, the two sets of minutiae should be aligned for the normalization of the location information. Usually a registration point is calculated for each fingerprint. The registration point is a landmark which corresponds to the same point on the real fingerprint. Singularity points are good candidates to be used as registration points. A single registration point provides translation invariance. It is also possible to make a better alignment including rotation and scale using the minutiae or the ridges on the fingerprint.

Introduction to Fingerprint Recognition

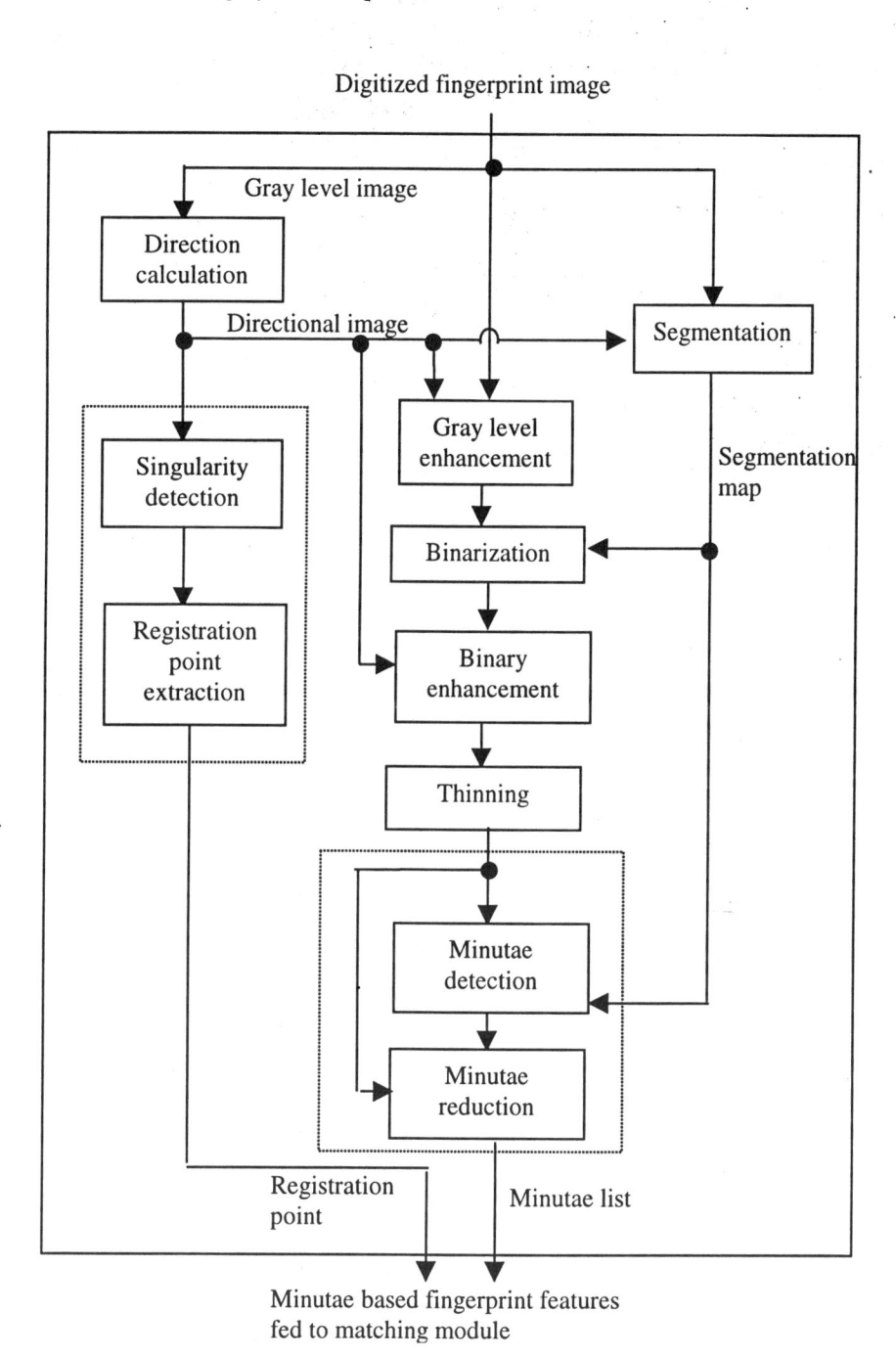

Figure 7. Block diagram for extraction of minutae-based fingerprint features.

(a) Original fingerprint image

(b) Enhanced fingerprint image

(c) Binarized fingerprint image

(d) Thinned fingerprint image

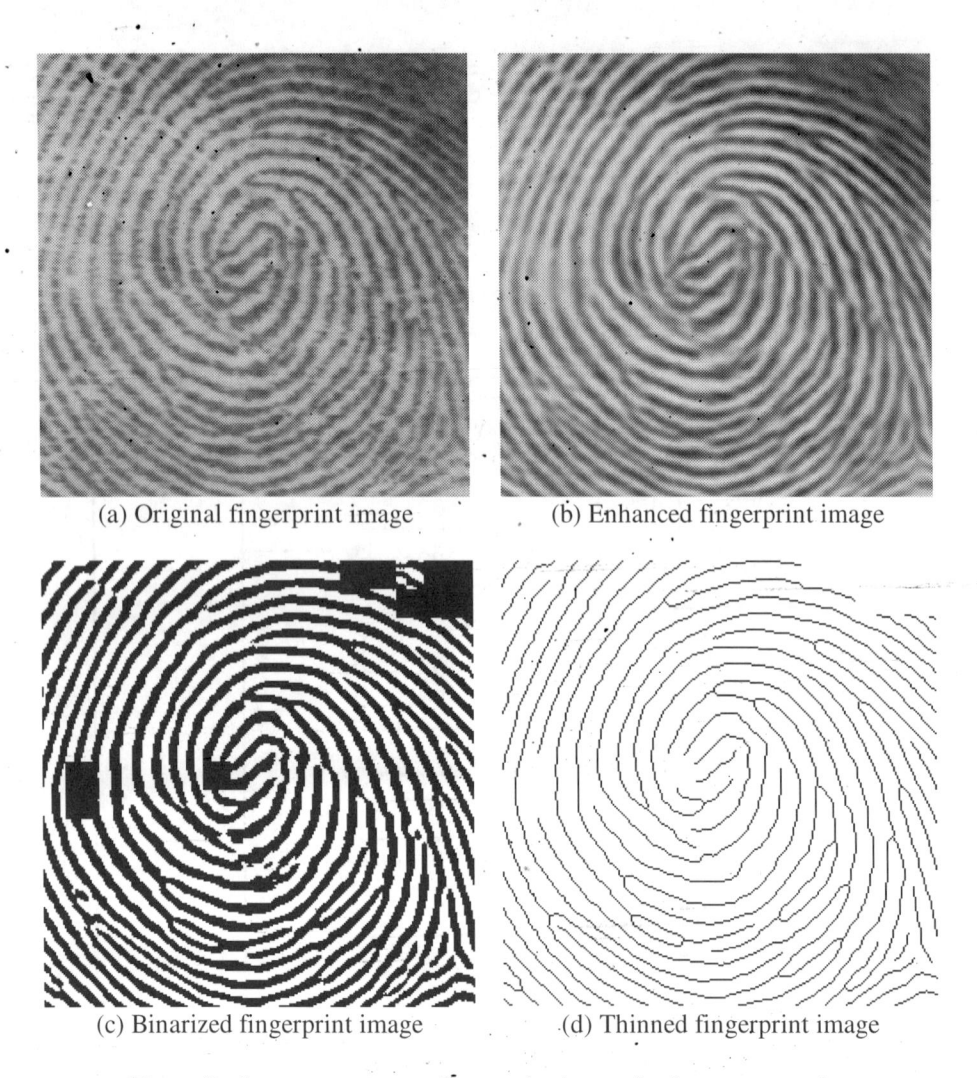

Figure 8. Some processes on fingerprint image for feature extraction

3.1 Direction Calculation

A direction map (see Figure 6) is a matrix of direction vectors representing the ridge or valley orientation at each location on the fingerprint image. Determining an orientation at a point needs examination of a large neighborhood around that point. For that reason, assigning orientation at each pixel requires a high computational capacity. Instead, the image is partitioned into square blocks of constant size and an average orientation is calculated for each block. It is also possible to use overlapping blocks. A local orientation is calculated for each point in the block. The vector having the same magnitude but orthogonal to the gradient represents the local orientation for that point. Then these vectors are averaged over all the pixels in the block. The direction of the resulting vector is the orientation for the block and the magnitude is a certainty measure for the calculated orientation. The resultant direction map is a low-resolution global representation of the fingerprint. This is the simplest and fastest algorithm for direction map calculation [31], [50], [56]. Another approach is using integrals over small line segments of different orientations passing through the point that the orientation is being calculated [20], [54]. A novel method is given in [53]. The fingerprint image is pre-filtered by equally spaced orientation selective band pass filters; then the orientation at the specified point is calculated using the outputs of each filter. The direction map is a highly utilized representation. It is used in enhancement, singular point detection, and even in binarization.

3.2 Segmentation

Segmentation is the process of discriminating the fingerprint area in the image. The image mostly consists of a uniform background and the fingerprint impression. In segmentation, areas corresponding to the background are determined and discarded from the rest of the processing. The image is partitioned into blocks and each block is then labeled as background or fingerprint area. The variance in the block, the certainty associated with the orientation in the block or the directional histogram in the block are used in the decision [20], [38], [39].

3.3 Singularity Detection

Singularity points, i.e., the core and delta points are located using the direction map. In [11] and [54], some structural heuristic is used for detection. But the most robust way is modeling the direction map as a direction field and calculating the *Poincare Index* which is the integral of the rate of change of orientation on a close contour [56], [33]. Ordinary points on the direction field have a Poincare Index of zero. Core and delta points have a Poincare Index of 1/2 and −1/2 respectively. One problem with this method is that fingerprints of the Arch type do not have any singularity in terms of Poincare Index so structural heuristic is used to locate the core point.

3.4 Gray-Level Enhancement

Enhancement of the image may be a simple low-pass filtering to eliminate noise. Enhancement of the fingerprint image may affect the performance of the system to a great extent. For that reason more complex algorithms are available.

In [44], oriented filter masks are designed. The parameters of the masks are average ridge and valley width and the direction. The image is convolved with these masks but the shapes of the masks are updated using the orientation information in the direction map during convolution. A similar approach has been tried in [14]. The orientation selective filters are constructed using a kind of Gabor filter, which is the product of a Gaussian and a plane wave.

In [53], the output of a set of orientation selective filters is used. In the enhanced image, each pixel value is taken as the value of the corresponding pixel in the filter corresponding to the dominant direction. A new enhancement method is reported in [31] and [32]. The idea is to use the projection along a line segment perpendicular to the orientation at a given location on the image. The ridges correspond to local maxima of the projection and valleys correspond to local minima of the projection. A special filter is designed to enhance the local maxima and minima on these profiles. The local orientation is again a parameter of the filter. In [55], a fuzzy set concept on grayscale image is used for edge detection and ridge extraction.

3.5 Binarization and Binary Image Enhancement

The histograms of fingerprint images are not bimodal. For that reason a global threshold cannot be used in binarization but local thresholding may be applied. The gray-level image enhancement is a must in such an approach.

A simple binarization scheme is given in [57]. Integrals over small line segments of different orientations are calculated. The lines are represented by a mask and correspond to slits on this mask. Using the minimum and maximum integral values and the gray scale value of the pixel, the binary value is determined. The same mask is also used in direction map construction (see Section 3.1). Again gray-level image enhancement is necessary. In [11], edges of the gray-scale image and the image itself are used in binarization. They use adaptive windows that run over the edge map and the gray-scale image. Using some features calculated over the windows the binary image is constructed. In [31], [32], and [49], projections along lines perpendicular to the ridge orientation are used. The peaks of the projections are chosen, as they correspond to the ridges. This scheme results in a fairly smooth binary image. In [43], iterative application of Laplacian operator is proposed for binarization and enhancement.

In most of the algorithms, the binary image obtained contains holes or breaks along the ridges. The holes are detected and filled [11] or the image is smoothed using morphological masks [16]. Any simple image-processing algorithm can be applied at this point. In [27], a complex algorithm which makes use of the direction map is proposed. The algorithm consists of an adaptive ridge width equalization stage, a ridge enhancement stage that makes use of the direction map, and a noise removal stage.

3.6 Minutiae Detection

Minutiae detection over the thinned image is a simple process. One seeks for bifurcations and ends on the thinned image by calculating zero crossings in the neighborhood of each pixel [45], [46]. Some attributes to be used in matching are also calculated at this stage. The criteria for choosing these attributes is invariance under possible distortions (see Section 2.1). Possible attributes can be listed as follows [2], [31], [32], [45], [50], [58]:

1) Angular Attributes: For ends the orientation of the corresponding ridge and for bifurcations the minimum angle between three branches and the orientation of the ridge that is not used in the formation of minimum angle.
2) Location with respect to registration point which is usually the core.
3) Number of ridge crossings along the line connecting the minutiae and the registration point.

4) The slope of the line connecting each minutiae to the registration point.
5) The minutiae type.

In [36], minutiae detection directly on the gray-scale image without binarization and thinning is proposed. The motivation for the study is the loss of information in binarization and thinning and unsatisfactory performance on low-quality fingerprint images. They perform ridge line following guided by the direction map. The local maxima of projections along lines orthogonal to the ridge directions are used for ridge line following. The orthogonal line is advanced in regular steps in the direction given by the direction map. At each step a local maxima is chosen representing a sample of the ridge curve. The tracing ends at minutiae or high curvature regions. The algorithm also handles ridge breaks.

3.7 Minutiae Reduction

An inevitable source of error in fingerprint recognition is false minutiae (see Figure 9). False minutiae are inevitable because of distortions such as scars, overinking, and sweat. Existence of false minutiae forces a matching process to handle false minutiae. It is not possible to handle excessive numbers of them; therefore after the minutiae detection phase the minutiae list is analyzed to eliminate false minutiae.

Minutiae reduction is based on intuitive heuristics [31], [32], [36], [50]. In simple approaches, they are eliminated using some distance criteria, e.g., minutiae which are too close to each other are discarded, or minutiae which are too close to the segmented regions, i.e., the regions corresponding to the background or bad areas in the fingerprint, are discarded. Also connectivity may be used for minutiae elimination. Minutiae that are connected by a short line segment are discarded. The problem with these simple rules is that it is possible to discard true minutiae because of false minutiae. For that reason more complex algorithms are available.

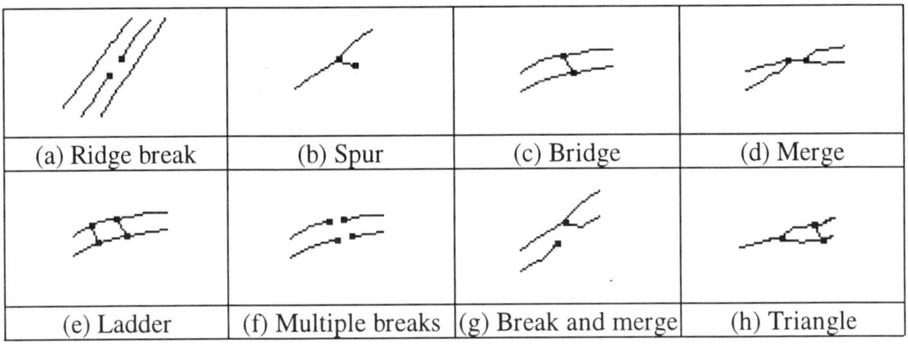

Introduction to Fingerprint Recognition 25

(a) Ridge break	(b) Spur	(c) Bridge	(d) Merge
(e) Ladder	(f) Multiple breaks	(g) Break and merge	(h) Triangle

Figure 9. Examples of false minutiae which are shown by dots in the figure.

In [27], the sources of false minutiae are classified as spurs, holes (a ridge curve forming a cycle), bridges, and breaks. Each of these structures is detected and corrected. To correct ridge breaks they make use of the thinned version of the negative image where a ridge break is observed as a bridge.

In [58], the false minutiae structures shown in Figure 9 are detected and corrected.

4 Performance Criteria

As in all recognition applications the construction of a test set is an important problem. There exist a number of databases that can be used for the development and testing of fingerprint recognition systems and are used in the available references. The National Institute of Standards Special Database 9 (NIST-9) is a database that includes 1350 mated fingerprint card pairs. It is divided into multiple volumes. Each volume has three CDs. Each CD contains 900 images of card type1 and 900 images of card type 2. Fingerprints on card type 1 were scanned using a rolled method and fingerprints on card type 2 were scanned using a live-scan method. In total there are 5400 fingerprint images in a volume. They are 8-bit gray-scale images scanned at approximately 500-dpi resolution and of size 832 × 768. It has a natural distribution from National Crime and Information Center Classes, which is whorl = 27.9%, left loop = 33.8%, right loop = 31.7%, arch = 3.7%, and tented arch = 2.9%. The National Institute of Standards Special Database 4 (NIST-4) includes two scans from one finger of 2000 persons, making a total of 4000 fingerprint images. The images are of size 512×512, have 256 gray scales, were scanned at approximately 500-dpi resolution, and are all scanner images of rolled fingerprints. The classes are evenly distributed. MSU is a database of 700 fingerprint images of 70 persons. The images are of size 640 × 480, have 256 gray scales, and were scanned using a fingerprint scanner of Digital Biometrics.

In classification studies, the performance of the system is given by the classification accuracy or classification error of the system. The NIST-4 database was used in [33], [57], and [20] for classification. In [33], a rule-based system is presented for classification of 4000 fingerprints of NIST-4 database. For the five Henry classes, a classification accuracy of 85.4% was achieved with no reject and it increased to 86.4% with 10% reject rate. In [57], a neural network-based classifier is used for classifying the fingerprints into five Henry classes. A multilayer perceptron network was trained on 2000 fingerprint images of the NIST-4 database and then tested on the other 2000 images of the same fingers. The best reported classification accuracy were 83% with the equally spaced grid, 86% with unequally spaced grid, and 95.4% with 10% rejects. In [20], a hierarchical neural network structure is proposed and used to classify the fingerprints in an unsupervised manner. The first 500 images from the NIST-4 database were used to train the network and the other 500 images of the same fingers were used in testing. In the best-performing network, 36 classes having at most 29 elements were emerged. The network is used as an indexing mechanism to the database. In this mechanism, 20% search space corresponds to dividing the database into five equal sized classes. Classification accuracy of the network was 90.2% with 20% search space and no reject. 5400 fingerprints of NIST-9 database is also used in [33], for which classification accuracy was 87.6% with no reject and it was 90.1% with 10% reject rate for five Henry classes.

In [16], hexagonal fast Fourier transform is used to classify a special database of 40 fingerprints of size 512 × 512 and of 256 gray scales into whorls, loops, and arches. Their classification accuracy was 85% with zero rejects. In [41], a fuzzy multilayer perceptron is used for the classification of fingerprints into whorl, left loop, and right loop. They used nine fingerprint images of size 256 × 256 and of 16 gray levels, three each belonging to the three categories. Perturbing the gray level of the original images generated a total of 189 images, of which 66 were selected for training the network and the remainder were used in testing. The best reported classification accuracy was 83.7% with no reject.

In AFIS applications, the performance of the system is given by the distribution of the correct match in the list constructed by the system. The tests are very costly because of the large search space therefore performance results are mostly missing or limited to a small database in the systems reported in literature. In [50], a test database of 1800 images from NIST-9 database is used and is tested using 100 randomly selected samples from the database. Ten percent of these samples are rejected and 80% is found at first 10 in the output list. They report a 31.7 seconds feature extraction time and a 1.015 seconds matching processing time on a SPARC 2 computer. The registration part of their algorithm is very costly as it is based on Hough Transform.

In AFAS applications there are four possible outcomes: (1) an authorized person is accepted, (2) an authorized person is rejected, (3) an unauthorized person is rejected, and (4) an unauthorized person is accepted. The rate of cases 2 and 4, which are called False reject rate (FRR) and False Acceptance Rate (FAR), are standardized metrics of the identification accuracy of biometric systems [32]. The evaluation of these rates depends on the problem. For example, in some situations FAR may be catastrophic. For this reason a Region of Operating Curve (ROC) is used for evaluation. The ROC is a graph of FAR vs. FRR. Another metric is Equal Error Rate (EER) which is the average error observed when FAR and FRR are equal to each other. But for the most part, these rates depend heavily on the specific database as it is nearly impossible to build a large database that can be guaranteed to represent the real operating environment.

In [5], a neural network approach to matching is presented and a custom database consisting of 5 different copies of one finger of 20 persons is constructed. The images are sized 512 × 464, have 256 gray levels, and are scanned directly by a CCD camera. The network is trained using fingerprint images of 5 persons and tested on the remaining 20 persons. An EER of 0.25% and a point of 0% FAR and

4% FRR on ROC is reported. In [32], EER of 3.07% and 2.69% on MSU and live scan part of NIST-9 database, respectively, are reported.

5 Summary

This chapter has introduced the basic concepts of fingerprint identification and authentication system. Fingerprint, face, iris, retina, and voice serve the purpose of recognizing individuals in applications including law enforcement, banking, and security. The following chapters in this book discuss specific projects pertaining to the fingerprint and face recognition and analysis. The chapters on fingerprint recognition include this introductory chapter, fingerprint feature processing techniques, and poroscopy, fingerprint sub-classification using neural networks, Gabor filter-based fingerprint identification, minutiae extraction and filtering from gray-scale images, and ridge extraction. The chapters pertaining to the face recognition include introduction to face recognition, neural networks for face recognition, face unit radial basis function networks, face recognition from correspondence maps, recognizing faces by elastic bunch graph matching, facial expression synthesis using radial basis function networks, and facial expressions in human computer interface. It is obvious that computational intelligence techniques offer definite advantages in fingerprint and face recognition.

References

[1] – (1997), "Access Control Applications Using Optical Computing," online available: http://www.mytec.com.

[2] Akhan, M.B. and Emiroglu, I. (1995), "A Flexible Matching Algorithm for Fingerprint Identification System," *Proc. The Tenth International Symposium on Computer and Information Sciences*, Izmir, Turkey, pp. 806-815.

[3] Alvarez, L., Lions, P.L., and Morel, J.M. (1992), "Image Selective Smoothing and Edge Detection by Nonlinear Diffusion II," *Siam J. Numer. Anal.*, 29(3), pp. 845-866.

[4] Arcelli, C. and Baja, G.S.D. (1984), "A Width Independent Fast Thinning Algorithm," *IEEE Trans. Pattern Analysis Machine Intelligence*, 2, pp. 223-231.

[5] Baldi, P. and Chauvin, Y. (1993), "Neural Networks for Fingerprint Recognition," *Neural Computation*, 5, pp. 402-418.

[6] Blue, J.L., Candela, G.T., Grother, P.J., Chellapna, R., and Wilson, C.L. (1994), "Evaluation of Pattern Classifiers for Fingerprint and OCR Applications," *Pattern Recognition*, 27(4), pp. 485-501.

[7] Canny, J. (1986), "A computational Approach to Edge Detection," *IEEE Trans. Pattern Analysis and Machine Intelligence*, 8, pp. 679-698.

[8] Casadei, S. and Mitter, S. (1998), "Hierarchical Image Segmentation – Part 1: Detection of Regular Curves in a Vector Graph," *Int. J. of Computer Vision*, 27(1), pp. 71-100.

[9] Chong, M.M.S., Gay, R.K.L., Tan, H.N., and Liu, J. (1992), "Automatic Representation of Fingerprints for Data Compression by B-Spline Functions," *Pattern Recognition*, 25, pp. 1199-1210.

[10] Clarke, R. (1994), "Human Identification in Information Systems: Management Challenges and Public Policy Issues," *Info. Technol. People*, 7(4), pp. 6-37.

[11] Coetzee, L. and Botha, E.C. (1993), "Fingerprint Recognition in Low Quality Images," *Pattern Recognition*, 26, pp. 1441-1460.

[12] Eleccion, M. (1973), "Automatic Fingerprint Identification," *IEEE Spectrum*, 10, pp. 36-45.

[13] Erol, A. (1995), *A Neural Network with Unsupervised Pre-classification and PCA Feature Extraction for Character Recognition*, MSc Thesis, Department of Electrical and Electronics Engineering, Middle East Technical University, Ankara, Turkey.

[14] Erol, A. (1998), *Automated Fingerprint Recognition*, Ph.D. Thesis Proposal Report, Dept. of Electrical and Electronics Eng., Middle East Technical University, Ankara, Turkey.

[15] Federal Bureau of Investigation, (1984), *The Science of Fingerprints: Classification and Uses*, Washington DC.

[16] Fitz, A.P. and Green, R.J. (1996), "Fingerprint Classification Using a Hexagonal Fast Fourier Transform," *Pattern Recognition*, 29(10), pp. 1587-1597.

[17] Freeman, J.A. and Skapura, D.M. (1991), *Neural Networks: Algorithms, Applications, and Programming Techniques*, Addison-Wesley.

[18] Freeman, W.T. (1991), "The Design and Use of Steerable Filters," *IEEE Trans. Pattern Analysis and Machine Intelligence*, 13(9).

[19] Galton, F. (1892), *Finger Prints*, Macmillan, London.

[20] Halici, U. and Ongun, G. (1996), "Fingerprint Classification Through Self-Organizing Feature Maps Modified to Treat Uncertainties," *Proc. of the IEEE*, 84(10), pp. 1497-1512.

[21] Halici, U., Erol, A., and Ongun, G. (1998), "Industrial Applications of the Hierarchical Neural Networks: Character Recognition and Fingerprint Classification," *Industrial Applications of Neural Networks*, L.C. Jain (Ed.), OCR Press, USA.

[22] Halici, U. and Gelenbe, E. (1998), *Lecture Notes on Neurocomputers*, Middle East Technical University Online Courses, http://www.ii.metu.edu.tr/metuonline .

[23] Halici, U. and Erol, A. (1995), "A Hierarchical Neural Network for Optical Character Recognition," *Proc. International Conference on Artificial Neural Networks*, Paris, pp. 252-256.

[24] Haralick, R.M. (1984), "The Digital Step Edge from Zero Crossings of Second Directional Derivatives," *IEEE Trans. Pattern Analysis and Machine Intelligence*, 6(1).

[25] Henry, E.R. (1900), *Classification and Uses of Finger Prints*, Routledge, London.

[26] Hrechak, A.K. and McHugh, J.A. (1990), "Automated Fingerprint Recognition Using Structural Matching," *Pattern Recognition*, 23, pp. 893-904.

[27] Hung, D.C.D. (1993), "Enhancement and Feature Purification of Fingerprint Images," *Pattern Recognition*, 26(11), pp. 1661-1671.

[28] Isenor, D.K. and Zaky, S.G. (1986), "Fingerprint Identification Using Graph Matching," *Pattern Recognition*, 19(2), pp.113-122.

[29] Iverson, L.A. and Zucker, S.W. (1995), "Logical/Linear Operators for Image Curves," *IEEE Trans. Pattern Analysis Machine Intelligence*, 17(10).

[30] Jain, A.K. (1989), *Fundamentals of Digital Image Processing*, Prentice Hall, Englewood Clifs.

[31] Jain, A.K., Hong, L., and Bolle, R. (1997), "On-Line Fingerprint Verification," *IEEE Trans. on Pattern Analysis and Machine Intelligence*, 19(4), pp. 302-314.

[32] Jain, A.K., Hong, L., Pankanti, S., and Bolle, R. (1997), "An Identity-Authentication System Using Fingerprints," *Proc. of the IEEE*, 85(9), pp. 1365-1388.

[33] Karu, K. and Jain, A.K. (1996), "Fingerprint Classification," *Pattern Recognition*, 29(3), pp. 389-403.

[34] Kohonen, T. (1989), *Self-Organization and Associative Memory*, Springer Series in Information Sciences.

[35] Lee, H.C. and Gaenssley, R.E. (1991), *Advances in Fingerprint Technology*, New York, Elsevier.

[36] Maio, D. and Hanson, A.R. (1997), "Direct Gray-Scale Minutiae Detection in Fingerprints," *IEEE Trans. on Pattern Analysis and Machine Intelligence*, 19(1), pp. 27-40.

[37] Malleswara, R.T.Ch. (1976), "Feature Extraction for Fingerprint Classification," *Pattern Recognition*, 8, pp. 181-192.

[38] Mehtre, B.M. and Chatterjee, B. (1989), "Segmentation of Fingerprint Images – A Composite Method," *Pattern Recognition*, 22, pp. 381-385.

[39] Mehtre, B.M., Murthy, N.N., and Kapoor, S. (1987), "Segmentation of Fingerprint Images Using The Directional Image," *Pattern Recognition*, 20, pp. 429-435.

[40] Miller, B., (1994), "Vital Signs of Identity," *IEEE Spectrum*, 31(2), pp. 22-30.

[41] Mitra, S., Pal, S.K., and Kundu, M.K. (1994), "Fingerprint Classification Using a Fuzzy Multilayer Perceptron," *Neural Computing and Applications*, 2, pp. 227-223.

[42] Moaver, B. and Fu, K.S. (1976), "An Application of Stochastic Languages to Fingerprint Pattern Recognition," *Pattern Recognition*, 8, pp. 173-179.

[43] Moayer, B. and Fu, K.S. (1986), "A Tree System Approach for Fingerprint Recognition," *IEEE Transactions on Pattern Analysis and Machine Intelligence*, 9(3), pp. 376-387.

[44] O'Gorman, L. and Nickeron, J.V. (1989), "An Approach to Fingerprint Filter Design," *Pattern Recognition*, 22, pp. 29-38.

[45] Ongun, G. (1995), *An Automated Fingerprint Identification System based on Self-Organizing Feature Maps Classifier*, M.Sc. Thesis, Middle East Technical University, Turkey.

[46] Ongun, G., Halıcı, U., and Özkalaycı, E. (1995), "HALafis: An automated fingerprint identification system," *Proc. Bilisim 95*, Istanbul, pp. 95-100.

[47] Ongun, G., Halıcı, U., and Özkalaycı, E., "A Neural Network Based Fingerprint Identification System," *Proc. The Tenth International Symposium on Computer and Information Sciences*, Izmir, Turkey, pp. 715-722.

[48] Rao, T.C. (1976), "Feature Extraction for Fingerprint Classification," *Pattern Recognition*, 8, pp. 181-192.

[49] Ratha, N., Chen, S., and Jain, A.K. (1995), "Adaptive Flow Orientation Based Feature Extraction in Fingerprint Images," *Pattern Recognition*, 28(11), pp. 1657-1672.

[50] Ratha, N.K., Karu, K., Shaoyun, C., and Jain, A.K. (1996), "A Real-Time Matching System for Large Fingerprint Databases," *IEEE Trans. Pattern. Analysis and Machine Intelligence*, 18(8), pp. 799-813.

[51] Roddy, A.R. (1997), "Fingerprint Features-Statistical Analysis and System Performance Estimates," *Proc. of the IEEE*, 85(9), pp. 1390-1420.

[52] Sherlock, B.G. and Monro, D.M. (1993), "A Model for Interpreting Fingerprint Topology," *Pattern Recognition*, 26, pp. 1047-1055.

[53] Sherlock, B.G., Monro, P.M., and Millard, K. (1992), "Algorithm for Enhancing Fingerprint Images," *Electronic Letters*, 8(18), pp. 1720-1721.

[54] Srinivasan, V.S. and Murthy, N.N. (1992), "Detection of Singularity Points in Fingerprint Images," *Pattern Recognition*, 25, pp. 139-153.

[55] Verma, M.R., Majumdar, A.K., and Chatterje, B. (1987), "Edge Detection in Fingerprints," *Pattern Recognition*, 20, pp. 513-523.

[56] Vizcaya, P.R. and Gerhardt, L.A. (1996), "A Non-Linear Orientation Model for Global Description of Fingerprints," *Pattern Recognition*, 29(7), pp. 1221-1231.

[57] Wilson, G.L., Candela, G., Grother, P.J., Watson, C.I., and Wilkinson, R.A. (1993), "Neural Network Fingerprint Classification," *J. Artificial Neural Networks*, 1(2), pp. 1-25.

[58] Xiao, Q. and Raafat, H. (1991), "Fingerprint Image Postprocessing: A Combined Statistical and Structural Approach," *Pattern Recognition*, 24, pp. 985-992.

Chapter 2:

Fingerprint Feature Processing Techniques and Poroscopy

FINGERPRINT FEATURE PROCESSING TECHNIQUES AND POROSCOPY

A.R. Roddy
U.S. Department of Defense
U.S.A.

J.D. Stosz
Motorola Arizona
U.S.A.

In this chapter, we explore methods of intelligent fingerprint processing for fingerprint minutia and pore feature extraction and matching, as well as fingerprint image segment extraction and correlation matching. We will also present the efficacy of using fingerprint pores in addition to minutiae in feature matching routines in order to improve the performance of fingerprint authentication systems. System parameters relating to the minutia and pore features and processing will be quantified in such a way that they can be used to predict the performance of a system that utilizes both minutiae and pores for feature matching. This chapter will conclude with our proposal of a testing plan for fingerprint authentication systems.

1 Introduction

In this chapter, we explore several techniques which may be considered intelligent techniques in fingerprint authentication. We detail general fingerprint preprocessing, extraction, and matching techniques, but we assert that to make these processes intelligent, a fingerprint quality determination should be added to the general preprocessing, extraction and matching routine.

The matching routine that we discuss in this chapter adopts a hierarchical approach; we show that one can use three different matching algorithms to determine if two fingerprints match. First, we apply a quality-driven correlation approach, which matches segments of the fingerprint image; then we employ fingerprint minutia and pore matching processes. In addition, we advocate the proper weighting of fingerprint features (minutiae and pores) depending on their quality, reliability and uniqueness.

Reliability and uniqueness of features are the two dominant parameters which contribute to false reject rates (FRRs) and false accept rates (FARs) in automated fingerprint authentication. Quantifying these parameters, among others, enables us to propose a probabilistic model to determine the performance of fingerprint authentication systems. We conducted a detailed statistical analysis of pores, and researched studies on minutiae in order to quantify uniqueness. In addition, we performed studies to analyze and quantify the reliability of fingerprint features. The results of these statistical analyses are discussed in this chapter, as well as the probabilistic model of performance.

While most methods for testing fingerprint authentication systems revolve around determining the FRR and FAR, these error rates can be misleading. In addition to factors such as algorithm performance and scanner characteristics, fingerprint quality of individuals and of database populations, as well as conditions inherent in certain applications (e.g., environment, posture, frequency of use, etc.) govern the performance of fingerprint authentication. At the end of this chapter, we propose a plan for testing fingerprint authentication systems which will enable comparison of systems by attempting to normalize the testing conditions.

Sections 2, 3, 4, and 5 of this chapter discuss the processing techniques involved in fingerprint authentication. The primary functions required for fingerprint authentication systems are image acquisition, preprocessing, quality analysis, feature extraction and selection, template creation (enrollment), matching (identification or verification), and decision making. This chapter does not discuss image acquisition, but focuses on all the steps that take place after the fingerprint image is acquired. Each of the steps listed previously, from preprocessing through decision making, is discussed in the context of using pores as well as minutiae and segment/correlation matching for fingerprint authentication. In Section 4, the three matching algorithms are discussed: correlation matching, minutia matching, and pore matching. Section 5, which discusses fingerprint matching, also provides the statistical analysis of fingerprint features, including the identification of common parameters involved in any automated fingerprint matching routine. In Section 6, we address the shortcomings of existing testing procedures for fingerprint authentication systems, and propose new procedures for testing fingerprint authentication systems.

2 Preprocessing

The purpose of this chapter is not to describe a specific fingerprint matching system; instead, we will present some of the many components which can be pieced together to make an intelligent and effective system. In this section, we describe fingerprint processing techniques which are used to simplify minutia and pore feature extraction and reduce the number of false features detected. Later, in Section 3, we will present intelligent correlation matching techniques.

Gray scale to binary image conversion, binary to skeleton conversion, and skeleton processing techniques are described below. These processing steps, which are adapted from [1], are outlined in Figure 1.

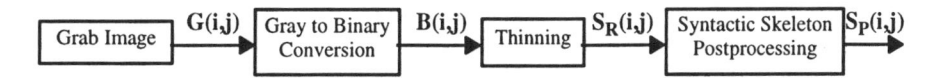

Figure 1. Image processing steps performed prior to feature extraction. The raw image, G, is processed to form the binary image, B, raw skeleton, S_R, and processed skeleton, S_P, which are used for extraction of fingerprint features including pores and minutiae.

2.1 Gray Scale to Binary Conversion

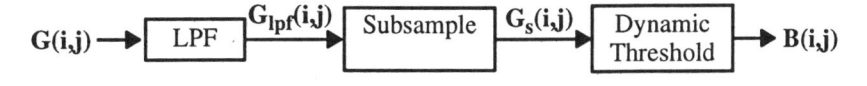

Figure 2. Gray scale to binary format conversion process.

A series of processing routines can be used to overcome undesired image degradation effects. (Image degradation will be discussed in detail in Section 3.3.1.) Initially, the gray scale image, G, is converted to binary format. Figure 2 is a block diagram representation of the gray to binary conversion process. First, G is low pass filtered. The resulting image, G_{lpf}, contains reduced noise at the cost of smoothing the high frequency regions of the print such as ridge/valley and ridge/pore transitions.

Next, if needed, the image is subsampled in the x-direction by the factor α to compensate for the difference in vertical and horizontal resolution introduced by the sensor:

$$G_s(i, j) = G_{lpf}(\alpha i, j) \quad \begin{aligned} &, \text{for } i = 0, 1, ..., X/\alpha \\ &, \text{for } j = 0, 1, ..., Y \end{aligned} \qquad (1)$$

where X and Y are the dimensions of the image. Finally, a dynamic thresholding routine, which can compensate for local lighting or finger pressure non-uniformity, converts the image to binary format, B. (See Figure 3).

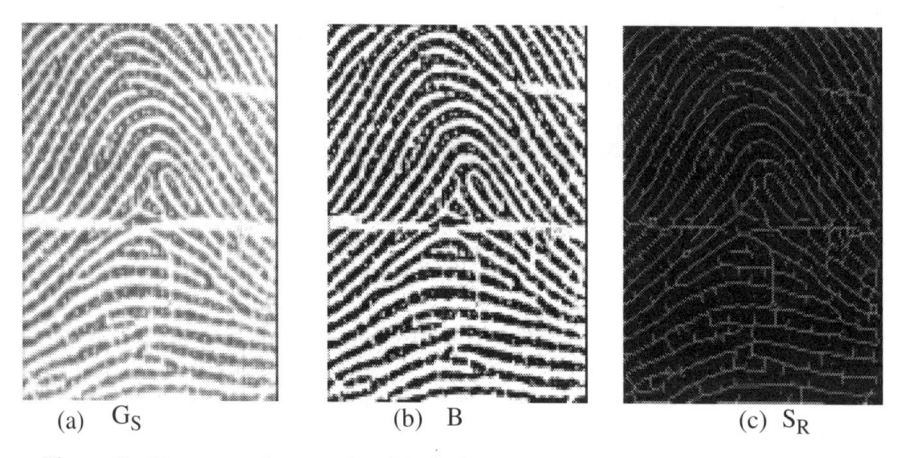

(a) G_S (b) B (c) S_R

Figure 3. Preprocessing results. (a) G_S is a smoothed, subsampled version of the original gray scale image, G, (b) B is the result of dynamic thresholding of G_S, (c) the result of thinning B to produce the raw skeleton image, S_R.

Adaptive thresholding is performed by segmenting G_S into X/A by Y/B subregions and then thresholding the subregions separately. X and Y are the dimensions of G_S, while $A \times B$ defines the size of the local analysis region. Thresholding begins with finding the mean pixel value, µ, within a subregion and then converting the pixels with gray level above µ, inside the subregion, to value one and all other pixels to value zero. Pixels with value one and zero are called white and black, respectively. The resulting binary image represents the original very well from a morphological perspective with the benefit of being much easier to process. Figure 3 shows examples of G_S and B.

$$\mu_{mn} = \left(\sum_{j=nB}^{(n+1)B-1} \sum_{i=mA}^{(m+1)A-1} G_S(i,j) \right) / (AB) \quad \begin{array}{l} \text{, for } m = 0, 1, ..., X/A \\ \text{, for } n = 0, 1, ..., Y/B \end{array} \quad (2)$$

2.2 Binary to Skeleton Processing

A skeleton image is produced by eroding the objects within a binary image until they are one pixel wide. It is important to maintain the object's connectivity and ensure that the topography of the original object is well represented by the skeleton. The advantage derived from using a skeleton image is that extraction of ridge features becomes a relatively straightforward procedure based on tracing line segments.

In the binary image, white areas represent both fingerprint valleys and pores while dark areas represent ridges. Our specific interest in the pores leads to the analysis, or thinning, of the white objects on the black background. An iterative, parallel thinning algorithm [3] is used for demonstration. The application of thinning rules to each pixel over a predetermined number of iterations results in the raw skeleton image, S_R, shown in Figure 3c. The neighbors of a central pixel, p_0, are defined in Figure 4, where $\Sigma(p_0)$ is the sum of nonzero neighbors of p_0 and $\tau(p_0)$ is the transition sum of the neighbors of p_0, which equals the number of $0 \Leftrightarrow 1$ transitions in the ordered set $p_1, p_2, p_3, p_4, p_5, p_6, p_7, p_8, p_1$.

p_8	p_7	p_6
p_1	p_0	p_5
p_2	p_3	p_4

Figure 4. The central pixel, p_0, and its neighbors.

The thinning rules are:

1) $1 < \Sigma(p_0) < 7$

2) $\tau(p_0) = 2$

3) $p_0 p_5 p_7 = 0$ or $\tau(p_7) \neq 2$

4) $p_3 p_5 p_7 = 0$ or $\tau(p_5) \neq 2$,

where the sum of the neighbors of element $S(p_0)$ is defined as:

$$\Sigma(p_0) = \sum_{n=1}^{8} S(p_n) \qquad (3)$$

and the transition sum of element $S(p_0)$ is:

$$\tau(p_0) = \sum_{n=1}^{9} S(p_n) - S(p_{(n+1)}) \qquad (4)$$

S is initially set equal to B, and after a sufficient number of thinning iterations, becomes S_R, the medial axis, raw skeleton representation of B. An ideal skeleton is exactly one pixel wide and centered within the valley or pore. From the skeleton, ends (ridge branch points) and branches (ridge end points) can easily be found by analyzing the neighboring pixels of a skeleton component. End points have only one neighbor and branch points have three neighbors, assuming the skeleton is 8-connected.

2.3 Skeleton Postprocessing

The raw skeleton image, S_R, may be severely degraded by artifacts resulting from pores or other sources. These degradations are substantially reduced by the skeleton postprocessing step shown in Figure 5. First, the quality of the skeleton is improved by analyzing and then extracting segments representing pores. This pore detection process will be discussed in detail in Section 3.1. Syntactic processing is used to "heal" undesirable artifacts arising from scars, wrinkles, oily or dry skin, etc., and transform the skeleton into a state from which valid minutia information can be extracted. Pore processing and cleaning are done independently of the healing routine.

Fingerprint Feature Processing Techniques and Poroscopy 43

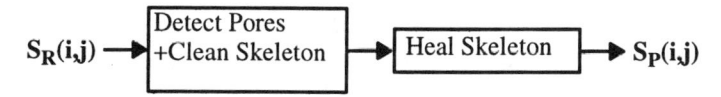

$S_R(i,j)$ → [Detect Pores +Clean Skeleton] → [Heal Skeleton] → $S_P(i,j)$

Figure 5. Syntactic skeleton processing methods are used to clean and heal the skeleton image as well as classify pores and minutia points. The resulting image is the processed skeleton, S_P.

Several commonly encountered skeleton structures are shown in Figure 6. Included in this figure are structures related to pores, minutia points, and undesirable conditions which can be corrected by skeleton postprocessing.

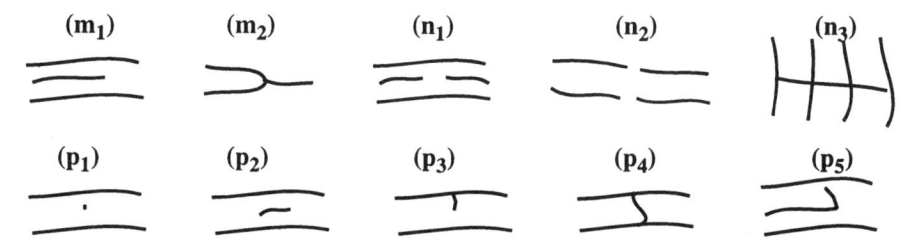

Figure 6. Basic Fingerprint Skeleton Structures. Note: skeleton line segments represent valleys and pores, not ridges; m_1 and m_2 are a valid ridge branch point and end point respectively; n_1-n_3 are 'problem' structures which look like valid minutiae; n_1 is a break in a valley which could be caused by dirt, noise, non-ideal thresholding, etc.; n_2 represents an area of the print in which valley information has been literally washed away – a typical effect caused by an excessively oily finger for which valleys are filled with moisture; n_3 represents a series of ridges with a wrinkle crossing through them; p_1-p_5 are typical artifacts due to the presence of pores; p_1 is a single pixel, p_2 is a short isolated line segment; p_3 is a short line segment connected to a valley; p_4 is a short line segment connecting two valleys, and p_5 is a short line segment appended to a valley end point.

2.4 Healing the Skeleton

Cleaning (removal of pores from the skeleton image) is based on the analysis of a single connected skeleton object. If an object conforms to a set of constraints, it is defined as a pore and then classified and removed. In contrast, healing requires a higher level of analysis, since generally, multiple neighboring skeleton segments are disrupted by a degradation process. These effects are illustrated by structures n_2 and n_3 in Figure 6. For example, in order to determine that structure n_3 is a result of a wrinkle, and that each component is not by itself a skeleton branch point or pore, the analysis of the whole disrupted local region is required. Similarly, healing of structure n_2 requires analysis of multiple skeleton segments in order to determine that the segments should actually be joined together.

Healing of the structures n_1, n_2, or n_3 is accomplished by finding the position and orientation of all branch and end points of the cleaned skeleton; see Figure 7b. In addition, for each skeleton minutia point, the position of the nearest several neighbors is determined. By analyzing this combination of information, a certain degree of healing of the skeleton can be done.

2.5 Healing Broken Skeleton Segments

Two disconnected skeleton line segments can be connected, or healed, if they satisfy a set of constraints:

1. The distance between end points e_0 and e_1 (see n_2 in Figure 6), must be less than a maximum distance threshold value.

2. The orientation of end points e_0 and e_1 is such that the line segments used to determine their orientation must be pointing at each other within a maximum difference angle threshold value.

If these two conditions are satisfied, the line segments are connected together by adding a straight line of pixels between points e_0 and e_1 in the skeleton.

2.6 Healing Wrinkles

Differentiating between the valid pore structure, p_4, of Figure 6 and the structure n_3, which represents a wrinkle, can be done by analyzing information on neighboring branch points. The horizontal line segments can be removed from structure n_3 if the following conditions exist:

1. A branch point, b_0, has at least two neighboring branch points, b_1 and b_2, which are each no further away than a maximum distance threshold value.

2. The orientation of one branch of b_1 must be 180 degrees, plus or minus a threshold maximum angle difference, different from the orientation of one branch of both b_0 and b_2.

3. These branches must also point at each other or be closely connected on common line segments.

3 Feature Extraction

Features are extracted in both the enrollment process and in the matching process of fingerprint authentication. Many of the feature extraction techniques which are discussed in the following sections can be considered preprocessing routines of the enrollment or matching process. However, we have chosen to discuss these routines in the context of extracting pores, minutiae, and regions of interest. The concepts of enrollment and matching processes will be discussed in later sections.

Section 3 discusses the extraction of minutiae and pores as identifying features, as well as segment extraction, where regions of interest in the image are extracted for correlation matching. Section 3 concludes with a discussion of factors which can impede successful feature extraction, such as image degradation.

3.1 Minutia and Pore Extraction Techniques

The techniques for extracting minutiae and pores are closely interwoven; thus, minutia extraction will be discussed in the context of pore extraction, and vice versa, in the sections to follow. The primary characteristics of both pores and minutiae are their locations in a given fingerprint. Minutia type and orientation, and the size and shape of pores are valid identifying characteristics, but they are secondary to location.

3.1.1 Cleaning – Pore Detection

The method used to extract pores as fingerprint features is critical to matching routines utilizing pores. The pore's position, size and shape are features making it distinct from other objects in an image. As mentioned in Section 2, pore detection is one of the steps in the "skeleton postprocessing" method (Figure 5).

The skeleton cleaning routine discussed in Section 2 determines which components of the skeleton represent pores. Once this information is known, the position of each pore is stored, and then all pixels related to the individual pores are pruned from the skeleton. The result is a list of pore locations and a simpler skeleton image, S_C. Figure 7a is an overview of the main steps involved in the cleaning process.

3.1.2 Determining End and Branch Point Locations

Pore detection can be facilitated by first determining and storing the location of all skeleton end and branch points. It should be noted that "skeleton end" and "skeleton branch" points are arbitrary terms that do not necessarily correspond to actual print minutia points due to degradations and noise. An end point is defined as any white pixel with either one or no neighbors, and a branch point has exactly three neighbors. All other skeleton components have exactly two neighbors and are referred to as connecting points.

There are N_E detected end points with locations denoted by $(x_{E,i}, y_{E,i})$ for $i = 0,1, ... N_E-1$, and N_B branch points with locations $(x_{B,i}, y_{B,i})$ for $i = 0,1, ... N_B-1$. The detected end and branch point locations are used to develop a minutia map in which pertinent minutia information can be stored for later use.

3.1.3 Skeleton Tracking and Pore Detection

Each end point is used as a starting location from which to track the skeleton. Tracking involves analyzing the current element, storing its position as (x_i, y_i) in the path variable P, and determining the location of the next element in the path. Tracking advances one element at a time until a stop condition occurs. The stop conditions for end point tracking are:

(1) another end point is detected, or
(2) a branch point is detected, or
(3) the path length, L_{Path}, exceeds a maximum allowed value.

Fingerprint Feature Processing Techniques and Poroscopy   47

Conditions (1) and (2) imply that the tracked segment is a pore. Under condition (1), the pore's position, (x_p, y_p), is defined by Equation 5 as the mean value of the coordinates of the elements forming the path of the tracked line segment.

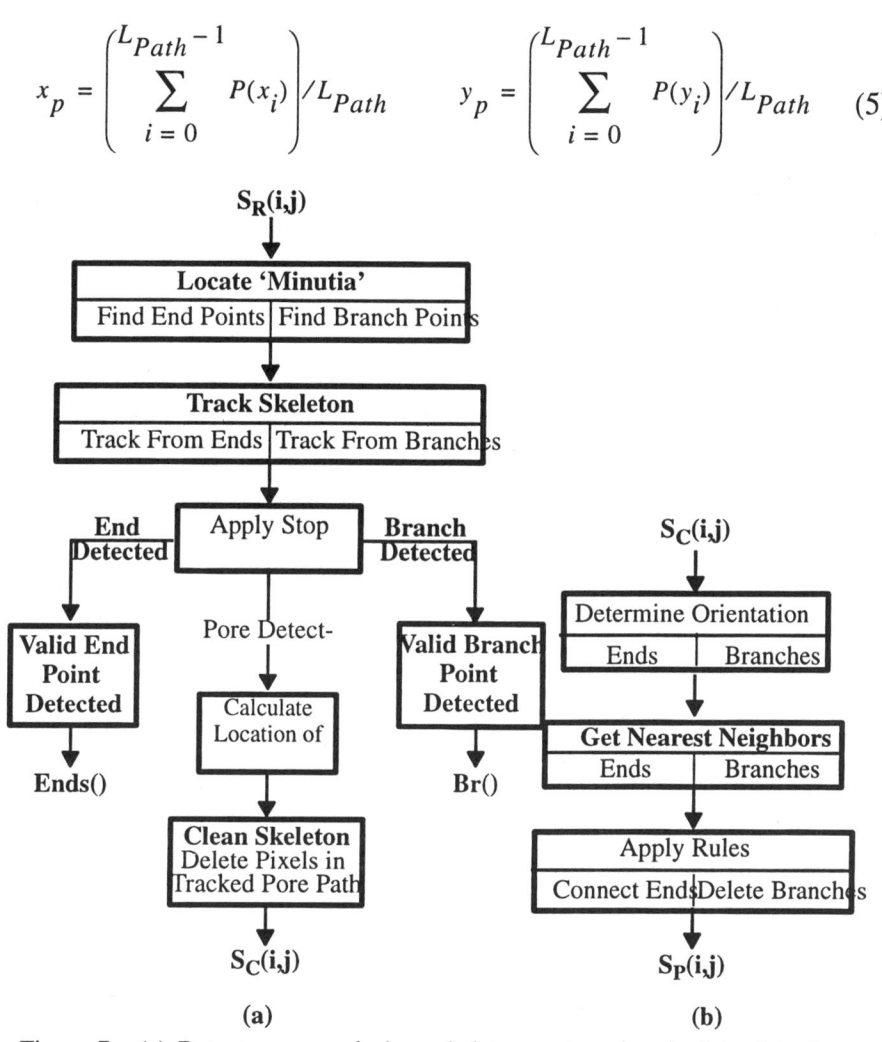

$$x_p = \left(\sum_{i=0}^{L_{Path}-1} P(x_i) \right) / L_{Path} \qquad y_p = \left(\sum_{i=0}^{L_{Path}-1} P(y_i) \right) / L_{Path} \qquad (5)$$

Figure 7. (a) Detect pores and clean skeleton – steps involved in detecting skeleton line segments relating to pores, classifying the pores, and removing pores which cleans the skeleton. (b) Heal skeleton – involves detection and correction of skeleton artifacts resulting from scars, wrinkles, or very moist fingers.

Under condition (2), the pore's position, (x, y), is defined by Equation 6 as the first element (starting end point) of the path.

$$x = P(x_0) \qquad y = P(y_0) \tag{6}$$

Condition (3) implies that the tracked segment represents a valid skeleton end point. In this case, the end point location is already known to be $P(x_0, y_0)$ and the orientation is defined to be:

$$\theta = \operatorname{atan}\left(\frac{y_p - P(y_0)}{x_p - P(x_0)}\right) \tag{7}$$

where x_p and y_p are the mean coordinate values of the elements in the tracked path as defined by Equation 5.

The tracking and cleaning procedure continues until all end points have been classified. Then, an analogous procedure of tracking starting at each branch point is done. The stop tracking conditions for branch point tracking are:

> (4) another branch point is detected, or
> (5) an end point is detected, or
> (6) the path length (of all tracked branches) is greater than the maximum allowed path length.

If condition (4) is satisfied, a pore is detected and its position is again the center of mass of the line segment connecting the two branch points. Condition (5) should have already been addressed by end point tracking. Condition (6) implies the presence of a valid skeleton branch point. The location is already known and the orientation is defined as the angle of the branch which is most separated in angle from the other two branches using Equation 7. Tracking, classification, and storage of end points, branch points, and pores produces both a minutia and a pore map which hold the information needed for feature extraction.

Note that if the fingerprint is of high quality then the reliability of end and branch point secondary features (such as orientation) may be sufficient for these features to be used in matching.

3.1.4 Pore Pruning and Minutia Classification

Once the location of a pore has been established, the elements contained in its path are pruned from the skeleton by converting them to

Fingerprint Feature Processing Techniques and Poroscopy 49

black pixels. At the same time, any branch or end points associated with the pore are deleted. Therefore, after all the cleaning has been done, the remaining branch and end points correspond to valid print features. Their location, type, and orientation represent the minutia map of the print. Similarly, the location of detected pores defines the pore map of the print. The results of the pore detection and cleaning process are demonstrated in Figure 8, which shows an improved, cleaned skeleton. End-point tracking enables the pruning of pores and classification of the structures m_1, p_1, p_2, p_3, and p_5 from Figure 6, corresponding to a valid end point and pores, respectively. Branch point tracking enables the pruning of pores and the classification of structures m_2 and p_4, from Figure 6, which are a valid branch point and pore, respectively.

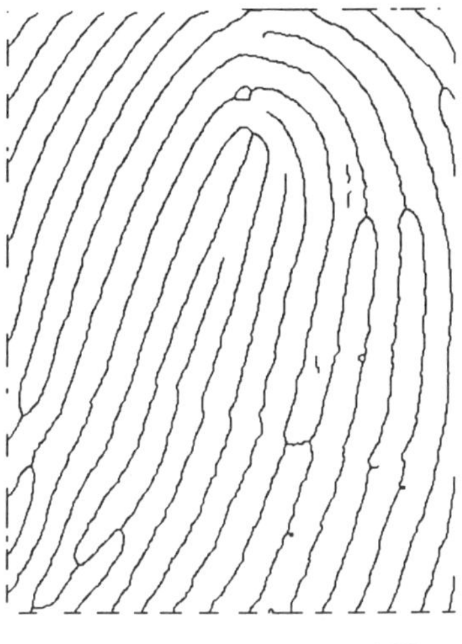

Figure 8. Pore detection and cleaning. The ability to reliably detect the position of pores and clean the skeleton is demonstrated here. Note: Pore detection and cleaning of the above images was based on end point tracking only – no branch point tracking was done.

3.2 Segment Extraction for Correlation Matching

For correlation based matching, it is impractical to use full images because of the effects of translation and rotation of the live-scan image relative to the enrolled print. In addition, plasticity of the finger tends to cause low correlation scores. A much better technique is to extract a set of small image segments and store these and their relative positions during enrollment. During matching, the segments are individually correlated against the full live-scan image resulting in a position and score at the best matching position. In fact, to allow for a more flexible algorithm, several positions and scores can be stored for each segment matched. Then, a verification decision can be made based on the relative position and maximum correlation score of the segments.

Segment extraction can either be performed manually or automatically. Only segments containing unique ridge structure should be selected. The segments should be large enough to include several ridges, but not large enough for plasticity of the finger to cause significant distortion. When incorporating correlation, minutiae, and pores into a matching algorithm, the segments can function as local origins from which to determine the position of the minutiae and pores. Automatic selection of segments can be done by performing auto- and cross-correlation analyses of potential segments. Segments which have high reliability (good autocorrelation with multiple images of the same print) and good distinction properties (cross-correlation analysis) should be selected.

The result of correlating a small segment with the live-scan image is both a maximum score and an associated position. The constellation of positions resulting from matching several segments is usually more effective for matching than the correlation scores. Efficient algorithms can be designed which rely on a combination of both the segment matching scores and positions. These algorithms can progressively match more segments as needed, depending on the quality of the print being matched.

For example, if after correlating three segments, the matching scores are very high, then identity can be confirmed. If the scores are somewhat lower but the relative locations of the segments match well, then a match results. Finally, if these conditions are not satisfied, then more segments can be correlated and another set of decisions made.

3.3 Poor Fingerprint Quality and Methods of Compensating

Large variations in the quality and the attributes of images of the same finger are common. Quality is defined by the amount of noise or degradation in the image, and attributes are defined as characteristics of the print such as wrinkles, scars, and even features such as pores. For example, finger dryness is a condition which leads to very noisy images, and oily or wet fingers produce prints in which the ridge flow is disrupted. Wrinkles are an examples of attributes which vary from image to image; they may be distinct in one image and not even visible in another image of the same finger. Fingerprint quality determines the ability to consistently extract features (for pore and minutia processing) and the amount of noise in a segment (for correlation matching). Fingerprint quality is closely related to the reliability of the fingerprint features. The following sections will discuss various quality problems in feature extraction (which also apply to fingerprint matching), such as image degradation, skin condition, and the physiological reliability of fingerprint features.

3.3.1 Image Degradation

Wrinkles, scars, and excessively worn prints result in images containing many "false minutia" structures. Also, non-ideal characteristics of an individual's finger, such as relatively oily or dry skin, can interfere with prism-based image acquisition and produce false minutia or significant "cloud-like" noise (see Figure 9). Finally, non-uniform image illumination, dirt, or latent prints on a prism produce unwanted effects and noise which must be removed in order to perform consistent feature extraction.

3.3.2 Skin Condition

The overall quality of the fingerprint depends greatly on the condition of the skin. Dry skin tends to cause inconsistent contact of the finger ridges with the scanner's platen surface, causing broken ridges and significant numbers of light pixels replacing ridge structure (dark pixels). For very oily or wet skin, the valleys tend to fill up with moisture, causing them to appear dark in the image similar to ridge structure (see Figures 10 and 11). Both of these conditions contribute noise and degradation to the image and cause an increased number of feature detection errors (and reduced performance). Figure 12 shows the expected distribution of skin condition over the general population. Based on experience, the number of users with dry skin outnumbers those with oily skin by a significant margin.

Wet finger

Dry finger

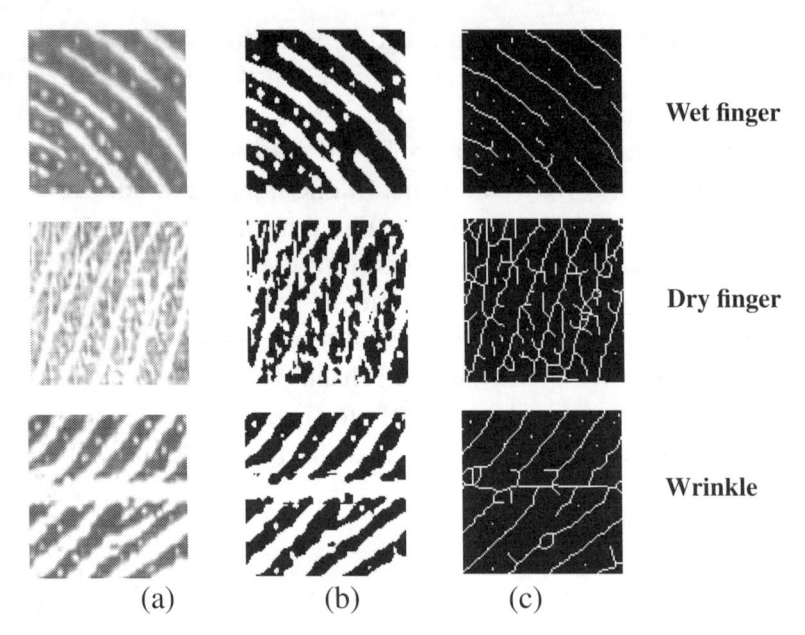

Wrinkle

(a) (b) (c)

Figure 9. Effects of various sources of degradation on the (a) gray scale, (b) binary, and (c) raw skeleton images. Wet and dry fingers suffer from randomly positioned "false minutia" skeleton objects. Wrinkles and scars create more structured patterns of false minutiae.

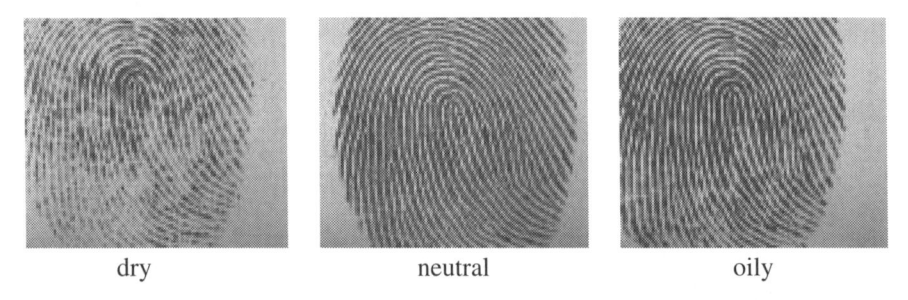

dry   neutral   oily

Figure 10. Variation in skin condition for the same finger.

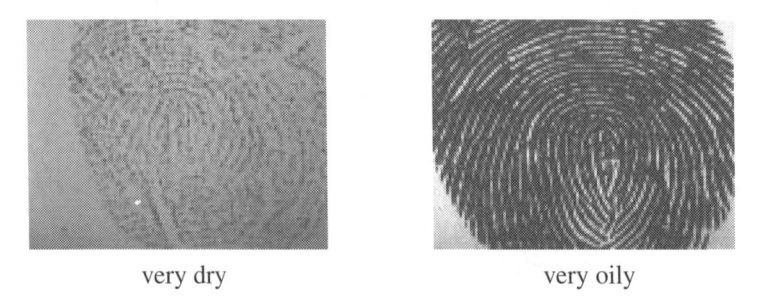

very dry   very oily

Figure 11. Indistinguishable ridge structure.

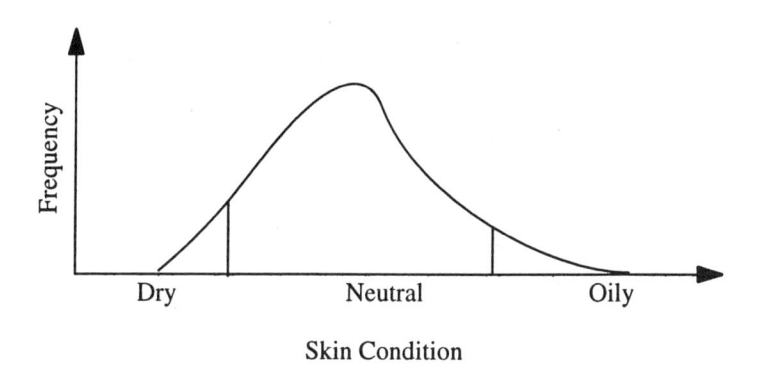

Figure 12. Expected distribution of skin condition over general population.

3.3.3 Quality as a Function of Position on the Finger

The quality of the fingerprint is a function of the position within the print as well as the amount of noise and degradation in the image. As seen in Figure 13, the fingerprint can be divided into a number of small segments which can then be rated for quality (in terms of skin condition). The higher the signal to noise ratio, the better the quality of the fingerprint. Very dry and very oily prints both have significant amounts of noise and degradation, and are of poor quality. Neutral and oily skin tend to produce the highest quality prints. A direct measurement of the signal to noise ratio for a fingerprint is not practical. The metric used to establish the quality of the print is the number of feature (minutia) detection errors. The number of missed detects (unreliable features) and false features detected can be determined by analysis of the results of minutia detection and rating routines run on various prints. The reliability of detection of a minutia is defined as the probability that a valid minutia point on the print is detected by the minutia extraction procedure. Falsely detected minutia correspond to detection of a minutia point where one does not actually exist. For a correlation-based matching scheme, a measure of the fingerprint quality may be determined through analysis of the distribution of matching scores for false reject type testing, or autocorrelation processing.

D	D	N	D	D	D
D	O	D	D	D	N
D	O	D	D	D	N
O	D	D	N	N	O
D	D	D	N	N	N
D	D	D	N	D	D

Figure 13. Print quality as a function of position on the fingerprint.
(D – Dry, N – Neutral, O – Oily)

Figure 14 illustrates the expected trend of increased detection errors as the quality of the fingerprint decreases. The link between skin condition and fingerprint quality is such that prints which are not too dry or too oily produce the best quality prints while excessively dry or oily skin produces very poor-quality prints. As seen in Figure 15, the quality rating of the print decreases as the skin tends towards being either dry or oily. Note that other factors can certainly influence the quality of the print such as worn or thin ridges, scars, wrinkles, and creases.

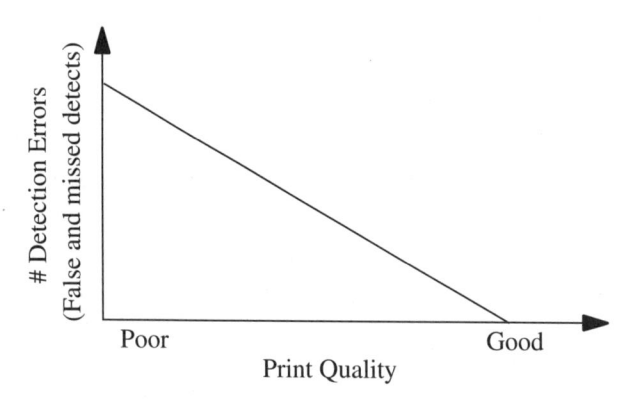

Figure 14. Expected trend of detection errors.

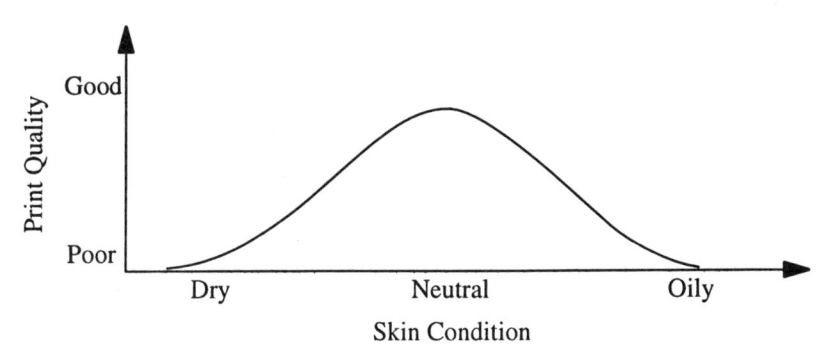

Figure 15. Print quality vs. skin condition.

3.3.4 Fingerprint Quality Determination

Optimally, an analysis of the quality and attributes of a print can be added to the preprocessing routine discussed in Section 2 in order to customize the preprocessing techniques, feature extraction, and matching algorithms in order to best manipulate a given print. In cases where print degradation, skin condition, wrinkles, and scars contribute noise to the print image, a front-end quality determination can decrease the number of errors in feature extraction and thus, matching. Figure 16 demonstrates where such a fingerprint quality determination could fit in the preprocessing routine discussed in Section 2.

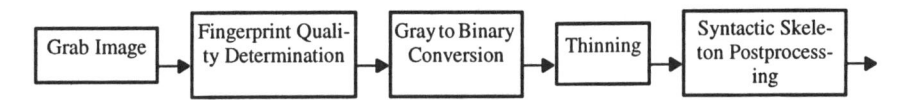

Figure 16. Fingerprint quality analysis can be added to the preprocessing routine discussed in Section 2. Note that, although this figure shows the fingerprint quality determination occurring at the front end of the preprocessing routine, it does not necessarily have to be a front-end process.

As discussed in the previous sections, a large variation in attributes and quality of samples of prints taken from a given finger exists. Also, a broad spectrum of print quality and attributes exists when all fingerprints from the entire population are considered. Thus, processing all prints in exactly the same way will not produce the best results. Experience has determined that certain processing techniques may be required for a given class of images and that an entirely different set of processing (or matching features – pores, minutiae, and correlation segments) techniques should be used for a different class of images.

Similarly, it is expected that the reliability of the fingerprint features such as ridge characteristics (minutiae) and the pores will vary depending on the type or class of print. Therefore, in addition to quality/type dependent processing, the matching routines used for various classes of prints should be specific to the quality of that class of prints. Later in this chapter, we will describe three independent matching techniques which are different based on their nature or in the features on which they depend. One technique involves correlation matching and the other techniques rely on minutia and pore features. Each of these techniques has advantages and disadvantages when all factors are considered; therefore, their use should depend on the front end print quality classification stage.

Two ways to determine fingerprint quality are described: (1) statistical and (2) skeleton processing based. First, a grid is effectively overlaid on the fingerprint image so that small segments can be classified according to their relative or absolute quality. Once quality is determined for each sub-segment of print, the results can be used either to weight or eliminate features from use in enrollment or matching, or even to determine the type of matching which will be performed (correlation, minutia, or pore based).

Moment (mean and variance) analysis of the gray-scale image is useful for distinguishing print quality. Relatively poor quality segments have different statistics from good quality prints. (A degree of quality can be assigned on a scale of 1 to 10, for instance). This scale factor can be used to weight features. The features may also be weighted according to two other factors – reliability and uniqueness. Note that segment (correlation) matching can also use weighting – good-quality segments can be weighted more. In fact, when performing automated extraction of segments, poor quality sections of the print may actually be eliminated so that they are not chosen or even processed further.

Analysis of the amount of debris resulting from skeleton processing is a good indication of fingerprint quality. The quantity of noise (small segments which wind up being "cleaned" or "healed") can be used to estimate the quality of the print (or segment of print). The quality of the segment can be directly related to the density of the "noise" objects.

During matching, low-quality sections of print, and the features contained within them, can be eliminated from further processing, feature extraction, or even matching.

4 Enrollment

The feature extraction processes and issues discussed in Section 3 are intrinsic to the enrollment process of fingerprint authentication, just as they are essential to the matching phase of authentication. In the enrollment process, the extracted features and/or segments are saved to a template file. This template file is, in essence, a vector representation of a user's fingerprint and contains pertinent authentication information. Figure 16 shows a block diagram of the enrollment procedure, which consists of fingerprint image acquisition, preprocessing, feature extraction, and creation of the user-specific template file.

Figure 17 gives a graphical description of the feature selection, extraction, and storage components of the enrollment procedure.

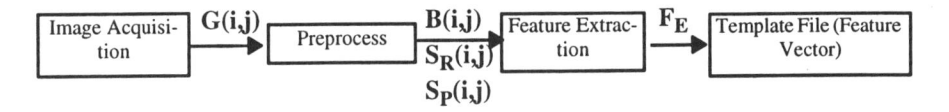

Image Acquisition $\xrightarrow{G(i,j)}$ Preprocess $\xrightarrow{\substack{B(i,j) \\ S_R(i,j) \\ S_P(i,j)}}$ Feature Extraction $\xrightarrow{F_E}$ Template File (Feature Vector)

Figure 17. Enrollment – main steps.

5 Fingerprint Matching

Verification is a form of authentication involving a one-to-one match between a live scan and the user's template. The verification steps are shown in Figure 19. A user provides a template name, T, to the system; the information is retrieved, and then matched against a live scan of the user's print. Verification may involve a comparison of the template's binary data segments, minutia characteristics, and/or detected pore features to the processed live scan image (see Figure 20a for details). Enrollment and verification require similar processing steps, including gray scale to binary and binary to skeleton conversion, quality analysis and skeleton processing.

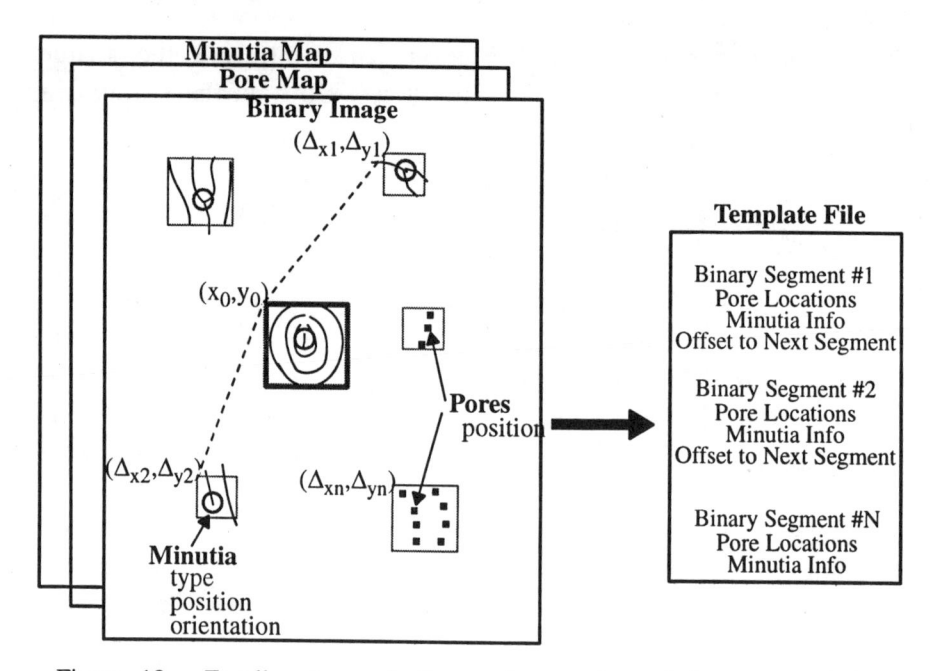

Figure 18. Enrollment – selection and extraction of features used for authentication. The template consists of three primary types of data: binary image segments, detected pore locations, and minutia information. As a results of preprocessing, the pore map and minutia map contain all of the pertinent information about pores and minutia points which is readily saved to the template file for a small region of interest once it is chosen.

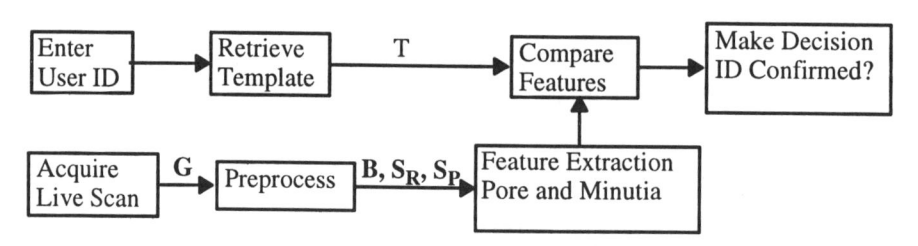

Figure 19. Main verification steps.

Matching can be done using correlation techniques, minutia point comparison, or using some combination of minutia, pore, and segment matching. Pore based matching will require either some minutia or segment extraction to create local points of reference from which the pore locations can be measured. Pore locations must be measured locally because there are many of them, and errors in identifying a given pore increase as the distance from the origin increases due to finger plasticity.

The matching algorithm can be made to rely on a quality rating of the live-scan print just as the feature extraction process (enrollment stage) can use quality ratings to improve performance. For instance, very poor quality segments of the fingerprint can be eliminated from further processing or matching. Also, the overall quality of the print can determine the preferred type of matching (and associated processing) which is to be used. For example, good-quality prints can rely on minutia matching alone; poor-quality prints may default to correlation-based matching; and prints which contain small high-quality segments may rely on a combination of minutia, pore, and correlation matching. Another method of matching might always perform minutia, pore, and segment matching and then analyze the resulting scores to determine the verification decision.

5.1 Correlation Matching

When performing correlation-based matching, a segment matching score, $S_{S,i}$, results for each image segment, i, used in the template. The image segments can be in either binary or gray-scale format. $S_{S,i}$ is weighted according to the quality rating of the segment and live-scan prints. An overall score, which combines the individual segment matching scores, is defined by (8) and used in the verification decision:

$$S_S = \left(\sum_{i=0}^{N_S - 1} S_{S,i} \right) / N_S \qquad (8)$$

where N_S is the number of image segments used.

S_S is normalized in the range of [0,1] with a score of 1.0 corresponding to a perfect pattern match. S_S represents an average measure of how well the template segments match the live-scan print and is independent of minutiae and only slightly dependent on the presence of pores within the segment.

The more powerful discriminator (than scores) resulting from correlation matching is the position information. When a number of small segments are allowed to translate across and down the image and the position of best match is recorded for each segment, the resulting constellation of points can be used to match prints. A verification decision is made based on how well the constellation resulting from registering the stored enrollment segments matches the actual relative locations of the segments determined at the time of enrollment. A score can be defined (S_L) based on the accuracy of the match between the enrolled and live scan information.

5.2 Multilevel Verification

In general, when using a combination of matching techniques, the result is a set of individual matching scores. A segment matching score, S_S, a segment position matching score, S_L, a minutia match score, S_M, and a pore match score, S_P. A user's identity is confirmed based on whether one or a combination of the following conditions is true:

$$S_S \geq T_S,$$
$$S_L \geq T_L,$$
$$S_P \geq T_P,$$
$$S_M \geq T_M,$$

where T_S, T_L, T_P, and T_M are the predefined threshold values for the segment (score and location), pore, and minutia matching scores respectively. Note that depending on the specific implementation of the algorithm, any of the above scores or a combination of the above scores can be used. By adjusting the threshold values, the error rates can be designed to match the desired performance of the system.

Fingerprint Feature Processing Techniques and Poroscopy

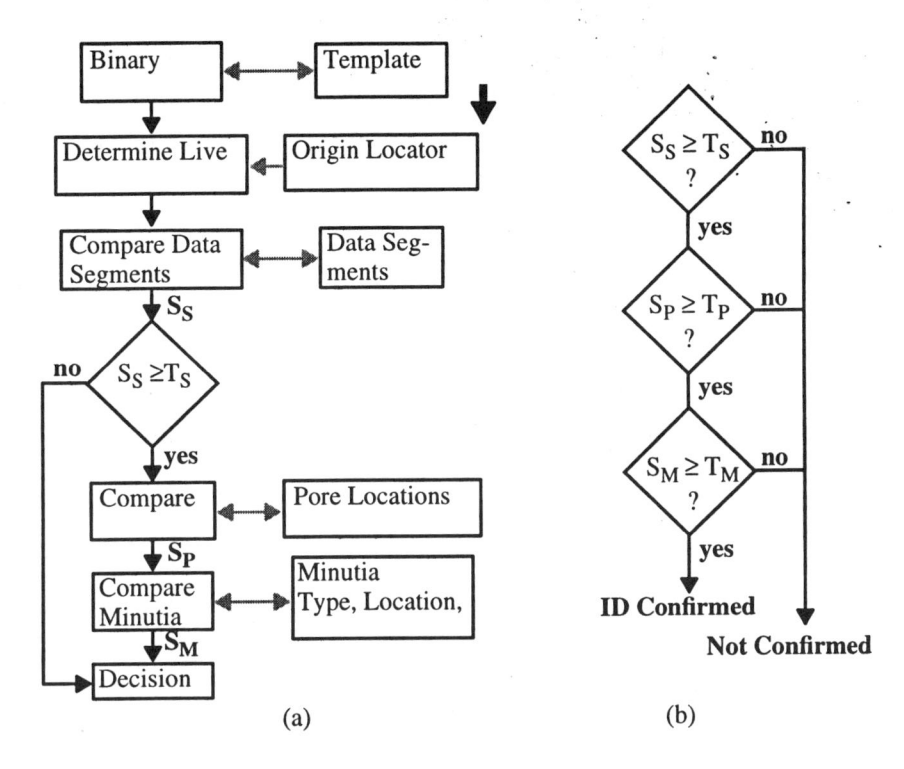

Figure 20. Multilevel verification – (a) main feature matching steps where live scan parameters and verification procedures are on the left side of the block diagram and stored template information needed for verification is located on the right side and (b) decision process involving data segment, pore, and minutia matching results.

5.3 Pore and Minutia Matching

The type, position, and orientation of minutia points and the location of pores (which are measured relative to minutia or correlation segments), are stored in the template file during enrollment. Preprocessing of the live-scan image results in the detection and classification of its minutia points and pores. The location of features from the same finger will be in alignment except for variations in position due to rotation and finger plasticity. The features can then be compared to check for a match.

The template features related to pores arise only from pores located close to image segments or minutia which were extracted from the binary image during enrollment and used as local origins. The first step in verification is alignment and scoring of these segments (or minutia) with the live-scan image. Once alignment has been done, pore matching consists of determining whether live-scan pores exist at specific positions stored in the template file. The pore-matching score, S_P, is the ratio of pores registered in the enrollment session to the number of pores confirmed during verification:

$$S_P = \left(\sum_{i=0}^{N_s-1} N_{MP,i}\right) / \left(\sum_{i=0}^{N_s-1} N_{P,i}\right) \qquad (9)$$

where $N_{P,i}$ is the number of pores detected in template segment, i, and $N_{MP,i}$ is the number of matching pores in segment i. Figure 21 provides an example of matching based on minutiae and pores for two segments from different fingers and also for two segments from the same finger.

The type, location, and orientation of the minutia points stored in the template can be compared to minutia information from the live scan. A individual minutia point match is considered to be made if the type and orientation of the template minutia matches a live-scan minutia in the correct location. Type matching is a binary decision (end point or bifurcation), whereas location and orientation allow for some variation. The constraint on location is simply that a minutia point from the live scan must be located within the image area corresponding to the template binary segment. The orientation constraint is that the detected minutia angle of rotation must be within 45° of the corresponding template minutia orientation value. The minutia matching score, S_M, is the ratio of correctly matched minutia points to the total number stored in the template file:

$$S_M = \left(\sum_{i=0}^{N_s-1} N_{MM,i}\right) / \left(\sum_{i=0}^{N_s-1} N_{M,i}\right) \qquad (10)$$

where $N_{M,i}$ is the number of minutia classified in template segment, i, and $N_{MM,i}$ is the number of matching minutia in the corresponding live scan region. Additionally, these features (and scores) can be weighted according to quality and uniqueness (value) of the individual features.

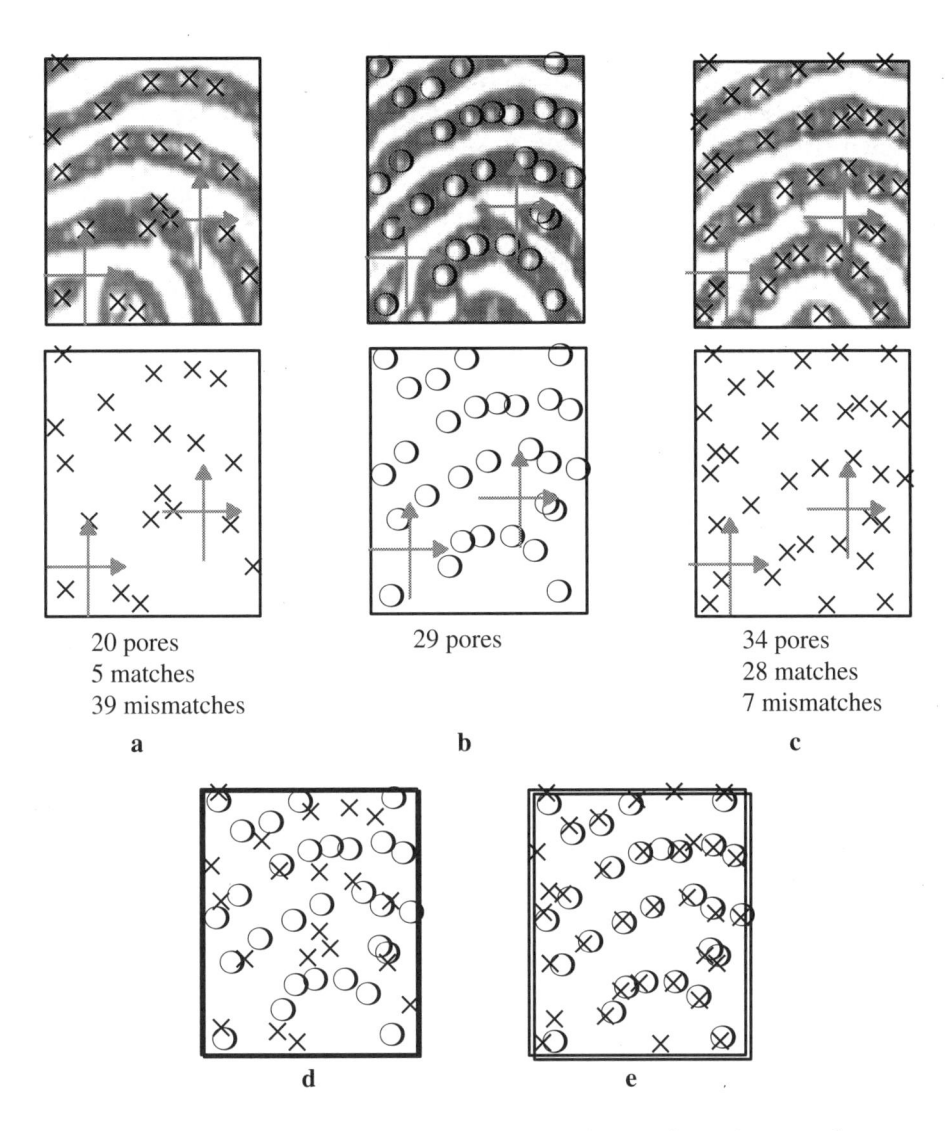

Fingerprint Feature Processing Techniques and Poroscopy

20 pores
5 matches
39 mismatches
a

29 pores
b

34 pores
28 matches
7 mismatches
c

d

e

Figure 21. Pore-based matching example: **b** and **c** are from the same finger and **a** is from a different finger. There are two very similar end-point minutiae in both print segments. In fact, it is likely that the different print segments would match based on minutia comparison alone. If the minutiae are used to align the prints, the pore information matches for the center and right images (**e**) but does not match for the center and left images (**d**).

5.4 Error Rates

Currently, the performance of fingerprint authentication systems is gauged mostly by error rate, which derive from (among other things) an algorithm's ability to match fingerprints correctly. Errors in a fingerprint authentication system can be one of two types. A false accept occurs when an unauthorized user is identified as an authorized user; a false reject occurs when an authorized user is not recognized as such. To describe the performance of a system, both the false accept rate (FAR) and false reject rate (FRR) must be determined.

5.5 Factors Affecting the Matching Routine and the Error Rates

5.5.1 Scanning Resolution

Some parameters become critical to the matching routine. For instance, the resolution at which the fingerprints are scanned determines the accuracy of feature location measurements. Inherently, there may be only one pore in a given 1 mm × 1 mm section of print, and at 1000 ppi (pixels per inch), this section is represented by approximately 40×40 (1600) pixels. In comparison, at a scanning resolution of 500 ppi, the same segment is represented by 20×20 (400) pixels. Therefore, the probability of another 1 mm^2 segment of print matching with respect to pore position is either 1/1600 or 1/400 depending on the scanning resolution. It can be seen that the false accept error rate will be reduced at a higher scanning resolution at the cost of an increased false reject error rate.

5.5.2 Feature or Search Area

The scanning-resolution issue can be made invariant by defining an absolute area to be associated with each feature (a feature area or search area). For instance, the location of a pore in the enrolled print segment may be determined to be (x,y), but in the corresponding live-scan segment, its location may be shifted some distance Δ as a result of rotation, plasticity, or other distortion. For the purpose of matching these two segments, a search area Σ, ($[x-\Delta x, x+\Delta x],[y-\Delta y, y+\Delta y]$) as seen in Figure 22, can be defined such that if the feature is within Σ, the features match with respect to position.

Fingerprint Feature Processing Techniques and Poroscopy

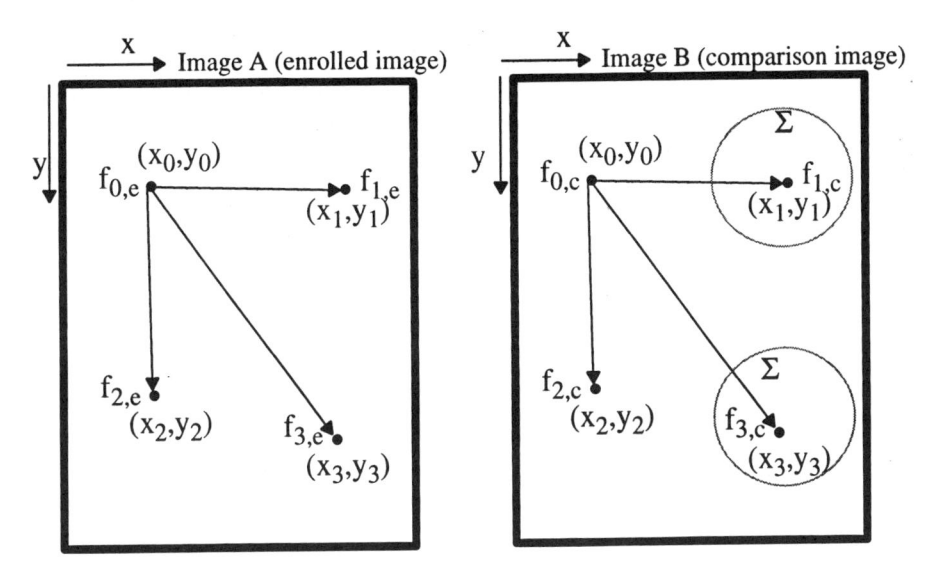

Figure 22. Basic matching principles. The concept of a search area represented by Σ is shown. Also, measurement is based on relative positions as opposed to relying on an absolute coordinate system. Feature $f_{0,e}$, which corresponds to $f_{0,c}$, is a feature (local origin) which is used to establish relative positions of other features.

The size of Σ is a parameter that influences the performance of the system (decreasing Σ produces a decreasing FAR and increasing FRR; increasing Σ produces increasing FAR, decreasing FRR). Practically, Σ should be large enough to account for effects such as plasticity of the finger and deviations in feature position due to variations in the data and effects of the processing algorithms, but not big enough for areas associated with distinct features to overlap. In a forensic comparison, plasticity and distortion of the finger are accounted for by human processing, but in an automated process, tolerances such as Σ must be incorporated to accommodate these inherent variations.

5.5.3 Plasticity, Scale, Translation, and Rotation

Plasticity of the finger is a trait which may cause substantial variation of the relative position and orientation of fingerprint features on a global scale. The degree of variation is relatively small when observing local regions of the print, which is the reason for using small image segments. Also, by allowing for variation in the size of the search area Σ for matching the segments, small changes in the relative position of the print features are overcome. When measuring the position of small high-density features such as pores, a local origin should be established. A minutia point can be used to establish a local origin.

The problem of image rotation must also be addressed. Overall image rotation can result from placement of the finger at an angle which differs from the enrolled print, or local regions can be rotated relative to one another due to the plasticity of the finger. Rotation causes a variation in the relative position and orientation of the print features. This problem can be compensated for by rotating the template's binary segments over the range of ±10°, in 2° increments, and then comparing each of the rotated states to the corresponding local area of the live-scan image. The state which matches best provides an estimated angle of rotation of the live-scan segment relative to the template segment. The angle of rotation is then used to modify the stored offset value which estimates the position of the next image segment. These modifications are intended to correct for uniform global print rotation. Once the relative angle of rotation between the stored template segments and the corresponding live-scan segments is known, the expected location and orientation of the minutia points and the location of the pores can also be modified.

5.5.4 Reliability

A critical factor when considering the performance of a fingerprint-matching system is reliability. Within the scope of this chapter, overall reliability is broken down into two components: inherent reliability and algorithm (processing) reliability. Inherent reliability refers to the physiological dependability of pores, which can be considered quantitatively to be the probability that a known pore will be visible in a particular live-scan print. Pores do not always appear on print images; factors such as temperature and skin condition can conspire to alter or suppress altogether the physical appearance of a given pore. (See Figure 23.) A multiple-fingerprint enrollment can be used to judge which features are reliable, and features can be weighted according to their reliability.

Fingerprint Feature Processing Techniques and Poroscopy 67

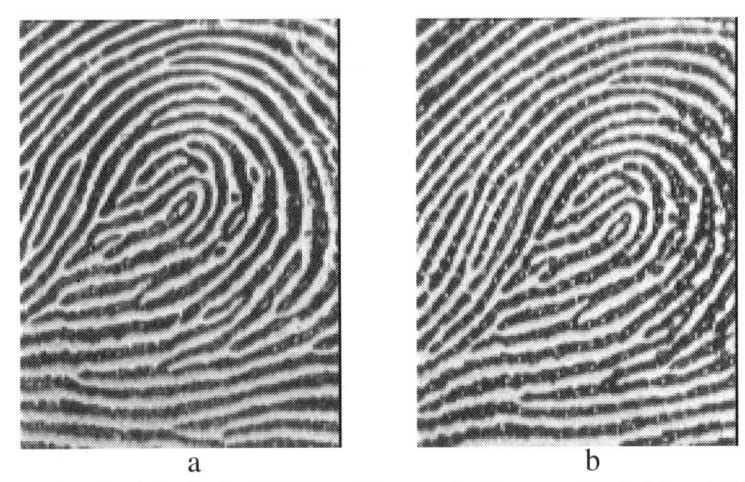

a b

Figure 23. Physiological reliability of fingerprint features. Both (a) and (b) are images from the same fingerprint; however, (a) shows that environmental factors have suppressed many of the pores evident in (b).

Algorithm, or processing, reliability must also be taken into account. Depending on the quality of the image, automated processing and detection algorithms make errors. There are two errors that the feature detection algorithm can make: a missed detect and an incorrect (false) detect. A missed detect occurs when a feature (pore or minutia) is discernible in an image, yet is not picked up by the detection algorithm. A false detect occurs when the algorithm mistakenly marks a feature, when in fact, no feature is present. The degree of noise and degradation in the image influences the quantity of errors. The probability of incorrect detection, p_{fd}, and missed detection, p_{md}, are parameters on which the performance of the system depends. A high p_{fd} or p_{md} will tend to increase the FRR but have little effect on the FAR.

6 Statistics of Features and Performance

Now that the different processes inherent in fingerprint authentication have been discussed, it is useful to consider the statistics of fingerprint features and to tie statistics and processes together to construct a model that will determine system performance. Section 6 discusses studies which we performed [4] to determine statistics of fingerprint features. Some of the statistics were derived from models, and some were derived from real data. Section 6.1 is devoted to determining uniqueness estimates which are related to the false accept error rate. Section 6.2 will discuss how reliability affects the false reject rate in fingerprint authentication. Section 6.3 will propose a probabilistic model of system performance that will tie together uniqueness, reliability, and the concepts discussed in the previous sections.

6.1 Uniqueness of Features and Configurations

The uniqueness of a set of minutiae or pores is defined as the probability of occurrence of the set. Therefore, the probability of an impostor match is directly related to the uniqueness. The more unique a set of features, the less likely an impostor is to match the set. Obviously, increasing the number of features used to represent the print will increase the uniqueness of the feature set, but the frequency of occurrence of particular arrangement of features will vary also.

A key issue in evaluating the uniqueness of a set of features is whether they are independent. Another issue is the feature distribution, for features may be regularly spaced, or the feature density (features per mm^2) may be non-uniform over the fingerprint. Addressing these issues is essential in determining the theoretical false accept rate of the system. For example, a particular combination of minutiae (such as the two end points of a ridge island) may occur frequently. In this case, an imposter is more likely to match both minutia points because of their close association, or dependence, than he/she would be to match two independent features. Therefore, features that exhibit dependence are less valuable than independent features.

6.1.1 Minutia Configurations

In this section, the methods used by Osterburg, et al. [5], to determine the uniqueness of a set or configuration of minutiae will be reviewed. The *P(configuration of features)* defines the uniqueness of a configuration of minutiae, in a given area of print, which is equivalent to the probability of two different print segments matching. Osterburg's results can be used to estimate the FAR of a minutia-based matching technique, although his model does not include some very important parameters inherent to automated-matching systems, such as feature reliability, detection errors, and search area variability. Therefore, the

Osterburg model is just a starting point for determining the theoretical performance (which includes both FAR and FRR estimates) of practical fingerprint matching systems.

By examining 1 mm × 1 mm segments, or cells, of fingerprints (see Figure 24 for a perspective on scales used in fingerprint processing), Osterburg determined the frequency of occurrence of 13 possible outcomes based on Galton fingerprint features. The results are provided in Table 1. The set of Galton features includes ridge endings, bifurcations, islands, dots, bridges, spurs, enclosures, double bifurcations, deltas, and trifurcations. Osterburg included three other outcomes: a broken ridge, an empty cell, or some other multiple combination of features.

The underlying assumption made by Osterburg is that the content of each cell is a random variable which is independent of all other cells. The implication is that any configuration of the same set of features has the same probability of occurrence meaning, for instance, that a tightly clustered pack of minutia is just as likely as the same set of minutiae being distributed uniformly over the print. Although the Osterburg study gives meaningful results, empirically the independence assumption is not valid because some configurations of Galton features are much less likely than others.

Table 1. Osterburg probability of feature occurrence.

Feature	Frequency	Probability of Occurrence
Empty cell	6584	0.766
End point	715	0.0832
Branch point	328	0.0382
Island	152	0.0177
Bridge	105	0.0122
Spur	64	0.00745
Dot	130	0.0151
Lake	55	0.00640
Trifurcation	5	0.000582
Double bifurcation	12	0.00140
Delta	17	0.00198
Broken ridge	119	0.0139
Other multiple occurrences	305	0.0355
Total	8591	1.00

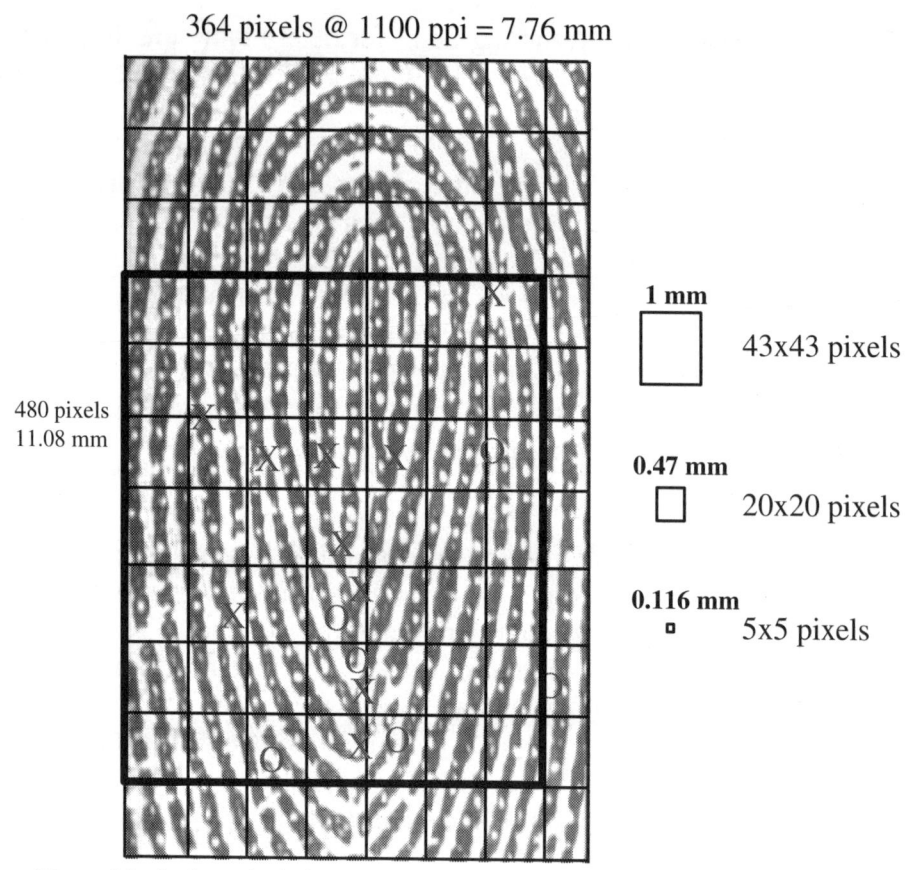

Figure 24. Scale and relative sizes used for fingerprint analysis. Inside the broad outline, the minutia end and branch points are marked with Os and Xs, respectively.

Based on the independence assumption, the individual feature probabilities are combined to yield the probability of a feature configuration:

$$P(\text{configuration of Galton features}) = p_0^{k_0} p_1^{k_1} \ldots p_{12}^{k_{12}} \qquad (11)$$

where p_i (for $i = 0, \ldots, 12$) is the probability that a given type of Galton feature, i, will occur in a cell, and k_i is the number of cells in which the feature occurs. The k_is sum to N, the number of cells.

As an example, the 7 mm × 7 mm block of data inside the broad box shown in Figure 24 has 4 cells containing one end point only, 8 cells containing one branch point only, 2 cells containing both a branch and an end point, and 35 empty cells (where only branch and end points have been considered for simplicity). Therefore, the probability of this configuration of 10 bifurcations and 6 end points (16 minutiae) in 49 mm² of print is:

$$P(\text{configuration}) = (0.0832)^4(0.0382)^8(0.0355)^2(0.8431)^{35} = 6.97 \times 10^{-22}.$$

Figure 25 demonstrates the need to expand the Osterburg model in order to determine practical values for the probability of matching. In Figure 25, **a** and **b** are two print segments that have the same set of features, but different configurations. These two prints would not match but would have the same probability of occurrence under the auspices of Osterburg model:

$$P(1^{st} \text{ configuration}) = P(2^{nd} \text{ configuration}) =$$
$$P(\text{end point})^{\text{\# cells with an end point}} P(\text{empty cell})^{\text{\# empty cells}} =$$
$$(0.0832)^4(0.766)^{21} = 1.78 \times 10^{-7}.$$

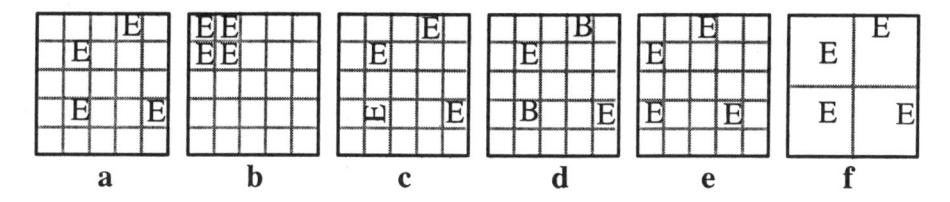

Figure 25. Osterburg minutia model and matching issues. In the Osterburg model, any configuration of the same set of features has the same probability of occurrence. The probability of configuration **a** or **b**, 4 end points and 12 blank grids in an area of 25 mm², is 1.78×10^{-7}. In a configuration, the relative position of the features is known as well as the type and orientation of the features. Note that the orientation of the minutia points is not the same in configurations **a** and **c** and although the configurations have the same probability of occurrence, they do not match. A common matching technique is to discard the type information; therefore, configuration **a** matches **d**, but the two have different probabilities of occurrence. Depending on the defined origin or reference point of the print, configurations **a** and **e** may match. Finally, the resolution at which feature locations are measured (or the dergee of allowed deviation in detected feature position) means that configurations **a** and **f** have a different probability of occurrence even though the minutiae are exactly the same (configuration uniqueness is a function of resolution).

The Osterburg model assumes that the 4 end points in this example can occur with equal probability within any 4 of the 25 cells. Thus, there are

$\binom{25}{4}$ = 12650 different configurations possible for the set of 4 end points in those 25 cells, each having the same probability of occurrence.

A comparison of **a** and **c** (in Figure 25) shows that although the feature types and locations match, the orientation of each feature is also crucial to the uniqueness of a configuration. The orientation is determined by the ridge flow direction, and the number of different orientations, N_O, is arbitrary (8 for instance). Whereas the probability of both configurations is the same under the Osterburg model, the different orientations actually distinguish the two configurations, thus:

P(configuration with orientation) =

P(configuration without regard to orientation) $\times (1/N_O)^{N_P}$ (12)

where N_P is the number of points in the configuration for which the orientation is determined.

In Figure 25, **d** demonstrates a situation where the location of all the features match **a**, yet the feature types do not match. These two configurations have different probabilities and do not match under the Osterburg model but will match for a routine which keys on geometry but is invariant with respect to type. The effect of translation is noted in **a** and **e**. Even though these configurations are different by definition, they may match depending on the choice of the reference point used to measure the minutia locations. Finally, the effect of the measurement accuracy (or resolution) can be seen comparing **a** with **f**. Even though the relative minutia positions and orientations are exactly the same, the cell size (feature area) is different resulting in a different probability of occurrence for the two configurations.

6.1.2 Pore Configurations

6.1.2.1 Pore Distribution

In order to describe the spatial regularity of pores, a deterministic model for pore distribution is proposed in which neighboring pores are separated by a constant distance d and are arranged in a lattice formation. This d represents the average distance between neighboring pores, where it is assumed that the pores are located in the center of a region containing only one sweat gland. The value of d is thus calculated from a live-scan image as:

$$d = (\text{area of ridges/number of pores})^{1/2} \quad (13)$$

For this extreme case, matching two images consists of simply lining up any pore on both images, since the rest of the pores would align themselves accordingly. The remaining pores in the image contribute no additional information; that is,

$$P(\text{all pores match}) = P(\text{one pore matches}).$$

The model is made stochastic by assuming that each pore position can deviate by a small random amount. The methods used in the next several sections to model intra-ridge pores are closely tied to the lattice model. The intra-ridge models represent a single ridge, whereas the lattice model can be thought of as being composed of multiple independent ridges covering an area.

6.1.2.2 Distribution of Distances Between Sequential Intra-Ridge Pores

In this section, a procedure for estimating the uniqueness of a sequence of pores based on measurements of real fingerprint data is summarized. To accomplish this task, pore locations along the ridges of live-scan prints were detected manually. Then, the distance between successive intra-ridge pores was calculated for individual prints. The plots of intra-ridge distance for individual fingerprints generally produced bimodal distributions with a dominant peak and an inferior peak resulting from missing or skipped pores. However, when all the data were combined, 3748 distance measurements resulted in a smoothed single mode distribution ($\mu_R = 0.377$ mm (16.955 pixels), $\sigma_R = 0.1820$ mm (8.1680 pixels)) with a significant upper tail, as seen in Figure 26.

The tail of the distribution begins at about 0.69 mm (30 pixels), and tapers off at 2.40 mm (104 pixels), the maximum observed distance between intra-ridge pores. The frequency value peaked at 0.30 mm (13 pixels), with a probability of occurrence of 0.0645 (at a measurement resolution of 1 pixel at 1100 ppi). This distance is defined as d_R, and its probability of occurrence is $P(d_R)$. In Figure 26, there are 104 bins represented, which is the maximum observed spacing between pores, meaning that each bin is the size of a pixel.

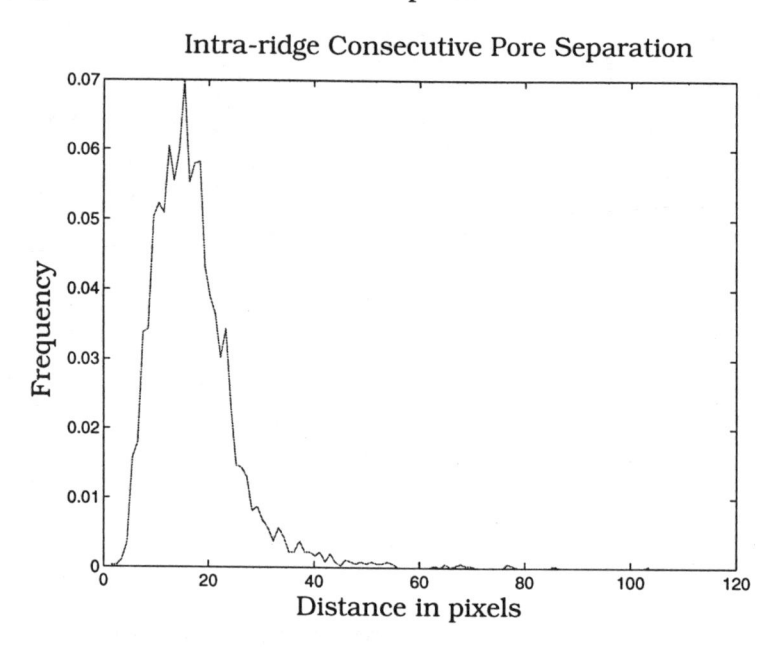

Figure 26. Plot of consecutive intra-ridge pore separation. The most frequent occurring separation is 13 pixels (0.3 mm).

Given this distribution, the probability of occurrence of any sequence of intra-ridge pores can be calculated by assuming the pores are independent. For this case, the only parameter of interest is the distance between consecutive pores. In addition, a lower bound for the uniqueness can be calculated by assuming that all the pores in a sequence are spaced by d_R (the most likely separation). Then any sequence of the same number of pores is guaranteed to be at least as unique as this bound.

Table 2 summarizes the uniqueness of sequences of intra-ridge pores with varying resolution. In this table, the results depend on the number of pores in the sequence and also on the measurement accuracy, or resolution, of a pore's position. As the measurement accuracy is decreased (fewer bins in the histogram), there are fewer distances possible between pores but the area under the distance probability density function remains constant, resulting in distances with higher probabilities. Therefore, d_R may not change but $P(d_R)$ will increase. The new values for $P(d_R)$ were determined by accumulating area of the normalized histogram around d_R symmetrically. From this table, with a resolution setting of $r=3$, the upper bound on the probability of occurrence is:

$$P_{\text{measured}}(\text{a sequence of 20 intra-ridge pores}) = 0.201^{20} = 1.16 \times 10^{-14}$$

for a sequence of 20 pores.

Table 2. Intra-ridge pore distances with varying resolution.

# pore	$r=1$ pix	$r=2$ pix	$r=3$ pix	$r=4$ pix	$r=5$ pix	$r=6$ pix	$r=7$ pix
	probability= (max probability at r) number of pores						
2	0.0645	0.1246	0.1848	0.2444	0.2879	0.3450	0.3839
3	0.0042	0.0155	0.0341	0.0597	0.0829	0.1190	0.1474
4	2.6821e-04	0.0019	0.0063	0.0146	0.0239	0.0411	0.0566
5	1.7297e-05	2.4123e-04	0.0012	0.0036	0.0069	0.0142	0.0217
6	1.1155e-06	3.0064e-05	2.1531e-04	8.7115e-04	0.0020	0.0049	0.0083
7	7.1936e-08	3.7467e-06	3.9781e-05	2.1287e-04	5.6934e-04	0.0017	0.0032
8	4.6391e-09	4.6694e-07	7.3500e-06	5.2016e-05	1.6391e-04	5.8215e-04	0.0012
9	2.9918e-10	5.8192e-08	1.3580e-06	1.2710e-05	4.7187e-05	2.0086e-04	4.7223e-04
10	1.9294e-11	7.2523e-09	2.5091e-07	3.1058e-06	1.3585e-05	6.9304e-05	1.8131e-04
11	1.2443e-12	9.0382e-10	4.6358e-08	7.5891e-07	3.9110e-06	2.3912e-05	6.9614e-05
12	8.0242e-14	1.1264e-10	8.5652e-09	1.8544e-07	1.1259e-06	8.2505e-06	2.6728e-05
13	5.1748e-15	1.4038e-11	1.5825e-09	4.5314e-08	3.2415e-07	2.8467e-06	1.0262e-05
14	3.3372e-16	1.7495e-12	2.9239e-10	1.1073e-08	9.3319e-08	9.8221e-07	3.9401e-06
15	2.1522e-17	2.1803e-13	5.4022e-11	2.7056e-09	2.6866e-08	3.3890e-07	1.5128e-06
16	1.3879e-18	2.7172e-14	9.9813e-12	6.6113e-10	7.7344e-09	1.1693e-07	5.8082e-07
17	8.9507e-20	3.3863e-15	1.8442e-12	1.6155e-10	2.2267e-09	4.0345e-08	2.2301e-07
18	5.7723e-21	4.2203e-16	3.4073e-13	3.9475e-11	6.4103e-10	1.3920e-08	8.5622e-08
19	3.7225e-22	5.2595e-17	6.2954e-14	9.6459e-12	1.8455e-10	4.8030e-09	3.2874e-08
20	**2.4007e-23**	**6.5547e-18**	**1.1632e-14**	**2.3570e-12**	**5.3130e-11**	**1.6572e-09**	**1.2622e-08**
21	1.5482e-24	8.1689e-19	2.1491e-15	5.7594e-13	1.5296e-11	5.7179e-10	4.8461e-09
22	9.9841e-26	1.0181e-19	3.9706e-16	1.4073e-13	4.4035e-12	1.9729e-10	1.8606e-09
23	6.4388e-27	1.2688e-20	7.3362e-17	3.4389e-14	1.2677e-12	6.8071e-11	7.1439e-10
24	4.1523e-28	1.5812e-21	1.3555e-17	8.4030e-15	3.6497e-13	2.3487e-11	2.7429e-10
25	2.6778e-29	1.9706e-22	2.5044e-18	2.0533e-15	1.0507e-13	8.1037e-12	1.0531e-10
26	1.7269e-30	2.4559e-23	4.6271e-19	5.0174e-16	3.0249e-14	2.7961e-12	4.0434e-11
27	1.1137e-31	3.0606e-24	8.5492e-20	1.2260e-16	8.7084e-15	9.6474e-13	1.5524e-11
28	7.1822e-33	3.8143e-25	1.5796e-20	2.9958e-17	2.5071e-15	3.3287e-13	5.9605e-12
29	4.6318e-34	4.7537e-26	2.9184e-21	7.3203e-18	7.2176e-16	1.1485e-13	2.2885e-12
30	2.9870e-35	5.9243e-27	5.3921e-22	1.7888e-18	2.0779e-16	3.9627e-14	8.7866e-13

A setting of 3 for the resolution parameter, r, is reasonable based on the fact that a typical pore has a diameter (assuming a circular shape) of about 5 pixels (115.5 µm) and the adjusted average spacing (compensating for skipped pores) between intra-ridge pores is 13 pixels (300.3 µm). Thus, setting r to 3 is equivalent to an allowed displacement of size 3 pixels (69.3 µm) in which to detect the pore. Therefore, the pore's position can vary slightly from its expected value, but too much variation will cause a mismatch. In addition, the "search area" is only big enough so that one pore is likely to be present and an adjacent pore is unlikely to overlap in this area.

6.1.2.3 Intra-Ridge Pore Distribution from Models

Assume that an individual pore occupies a unit area, or cell. Unit cells are bounded on the edge by the ridge border, and their length is defined as the average distance between intra-ridge pores. Unit cells are not a physiological feature; they are simply defined as the average area of print associated with a single pore. The location of the pore within a unit cell is a random variable and therefore, a sequence of intra-ridge pores is represented by a random vector for which the elements are independent and identically distributed. In addition, there is a finite probability of a pore not occurring in a unit cell (an empty unit cell). This condition is used to explain the occurrence of relatively long segments of ridge which have no pores.

As shown in Figure 27, the unit cells are divided into 9 sub-regions, $A_i,...,I_i$, for a given unit cell i. For this analysis, two models will be examined. In the first model, Model 1, the position of the pore is uniformly randomly distributed over the entire unit cell. This assumption eliminates the need for the 9 sub-regions. In the second model, Model 2, the probability that a pore occurs in the center of the unit cell, or ridge, is greater than that for the edge of the ridge. For this case, the pore can be located anywhere within the sub-division of the unit cell with equal probability.

The distribution of distances between consecutive intra-ridge pores in a sequence can be simulated using these two models and the results compared to the distribution obtained using the actual measured data.

6.1.2.4 Model 1

Model 1 assumes that pores are randomly located within a unit cell with rectangular dimensions. The width of these unit cells corresponds to the average width of a ridge and the length is the average adjusted spacing between intra-ridge pores (derived from measured data). In addition, the provision of a unit cell with no pore present (skipped pores) in Model 1 is used to account for the long tail of the distribution of the real data.

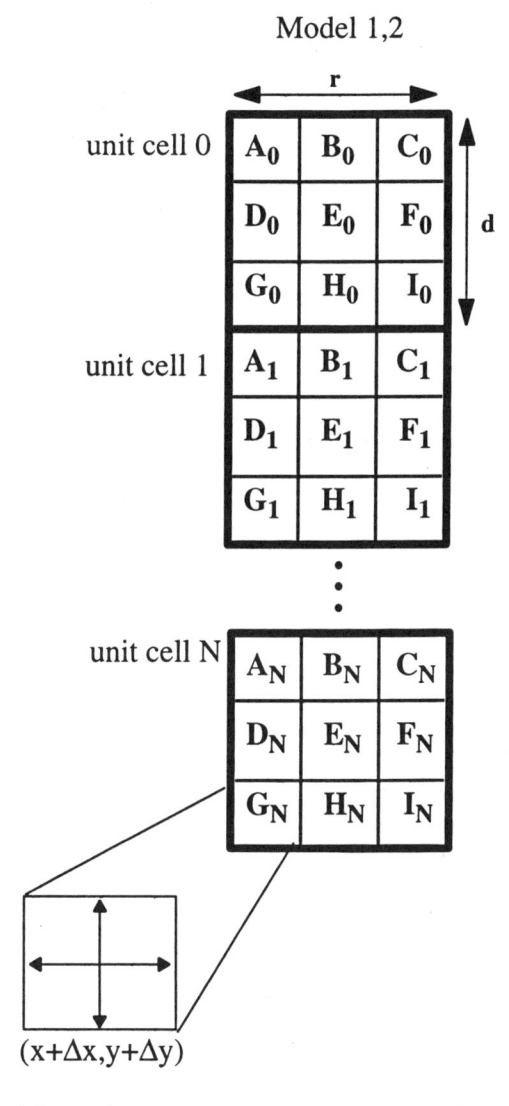

Figure 27. Models used to represent pore occurrence along a single ridge. These models can be used to predict the probability of occurrence of a sequence of N pores on the ridge, where **d** is the average distance between intra-ridge pores and **r** is the average ridge width.

The probability of a skipped pore is estimated using the distribution of distances derived from measured data and is approximated as 8.3%. Finally, Model 1 limits the minimum possible spacing between pores so that simulated pore positions are not unreasonably close together. The distribution of a set of simulated pore positions based on Model 1 is shown in Figure 28a and is very similar to the distribution measured from real data.

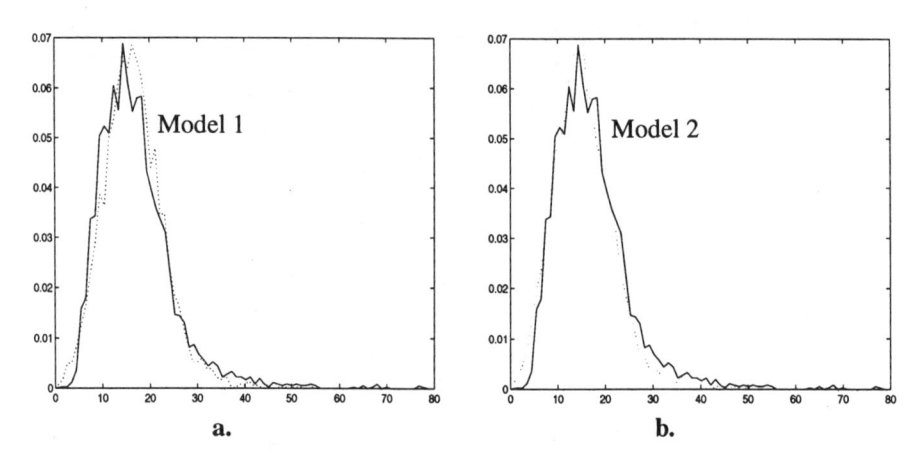

Figure 28. Distribution of intra-ridge pores. The measured distribution of spacing between consecutive intra-ridge pores is shown here as a solid line in both plots ($\mu = 16.96$, $\sigma = 8.16$) (where distance is measured in pixels). Model 1 is represented by the dotted graph in **a**. Model 1 also has a provision for missing pores, which can be used to account for the tail of the real data distribution as shown in the solid curve in **a**. For Model 1, the unit cell size was 21 pixels wide by 13 pixels long, the fraction of skipped pores was 0.083, and the minimum allowed pore separation was 1 pixel ($\mu = 16.7$ and $\sigma = 6.5$ for Model 1 with 50000 samples). Graph **a** shows the close fit between the real data and the Model 1 simulation. The second model, Model 2, includes the ability to force pores to be located on the center of the ridge with a higher probability than on the edges. With the probability of a pore occurring in the middle 1/3 of the ridge increased, the distribution is plotted in **b** along with the real data distribution for comparison.

6.1.2.5 Model 2

Model 2 is an extension of Model 1 and is meant to account for the fact that in real fingerprints, the pores tend to be located on the center of the ridge. The assumptions used in Model 2 are the same as those used in Model 1 except that the probability of the pore being located in the center of the ridge is higher than the probability of a pore occurring on the edge of the ridge. It is seen from Figure 28 that Model 2 simulates the real data distribution closely but there is not a significant difference between simulations of Models 1 and 2.

6.1.3 Ridge-Independent Pore Configurations

Knowing the probability of occurrence of a sequence of intra-ridge pores is of value for proving the efficacy of using pores for identification. However, in practice, it may be unnecessary or too difficult (for example with very noisy images) to associate pores with specific ridges. In these cases, extracting only the pore information while disregarding their ridge association is preferable. In this section, the value of ridge-independent configurations of pores will be examined. For this analysis, the configurations are made up of pores which exist in a local region of the fingerprint but may reside on several different ridges.

6.1.3.1 Binomial Distribution of Pores

The technique used by Osterburg, outlined in Section 6.1.1, for determining the uniqueness of a configuration of minutiae can be extended for use with pores. For this purpose, 67 different prints of varying quality were analyzed on a 5×5 pixel scale (see Figure 24 to get a feel for this size segment). Each image was divided into 5×5 pixel segments and the number of pores per segment was counted. In such a small area, there is little chance of more than one pore occurring; therefore, a binomial distribution should result. From the data, 93.3% of search areas contained no pores and 6.7% contained one pore.

Defining $\quad p_\alpha = P(\text{one pore in a 5×5 pixel cell}) = 0.067$

and $\quad q = 1 - p_\alpha = P(\text{no pores in the cell}) = 0.933$

and assuming that pores are independent, the probability of occurrence of a configuration of N_P pores in a region of N_C cells is given by:

$$P(N_P) = p_\alpha^{N_P} q^{(N_C - N_P)} \tag{14}$$

where there are

$$\binom{N_C}{N_P} = \frac{N_C!}{N_P!(N_C-N_P)!} \tag{15}$$

different configurations possible when N_P pores are present.

Assuming independence among the pores, the binomial distribution can be used to yield the probability of any particular configuration of pores occurring in any area of print. For example, a section of fingerprint which measures 20×20 pixels (0.462 mm × 0.462 mm) consists of an array of 16 grids of size 5×5 pixels each. Therefore, the probability of occurrence of a configuration with between 0 and 16 pores in that area can be calculated. Table 3 summarizes the results for this analysis. From the table, the most likely configuration of pores in a 20×20 pixel region has a single pore. The probability of occurrence of a certain configuration with one pore is 0.0237 and since there are 16 configurations of one pore possible, the likelihood that one pore is present (regardless of the configuration) is 0.379.

Table 3. Calculated probability of n pores occurring in a 4×4 grid area.

# pores n	# empty cells m	# configurations $\binom{N}{n}$	configuration probability $p^n q^m$	$P(n$ pores in 20×20 pixel region) = $\binom{N}{n} p^n q^m$
0	16	1.00000e+00	3.29690e-01	3.29690e-01
1	**15**	**1.60000e+01**	**2.36755e-02**	**3.78808e-01**
2	14	1.20000e+02	1.70017e-03	2.04021e-01
3	13	5.60000e+02	1.22092e-04	6.83713e-02
4	12	1.82000e+03	8.76756e-06	1.59570e-02
5	11	4.36800e+03	6.29611e-07	2.75014e-03
6	10	8.00800e+03	4.52132e-08	3.62067e-04
7	9	1.14400e+04	3.24682e-09	3.71436e-05
8	8	1.28700e+04	2.33159e-10	3.00075e-06
9	7	1.14400e+04	1.67434e-11	1.91545e-07
10	6	8.00800e+03	1.20237e-12	9.62857e-09
11	5	4.36800e+03	8.63438e-14	3.77150e-10
12	4	1.82000e+03	6.20046e-15	1.12848e-11
13	3	5.60000e+02	4.45264e-16	2.49348e-13
14	2	1.20000e+02	3.19750e-17	3.83700e-15
15	1	1.60000e+01	2.29617e-18	3.67387e-17
16	0	1.00000e+00	1.64891e-19	1.64891e-19

Fingerprint Feature Processing Techniques and Poroscopy

In practice, a larger segment of fingerprint is desirable. The results for a 40×40 pixel (0.924 mm × 0.924 mm) area, 4 times the area of the 20×20 pixel segment, are discussed next. For this case, the segment consists of 64 grids of size 5×5 pixels. A segment of this size is large enough to contain significant ridge structure while not exhibiting distortion due to finger plasticity sometimes present in larger areas of print. As seen in Table 4, the most likely number of pores in this area is 4, with a probability of 20.0%.

Table 4. Calculated probability of n pores in an 8×8 grid area

# pores n	# empty cells m	#configurations $\binom{N}{n}$	configuration probability $p^n q^m$	$P(n$ pores in 64×64 pixel region) = $\binom{N}{n} p^n q^m$	# pores n	# empty cells m	#configurations $\binom{N}{n}$	configuration probability $p^n q^m$	$P(n$ pores in 64×64 pixel region) = $\binom{N}{n} p^n q^m$
0	64	1.00000e+00	1.18148e-02	1.18148e-02	33	31	1.77709e+18	2.12226e-40	3.77145e-22
1	63	6.40000e+01	8.48435e-04	5.42998e-02	34	30	1.62029e+18	1.52403e-41	2.46936e-23
2	62	2.01600e+03	6.09272e-05	1.22829e-01	35	29	1.38882e+18	1.09442e-42	1.51996e-24
3	61	4.16640e+04	4.37527e-06	1.82291e-01	36	28	1.11877e+18	7.85921e-44	8.79265e-26
4	60	6.35376e+05	3.14194e-07	1.99631e-01	37	27	8.46637e+17	5.64381e-45	4.77825e-27
5	59	7.62451e+06	2.25627e-08	1.72030e-01	38	26	6.01558e+17	4.05289e-46	2.43805e-28
6	58	7.49744e+07	1.62026e-09	1.21478e-01	39	25	4.01039e+17	2.91044e-47	1.16720e-29
7	57	6.21216e+08	1.16353e-10	7.22803e-02	40	24	2.50649e+17	2.09003e-48	5.23863e-31
8	56	4.42617e+09	8.35546e-12	3.69827e-02	41	23	1.46721e+17	1.50088e-49	2.20211e-32
9	55	2.75406e+10	6.00017e-13	1.65248e-02	42	22	8.03474e+16	1.07780e-50	8.65984e-34
10	54	1.51473e+11	4.30880e-14	6.52668e-03	43	21	4.11080e+16	7.73982e-52	3.18169e-35
11	53	7.43596e+11	3.09421e-15	2.30084e-03	44	20	1.96197e+16	5.55807e-53	1.09048e-36
12	52	3.28421e+12	2.22200e-16	7.29751e-04	45	19	8.71988e+15	3.99133e-54	3.48039e-38
13	51	1.31369e+13	1.59564e-17	2.09618e-04	46	18	3.60169e+15	2.86623e-55	1.03233e-39
14	50	4.78557e+13	1.14585e-18	5.48357e-05	47	17	1.37937e+15	2.05828e-56	2.83912e-41
15	49	1.59519e+14	8.22854e-20	1.31261e-05	48	16	4.88527e+14	1.47808e-57	7.22080e-43
16	48	4.88527e+14	5.90902e-21	2.88672e-06	49	15	1.59519e+14	1.06143e-58	1.69318e-44
17	47	1.37937e+15	4.24335e-22	5.85315e-07	50	14	4.78557e+13	7.62225e-60	3.64768e-46
18	46	3.60169e+15	3.04721e-23	1.09751e-07	51	13	1.31369e+13	5.47364e-61	7.19064e-48
19	45	8.71988e+15	2.18824e-24	1.90812e-08	52	12	3.28421e+12	3.93070e-62	1.29092e-49
20	44	1.96197e+16	1.57141e-25	3.08306e-09	53	11	7.43596e+11	2.82269e-63	2.09894e-51
21	43	4.11080e+16	1.12845e-26	4.63882e-10	54	10	1.51473e+11	2.02701e-64	3.07038e-53
22	42	8.03474e+16	8.10354e-28	6.51099e-11	55	9	2.75406e+10	1.45562e-65	4.00887e-55
23	41	1.46721e+17	5.81926e-29	8.53810e-12	56	8	4.42617e+09	1.04530e-66	4.62668e-57
24	40	2.50649e+17	4.17889e-30	1.04744e-12	57	7	6.21216e+08	7.50646e-68	4.66314e-59
25	39	4.01039e+17	3.00092e-31	1.20348e-13	58	6	7.49744e+07	5.39049e-69	4.04149e-61
26	38	6.01558e+17	2.15500e-32	1.29636e-14	59	5	7.62451e+06	3.87099e-70	2.95144e-63
27	37	8.46637e+17	1.54754e-33	1.31020e-15	60	4	6.35376e+05	2.77981e-71	1.76622e-65
28	36	1.11877e+18	1.11131e-34	1.24330e-16	61	3	4.16640e+04	1.99622e-72	8.31704e-68
29	35	1.38882e+18	7.98044e-36	1.10834e-17	62	2	2.01600e+03	1.43351e-73	2.88996e-70
30	34	1.62029e+18	5.73086e-37	9.28565e-19	63	1	6.40000e+01	1.02942e-74	6.58831e-73
31	33	1.77709e+18	4.11541e-38	7.31345e-20	64	0	1.00000e+00	7.39243e-76	7.39243e-76
32	32	1.83262e+18	2.95533e-39	5.41601e-21					

As an example, any configuration of 4 pores has a probability of occurrence of 3.14×10^{-7}, and there will be 6.35×10^5 different configurations of 4 pores in a 64 grid area. For comparison, from the table, it is noted that the most likely configuration is the one with no pores, which occurs with a probability of 1.18%. For perspective, assuming the most likely sequence of pores occurs in a 40×40 pixel segment, the probability of a different fingerprint matching is 0.0118 based on comparison of pore location only with a resolution (or measurement accuracy) of 5×5 pixels.

In order to compare the results for intra-ridge sequences to ridge independent configurations, assume that 20 pores occur in an area of size 4 mm^2. This is a good approximation based on the density of pores being about 5 pores/mm^2. Again using cells of size 5×5 pixels, the area consists of 300 cells. Given this area, the probability of occurrence of a configuration of 20 ridge independent pores would be:

$$P_{RI}(\text{a configuration of 20 ridge independent pores}) = 1.23 \times 10^{-32}$$

which is 1.06×10^{-18} times smaller than the probability of an intra-ridge sequence with the same number of pores (assuming all pores are spaced by the most likely separation).

6.1.3.2 Measuring Configuration Probabilities

In the previous discussions, the underlying assumption of independence makes uniqueness calculations possible. In reality, though, the independence assumption is not accurate. There appears to be a definite influence on a pore's position depending on the relative positions of the neighboring pores. If the independence assumption is not valid, then the assumption that all possible configurations of N pores are equally likely is also not valid. In this case, it is desirable to determine the exact probability of occurrence of each possible configuration of N pores by finding the histogram or pdf of the configurations. This information could be used by the processing routine to ignore highly likely configurations and to search for very distinctive pore configurations providing inter-class separability. The expected outcome is a reduction in the number of false accept errors.

As alluded to before, a segment with an area of between 1 and 4 mm^2 should be optimal for processing pores. For this size area, the number of pixels (at 1100 ppi = 43.3 ppmm scanning resolution) is approximately 40×40 (1600) to 80×80 (6400). The resolution is effectively reduced by analyzing 5×5 pixel segments, leading to areas of 8×8 (64) or 16×16 (256) cells. In fact, to determine the histogram of all possible configurations is unreasonable even for the simplest case of low resolution and smaller area. There are 2^{64} possible configurations for the smaller area with low resolution and 2^{6400} configurations at high resolution and larger area. As an interesting note, about 4,000 km by 4,000 km of print area would be needed in order to fill each of the 2^{64} histogram bins with only one entry. Even with the enticement of free doughnuts, it is unlikely that the researchers could have gathered sufficient subjects to accumulate the data for this experiment.

These numbers are actually exaggerated, since only a small subset of the total number of theoretical configurations are really possible. Based on measurements, there were never more than 12 pores detected in a print area of size 40×40 pixels, and this event had the remote probability of 0.03%. The most probable number of pores to occur in this area was 4, with a probability of 22.3%. The complete distribution of measured data is shown in Figure 29. Even by applying such a realistic constraint on the pore density, the number of configurations is still enormous.

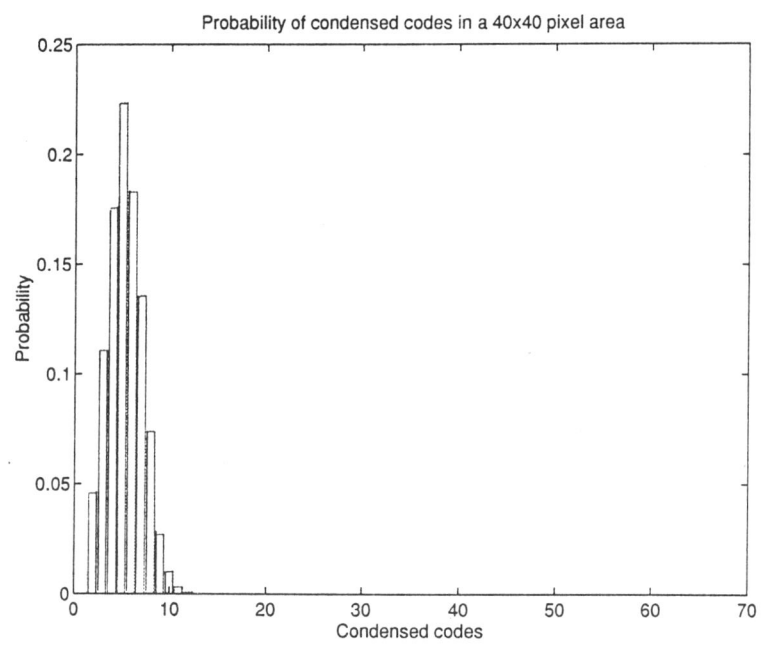

Figure 29. Measured density of pores (pores/mm^2) in a 40×40pixel area of print.

6.2 Reliability – FRR Analysis

Many factors that determine the FAR of a system were discussed in the previous sections; Section 6.2 is devoted to factors that can affect the FRR. Whereas the FAR analysis centers on the differences between an impostor and an authorized user, the FRR focuses on variability that occurs within an authorized user's fingerprints over time. These variations can be studied to determine the feature reliability, with regard to physiology and algorithm.

To ensure authentication of an individual with 100% accuracy, either the individual's live-scan fingerprint must be exactly the same as the enrolled print or, in the presence of noise and distortions, the features of the live-scan print corresponding to the enrolled print must be extracted without error. In the real world, noise and distortions are always present, and no automated process is perfect. In addition, the physiological reliability of pores falls short of 100%; therefore, the inherent and algorithm (or processing) reliability warrant further study.

6.2.1 Inherent Reliability

As discussed earlier in the chapter, the physiological reliability of pores (or inherent reliability, R_i) depends on environmental factors; temperature and skin condition can conspire to alter or suppress altogether the physical appearance of a given pore. Individual pores from 516 images of ten different fingerprints were analyzed to determine R_i with respect to both pore visibility (or detection) and size (both absolute and relative to other pores in the image). Clarity of the pore, image quality, skin condition, and pore density were also recorded. It should be noted that all of this data was collected manually (by eye) to prevent introducing algorithm errors.

The specific pores studied were visible on average in 91% of the images ($R_i = 0.91$). The least reliable pore was visible in only 75% of the fingerprint images. In this case, the reason for the low reliability is that during capture of these images, one individual altered his prints through a variety of means, e.g., gripping a cold soda can prior to image capture. Therefore, 75% can be estimated as a lower bound for R_i. Although pore size and shape are of significance, the most important

Fingerprint Feature Processing Techniques and Poroscopy

aspect of R_i is whether or not the pore is actually present (detectable). It is also important to remember that, although the lower bound for detection is 75%, that lower bound is for one pore, not a configuration of pores. If twenty pores are used to match prints, the 75% refers to the reliability of only one pore out of the twenty.

Although the characteristic skin condition, image quality, size, and shape consistency varied somewhat among the individuals' prints, there were several correlations between categories. For instance, the more neutral prints (with regard to skin condition) had the highest image quality. Furthermore, prints of lower image quality tended to correspond to a dry skin condition, and their pores were less consistent in shape. Finally, circular pores proved to be the most reliable with regard to shape.

6.2.2 Algorithm Reliability

The reliability, R_d, of the algorithm used in this study was also examined. Using high quality prints, the prevalence of missed detects and false detects was recorded and the causes for both types of errors were assessed. The detection algorithm missed 11% of the pores present ($R_d = 0.11$), and had a false detect occurrence of 1%. It was found that the predominant source of missed detections was the thresholding stage used in preprocessing to convert gray scale images to binary. These processing errors occurred in 4.8% of the detections. On the other side of the coin, most of the false detects were caused when the algorithm detected a pore in the middle of the valley. This physically impossible situation may have resulted when the curvature of the ridge implied that a pore was present on the edge of the ridge. The algorithm might interpret such a structure as a pore.

These reliability statistics apply to a single pore in a configuration. The probability of the algorithm missing a given pore is 11%, but the probability of missing a configuration of many pores is orders of magnitude smaller.

In further discussions, R_i will be defined as the probability of a feature appearing in a fingerprint image, and R_d the probability that the feature is properly detected by the algorithm. Therefore, R (total reliability), is defined as the probability that a feature appears and is properly detected.

6.3 Performance

Section 6.1 examined the uniqueness of configurations of features and Section 6.2 addressed feature reliability. In this section, uniqueness and reliability are conjoined to establish performance, which is defined in terms of the number of false reject and false accept errors the system produces. The number of false accept errors is related to the uniqueness of a configuration while the number of false reject errors depends on the reliability of the features. Some parameters which contribute to uniqueness have been discussed earlier: number of features, density of features, and feature area. Parameters critical to the reliability are the inherent feature reliability and the efficiency of the feature detection algorithm.

Consider comparing two fingerprint segments of equal size. Segment 1 is the enrolled segment and segment 2 is a subsequent live-scan which originated from either the same user or a different user. The comparison is based on feature location only and segment 1 contains n_1 features while segment 2 contains n_2 features. Assume that there were no errors in detecting features during enrollment (all real and no false features were detected). For the two possible sources of segment 2, it is necessary to determine the probability density function for the feature matching score.

Note that prior to this section, all references to probability of matching or uniqueness of a set of features related to the entire set of features. Every feature was required to match for the entire set to match. In this section, a more realistic approach is taken in which the number of features in both segments as well as the number that actually match are taken into account. A matching score provides the degree of matching between two segments with a range of a complete non-match to a complete match.

Some relevant parameters needed for the matching problem are:

n_c – the number of cells or feature areas within the segment of print being analyzed
n_1 – the true number of features in segment 1 minus the enrolled segment
n_2 – the number of features detected in segment 2 minus the live-scan segment
n_m – the number of matching features
n_{fd} – the number of features falsely detected

p_{md} – probability that a valid feature is not detected
$p_{cd} = 1 - p_{md} = R_d$ minus the probability that a valid feature is correctly detected
p_{fd} – probability of detecting an invalid, or false, feature at any location
$p_{cm} = 1 - p_{fd}$ – probability that a feature is not detected in an invalid location.

If segment 2 originates from the same user as segment 1, then reliability must be addressed:

R – reliability
R_i – inherent reliability of the feature
R_d – algorithm detection reliability
$R = R_i \times R_d$

In addition, two sources contribute to the number of matching features:

$n_m = n_{m,R} + n_{m,F}$
$n_{m,R}$ – the number of correct (valid feature) matches
$n_{m,F}$ – the number of false detects (invalid features) that match real features

Reliability is defined as the probability of detecting a valid feature in the correct position. R can range from 0 (totally unreliable features) to 1.0 (no missed detects). When segment 2 is from an impostor print, R is considered to be 0 and p_{fd} is set equal to the measured value for the probability of a pore in a grid cell. This situation simulates a randomly located set of independent features in segment 2.

A feature match is defined as the detection of a feature in the live-scan segment at a valid (or enrolled) feature location. A feature mismatch is defined as any feature in the live scan which does not match an enrolled feature. It is possible to make a false detection at the location of a valid feature which is unreliable. This situation results in an incorrect (false) match but is not a mismatch.

Define the feature matching score to be:

$$S_F = \frac{n_m}{n_1} - \frac{n_2 - n_m}{n_{2,max}} \qquad (16)$$

where $n_{2,max}$ is the maximum number of features allowed in segment 2.

The range of S_F is [-1,+1] where a score of -1 corresponds to the case where $n_m = 0$ and $n_2 = n_{2,max}$. This is the worst possible match; no features match and segment 2 contains the maximum number of detected features allowed. A score of +1 results when $n_1 = n_2 = n_m$, a perfect match with no mismatching features.

Given n_c, n_1, n_2, $n_{m,R}$, $n_{m,F}$, R, and p_{fd}, the probability of n_m pores matching can be determined by calculating the probability of the score resulting from each variation of the input parameters and then accumulating the probability of like scores. The result is the pdf of the matching score, S_F, which can be used to calculate false accept and false reject error rates.

First, assume that the two fingerprint segments originate from different fingers. Furthermore, for simplicity, assume that the features are independent. Given an enrolled feature set, the number of features in segment 2 and their positions are random. Therefore, the number of matching features is a random variable. Whether features detected in segment 2 are real or result from detection errors is transparent since the only concern is how many features match. Given n_c, n_1, and n_2 the probability of matching n_m features between segment 1 and segment 2 is:

Fingerprint Feature Processing Techniques and Poroscopy

$$P_{FAR}(n_m) = \left[\binom{n_1}{n_m} \frac{\left(\prod_{i=0}^{n_m-1} n_2 - i\right)\left(\prod_{i=0}^{n_1-n_m-1} n_c - n_2 - i\right)}{\left(\prod_{i=0}^{n_1-1} n_c - i\right)} \right] \left[\binom{n_c}{n_2} p_\alpha^{n_2} (1-p_\alpha)^{n_c - n_2} \right] \quad (17)$$

where the product terms are valid for $i > 0$. Or, by defining:

$$P_{FAR}(n_m) = P_{n_m} P_{n_2} \quad (18)$$

The first term in Equation 17 and Equation 18 is p_{nm}, the probability that n_m features from segment 1 match in segment 2. The second term is p_{n2}, the probability that there are n_2 features in segment 2, where p_α is the probability that there is a feature in a cell. In this situation, the second term incorporates p_{fd} into p_α and reliability is not an issue since the print segments originate from different fingers.

As an example, assume that for a given $n_c = 300$, $n_1 = 20$, and $n_2 = 25$, that there are $n_m = 10$ matching features. The probability of n_2 having 25 features is 0.0457 and the probability that 10 of them match those in segment 1 is 2.5175×10^{-7} giving an overall probability of 1.1505×10^{-8} of this situation occurring. The corresponding match score, S_F, from Equation 16 is 0.45 assuming that $n_{2,max}$ is equal to n_c.

Next, assume that the two print segments originate from the same finger. Furthermore, the rotation and positioning of the segments are assumed to be known exactly. In this case, the feature reliability and the number of false detects are of critical importance. A feature may or may not be correctly detected in the live-scan segment depending on its reliability, leading to a reduction in the number of features matched. In addition, there are random false detection errors in the live-scan image segment which, depending on their position, will match (improve the score) or mismatch (reduce the matching score). The probability of matching n_m features is:

$$p(n_m) = P_{n_m, R} P_{n_m, F} P_{n_2} \quad (19)$$

where $n_m = n_{m,R} + n_{m,F}$,

$$p_{n_{m,R}} = \left[\binom{n_1}{n_{m,R}} R^{n_{m,R}} (1-R)^{n_1 - n_{m,R}} \right] \quad (20)$$

is the probability of $n_{m,R}$ matches corresponding to real features which were correctly detected.

$$p_{n_{m,F}} = \left[\binom{n_1 - n_{m,R}}{n_{m,F}} \frac{\left(\prod_{i=0}^{n_{m,F}-1} n_2 - n_{m,R} - i \right) \left(\prod_{i=0}^{n_1 - n_{m,R} - n_{m,F} - 1} n_c - n_2 - i \right)}{\left(\prod_{i=0}^{n_1 - n_{m,R} - 1} n_c - n_{m,R} - i \right)} \right] \quad (21)$$

where $p_{nm,F}$ is the probability of $n_{m,F}$ matches which are matches of falsely detected features in the live-scan segment randomly occurring at valid feature locations.

The probability that there are n_2 features in segment 2, p_{n2}, is given as:

$$p_{n_2} = \left[\binom{n_c - n_{m,R}}{n_{fd}} p_{fd}^{n_{fd}} (1 - p_{fd})^{n_c - n_{m,R} - n_{fd}} \right] \quad (22)$$

In Equation 22, p_{fd} is used instead of p_α, which was used in Equation 17.

In Equation 19, by setting $R = 0$, $p_{nm,R} = 1$, $n_{m,R} = 0$ therefore, $n_m = n_{m,R}$, and $n_{fd} = n_2$. In addition, if p_{fd} is set equal to p_α, then Equation 19 reduces to Equation 18. This situation corresponds to totally unreliable features and a false detection rate which provides the same feature density as the measured feature density. The result is that segment 2 is equivalent to an imposter print segment.

If no false detects are allowed, then p_{fd} is 0 for Equation 19, and $n_{m,F} = 0$ and $n_{fd} = 0$, therefore, $n_2 = n_m = n_{m,R}$. For this case, Equation 21 and Equation 22 reduce to 1 and the result is:

Fingerprint Feature Processing Techniques and Poroscopy

$$p(n_m) = p_{n_{m,R}} = \left[\binom{n_1}{n_m} R^{n_m}(1-R)^{n_1-n_m} \right] \tag{23}$$

which is just the probability that n_m features are reliable given n_1 features to start.

If the reliability is 100%, then $n_{m,R} = n_1$, Equation 20 and Equation 21 both reduce to 1, and Equation 22 becomes:

$$p_{n_2} = \left[\binom{n_c - n_1}{n_{fd}} p_{fd}^{n_{fd}}(1 - p_{fd})^{n_c - n_1 - n_{fd}} \right] \tag{24}$$

which represents the probability of n_{fd} falsely detected features in segment 2 given that there are n_1 correctly detected ($n_2 = n_1 + n_{fd}$).

Equations 19-21 are used to determine the expected performance of a system. The parameters such as the number of features enrolled, the accuracy of measurement (feature area), the feature reliability, and the algorithm efficiency can all be evaluated. The plots in Figure 30 show how variations in the critical system parameters affect performance.

7 Testing

This section discusses levels of testing for fingerprint authentication systems; testing the algorithms will be the primary focus. There are two levels of testing: component level testing and system level testing. This section will define these different levels of testing and the terms associated with them.

7.1 Component Level Testing (Database Testing)

Component level testing consists of testing individual components of a fingerprint authentication system separately from the entire functioning system. At times, this is desirable as it may pinpoint troublesome errors caused in one area of the system's function. While this section will focus on the testing of the algorithm, it is useful to discuss all of the components that may be involved.

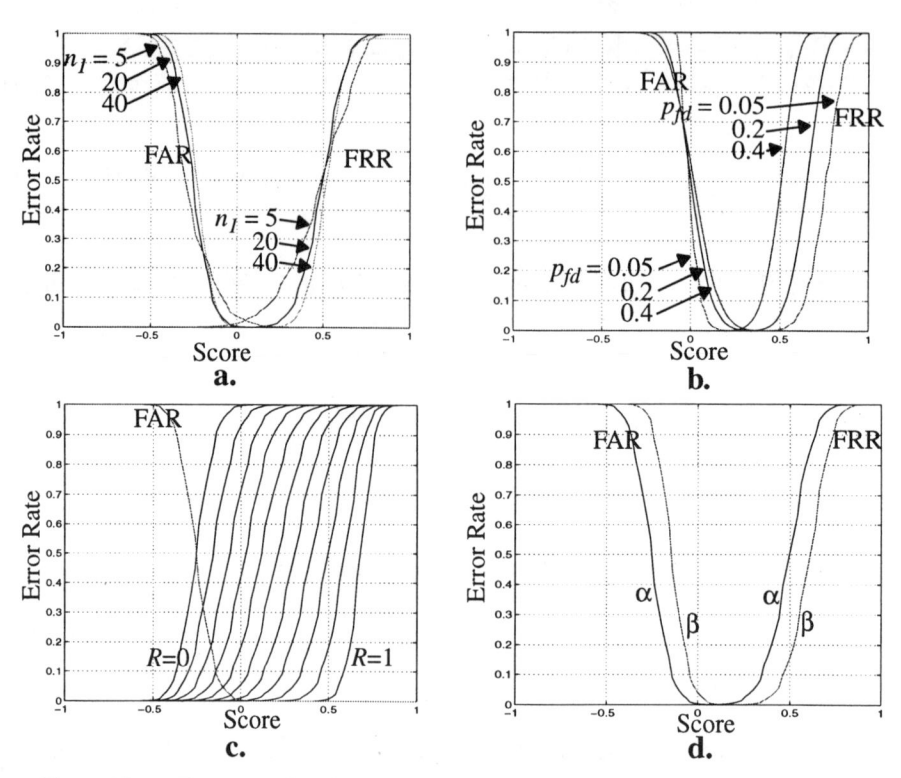

Figure 30. **a.** Error rate plots for variation in enrolled feature density. The parameters used in this simulation were: $n_c = 300$, $R = 0.8$ (for FRR and $R = 0$ for the FAR plots), $p_{fd} = 0.067$, n_2 ranges from [0,60], and n_1 is set to 5, 20, or 40. The Equal Error Rate (EER) is 2.43×10^{-2} for $n_1 = 5$, 6.2×10^{-4} for $n_1 = 20$, and 3.2×10^{-4} for $n_1 = 40$. **b.** Variation in feature detection error rate (fingerprint noise level). The parameters used in this simulation were: $n_c = 300$, $R = 0.8$ (for FRR and $R = 0$ for the FAR plots), $n_1 = 20$, n_2 ranges from [0,300], and p_{fd} is set to 0.05, 0.2, or 0.4. For a lower detection error rate (cleaner fingerprint image), an enrolled user will gain access more often (the FRR error rate curve is steeper and more towards the right). For high detection error rates (noisy images), an imposter will have a higher feature density and will have a greater chance of gaining access. The Equal Error Rate (EER) is 7.2×10^{-4} for $p_{fd} = 0.2$ and 9×10^{-3} for $p_{fd} = 0.4$. **c.** Error rate plots for variation in feature reliability. As R decreases, the performance of the system degrades. An FAR plot is shown for comparison of actual system performance as a function of the feature reliability. The parameters used in this simulation were: $n_c = 300$, $n_1 = 20$, $p_{fd} = 0.067$, n_2 ranges from [0,60], and R ranges from 0 to 1.0 in 0.1 increments ($R = 0$ for the FAR plot). **d.** Resolution or measurement accuracy curves. Parameters for α curves: $n_c = 300$, $n_1 = 20$, $R = 0.8$ (for FRR), $p_{fd} = 0.067$, n_2 ranges from [0,60]. For β curves: $n_c = 200$, $n_1 = 20$, $R = 0.8$ (for FRR), $p_{fd} = 0.1$, n_2 ranges from [0,60]. Both sets of parameters simulate a comparison of two segments with the same area and detection error rates. The α curves simulate a system using a more precise feature position determination and corresponding smaller search area than the β curves. Higher resolution will tend to make it more difficult for both a valid user and an imposter to match (but the EER may be the same as for lower resolution settings) as seen from the curves.

All fingerprint authentication systems consist of a number of common components, some of which are more tangible than others. Some systems include components that perform non-standard functions. A fingerprint authentication system may consist of some combination of the following components:

- application
- scanner
- algorithm (processing and matching software)
- liveness/anti-spoofing hardware
- tamper detection hardware
- other hardware - door unlock mechanism, user feedback device, PIN entry pad

There may also be features which are present in one system but not another which may influence the end users' decision, such as an ethernet connection or network ability.

The application component indicates the capacity in which the system will be used (e.g., a stand-alone unit, command line, graphical user interface for system administration functions, etc.). It may include functions for performing auditing, manipulation of databases of enrolled users, a protected enrollment capability, identification/-verification mode, virus checking, etc. The scanner component is the hardware used to acquire the fingerprint image. The algorithm component is the software which produces the template (enrollment algorithm) and performs the acquisition (scanner to algorithm link), processing, and matching of fingerprints. Some systems may have special hardware and software components responsible for detection of non-live and fake fingers. Tamper-detection techniques may be built into the scanner or associated hardware.

There are several other facets of a fingerprint biometric that will indicate the worth of the system; these warrant examination, but not necessarily testing. These may include the timing of the system (the length of time required to enroll, identify, or verify), template size (which is directly related to cost), the availability of an adjustable threshold to the user (so that performance can be customized), and whether or not an option for protected enrollment exists.

It is not always possible to isolate each component for testing purposes. When testing an algorithm, whether for comparison to another algorithm, or for an absolute performance rating, it is important to normalize any effects or artifacts of other components. For this reason, a standard database of images should be used to test algorithms.

Factors associated with database testing will be discussed, along with more details of component testing, in Section 7.4.

7.2 System Level Testing (Field Testing)

System level testing involves testing the entire fingerprint authentication system as a unit. Field testing is large scale system level testing in the environment required for a specific application. For example, if a fingerprint system is to be used for authentication in drawing money from an ATM (automatic transaction machine), a field test would require the fingerprint system to be installed on an ATM and placed under the same environmental constraints that it would endure in the field. Field testing is specific to the application, since different applications and their respective environments will evoke different sources of errors, etc.

System level testing can be done on a much smaller scale than the field testing described above. Preliminary testing of the system can also be performed to determine whether or not the system is functioning correctly, and whether it merits further testing (i.e., database testing or field testing).

A system which performs identification mode of operation is much more difficult to test than a system performing only verification. The results (especially the FAR but also the FRR) from a system performing identification are dependent on the size of the database (or number of enrolled users). Based on this fact, a verification system cannot be compared fairly to an identification system and the results of testing a system in identification mode must be presented very carefully.

7.3 Flow Chart:
A General Overview of a Biometric Test Plan

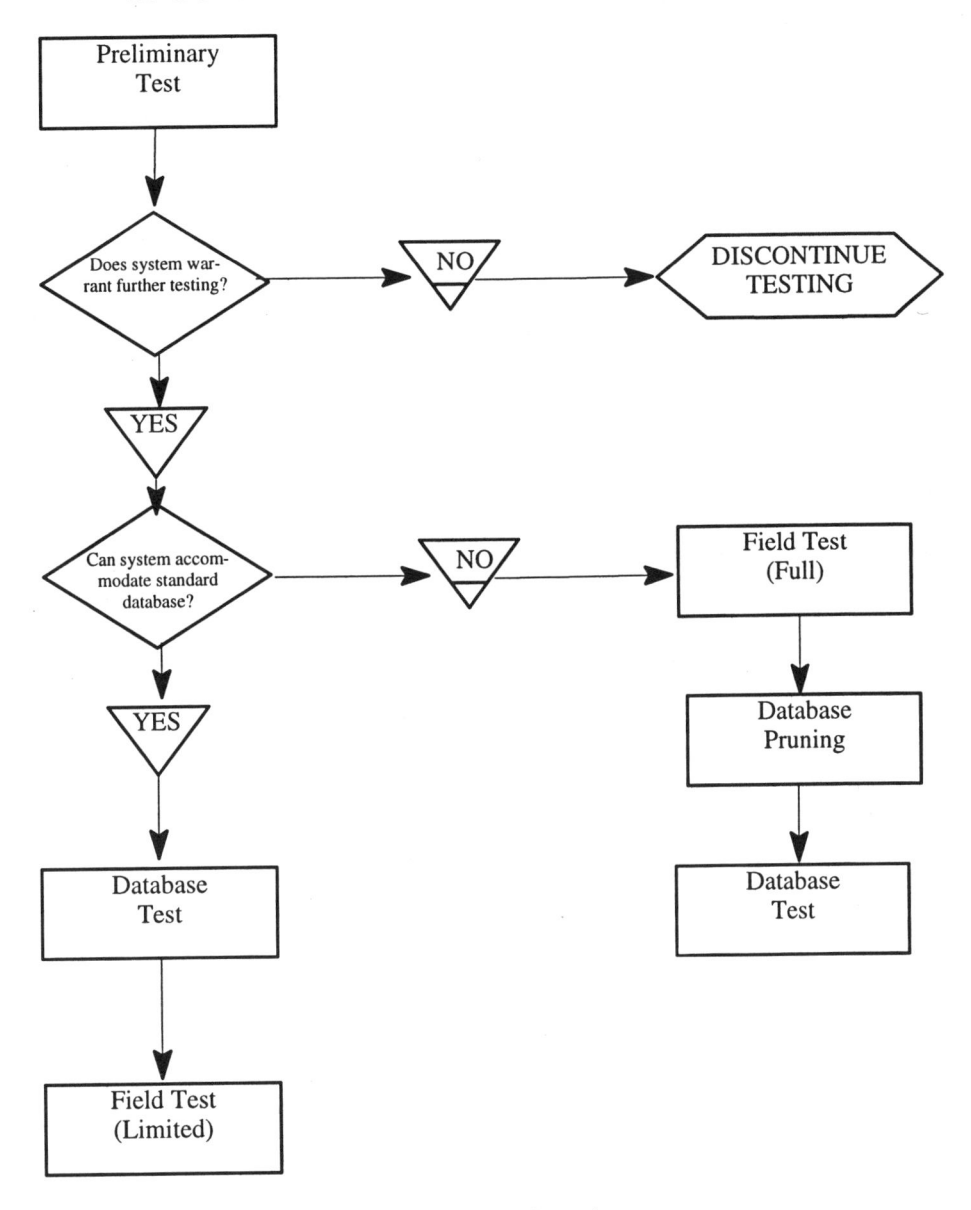

Figure 31. Test flow chart.

7.4 Detailed Description of Procedure

The flow chart in Section 7.3 diagrams the testing process for a fingerprint biometric system. This section will explain each step in detail. As can be seen from the flow chart, testing can be done with either a standard database or a non-standard database. The standard testing procedure, including preliminary testing, database testing, and field testing, will be discussed first, followed by the non-standard testing procedure.

7.4.1 Standard Testing Procedure

7.4.1.1 Step 1 – Preliminary Test (System Level)

Preliminary testing is performed to determine if the system is in working order and if it warrants further testing. This is a small-scale system-level test that yields only qualitative results (no error rates). The flow chart in Section 7.3 shows that after preliminary testing, either all testing is stopped, or the decision is made to continue on to the next phase of testing.

Preliminary testing should also include an indication as to whether or not the system can be examined at the component level (generally systems do not have this capability). A requirement for algorithm-component level testing is the ability to do a batch-mode verification with as input a list of templates/images and another list of images which are cross verified against the first list. The output is the match score or the verification decision.

A description of the procedure for preliminary testing follows: the fingerprint system should be set up, albeit in a less formal structure than rigorous field testing would require, in a manner that is representative of the application. It can be set up on a lab bench instead of, for example, in an ATM. A small set of users (volunteers) will be enrolled, and then asked to periodically verify or identify themselves using the system. A log should be kept to document parameters and resulting events. Some of the parameters and events that should be documented are:

- the ID number of the individual attempting to gain access,

- the ID number of the impostor, if the person attempting to gain access is not the authorized user.
- whether the attempt was successful or not
- match scores
- the amount of time required to enroll/verify (timing analysis)
- the amount of memory occupied by the template (storage analysis)
- comments/remarks about that attempt.

The qualitative results that are reported at the conclusion of preliminary testing should include a description of any problems that surfaced during testing, recommendations for better performance or better ergonomics, reasons for not continuing on to the next phase of testing, etc. In addition, this preliminary testing will indicate whether or not the application for which the biometric is intended is, in fact, feasible. In the end, a measure will be given of how reliable the system seems to be in terms of (1) reliably accepting authorized users, and (2) reliably rejecting impostors. This measure may be on a scale of 1 to 5, or on a graded scale from "unacceptable" to "acceptable."

7.4.1.2 Step 2 – Algorithm (Database) Testing – Component Level, Standard Database

Database testing is performed for three primary reasons: repeatability, FAR analysis, and to enable fair comparisons of different biometric components. This component level test is done in order to ascertain how well certain components of the system (in this case, the algorithm) perform; it serves as a baseline performance indicator. This is a larger scale test than the preliminary testing discussed previously, and quantitative results – component level FAR and FRR – will be determined. Component testing can enable the comparison of components across systems (i.e., from different vendors or even different biometrics).

As mentioned earlier, algorithm testing is done using a database of print images. This section will discuss testing with a standard database of live-scan prints, and is intended for systems which comply with parameters associated with the acquisition of the standard database.

The size and composition of this database is assumed to be known: the database should include a fair representation of fingerprints of varying skin condition or print quality (dry, neutral, oily). The demographics of the individuals whose prints are in the database should be recorded. (Bricklayers have different print quality than professional office workers).

	Highest possible quality print for the individual – conditioning of the finger allowed to improve quality	Multiple prints which represent the variation possible in each individual's fingerprint	
ID: User #1 User Demographics	Enrollment Prints Highest Quality	Live-scan prints first attempt	Live-scan prints second attempt
ID: User #2 User Demographics	Enrollment Prints Highest Quality	Live-scan prints first attempt	Live-scan prints second attempt
ID: User #3 User Demographics	Enrollment Prints Highest Quality	Live-scan prints first attempt	Live-scan prints second attempt
⋮	⋮	⋮	⋮
ID: User #N User Demographics	Enrollment Prints Highest Quality	Live-scan prints first attempt	Live-scan prints second attempt

Figure 32. Structure of Database.

All of the images in this database should be captured with the same scanning device, so that the effects of scanner parameters (contrast, artifacts, distortion, etc.) will not favor any particular algorithm. Database testing can be used to compare algorithms from different systems/vendors; using a standard database of images captured with the same scanner ensures that the same data is presented to the algorithms. Several reasons exist for not using the NIST databases (numbers 4 and 9). The number 4 database contains live-scan captures, while 9 contains inked prints that have been scanned in. Both of these databases have already gone through quality control to eliminate poor quality prints. However, poor quality prints exist in normal use, and eliminating these from the database will yield an artificially low FRR when testing is performed. Secondly, there are only two copies of each finger in both databases. In order to do a thorough FRR test, numerous copies of the same finger are required. In addition, the NIST databases were created with AFIS applications in mind – not biometrics. AFIS systems (mainly searching of large databases) have a host of user issues that are not germane to biometrics (primarily access control applications which require binary decisions – access approved/not approved), and vice versa.

The setup of the fingerprint system is of little consequence in this type of testing, since no users will be required (once the standard database is established). The images in the database can be enrolled in lieu of live user enrollment. All fingerprints belonging to an individual should be saved with information unique to that person, so that they may be easily traced to that individual. Each print in the database should be classified or labeled with its particular characteristics (quality of print, skin condition, etc.) and a description of the person (such as occupation, gender, age, etc.). Then testing can be done for each category of print in the database and on the database as a whole. The results (FRR-type testing) will depend on the category of print in the test. FAR-type testing would compare a particular category of prints against the rest of the database. The advantage of stating results as a function of print category is that results can be extrapolated or used to provide more accurate results for a particular application which may have a concentration of users from a particular category represented in the overall database (for example retirement benefits programs for elderly people compared to an ATM application). Additionally, the performance of each category of print will be known.

After all the images in the database are enrolled, they must be matched against each other. Trying to match Person A's prints to Person B's prints is considered an impostor attempt. Matching Person A's index print to their own index print is termed a matching print attempt. Matching Person A's index print to their thumb print (or a print from any different finger) is considered a non-matching print (or an impostor) attempt.

While establishing the database, the way that the data are captured is critical. It is very important to understand the concept of "independent trials" when performing both non-matching print and matching print testing. An example of two trials that are not independent is capturing sequential fingerprint images without the user removing his/her finger from the platen of the scanner. A recommendation for database capture is that the data represent as well as possible the full range of expected variations in the individual user's print. Generally, capturing data over a period of time (or seasons) is a good practice. The database may also be established so that either a one or a two trial access attempt can be simulated. In this case, the user would present two prints per session (not necessarily independent trials) in such a way that it is assumed that the first attempt is denied access.

The output of database testing should be either a match/no match decision or a matching score, which can be used to obtain error rates. The algorithm component level FRR and FAR can be obtained from these match scores; however, the system level FRR cannot be obtained from them. The system level FAR is assumed to be the same as the component level FAR. The error rates obtained from this testing should be stated as a function of print quality classification; i.e., results should be given separately for dry prints, neutral prints, and oily prints, as well as an overall error rate. This helps extrapolate error rates between different types of fingerprint quality or skin condition.

Although this discussion focuses mainly on algorithm testing, it is important to note that all components of a system can be tested in a similar manner. However, there are components which may be available in some systems, yet not in others. For example, some fingerprint recognition systems may contain a component that indicates when a device has been tampered with; other systems may not contain this component. If the two systems are to be compared, the existence of the additional component must be taken into account. However, it should not be assumed that the absence of a component makes a system less desirable; in fact, additional components can work to increase the FRR of systems.

7.4.1.3 Step 3 – Limited Field Testing (System Level)

The most valuable information for the vendor or customer is knowing what the system level performance is for a particular application. Therefore, the most important step in testing a biometric system is the field test, for this is when a system is called to perform as the sum of its components, with another unreliable factor – the user. However, recall that system level testing cannot pinpoint which component is causing problems.

Although database testing indicates the performance of the algorithms and the scanner, most databases only consist of valid data; no "garbage images" are included. Garbage can include instances when the scanner captured an image although no fingerprint was present, whether due to poor scanner calibration or a dirty scanner platen. They can include images where the fingerprint was not correctly placed (excessive translation, too much pressure, not enough pressure), or images where the fingerprint was rotated by a significant amount. Although these images are usually pruned from databases, they reflect real attempts to gain access by valid users.

In field testing, the biometric system must be set up appropriately for the application for which it is intended; if it is to be used in an ATM, it should be installed in an ATM and used to withdraw money. If it is to be used on a door for physical access control, it should be installed on a door and used in concert with a locking/unlocking device. Field testing for a standard biometric system (one that complies with the standard database parameters) is considered "limited" because no access attempt images need to be stored. No further processing of data is required after enrollment; only on-line processing (accumulation of the number of false rejects and false accepts – if performed) is done.

A log should document the following:

- the ID number of the individual attempting to gain access
- the ID number of the impostor, if the person attempting to gain access is not the authorized user
- whether the attempt was successful or not
- match scores
- comments/remarks about that attempt.

The output of an attempt in a field test is binary. Either the person is granted access or they are not, depending on whether the fingerprints presented matched the template. The system level FRR can be obtained from this test. These error rates should be the sum of the algorithm component level errors, plus the errors induced by garbage images and other component errors (due to scanner, poor presentation of print, etc.). FAR analysis is not required in field testing, since it is assumed that the FAR can be determined from database testing. In fact, it would be impractical to perform an FAR analysis in a field test for which it is desired to compare multiple imposter prints against all of the templates (enrolled valid users) in the database.

7.4.2 Non-Standard Testing Procedure

The following discussion is devoted to testing biometrics systems using a non-standard database. These systems do not comply with parameters associated with the acquisition of the standard database; i.e., the scanner used in such a system may be operating at a higher resolution than the scanner used to build the standard database. Another example would be a print captured at 500 dpi from an IC-type scanner as opposed to an optical scanner. Whereas the standard method of testing suggests conducting preliminary testing, followed by database and then field testing, the non-standard procedure suggests preliminary testing, followed by field testing, and then database testing. The database used in the last step is obtained during the field testing. The advantage of producing a system specific database is repeating component level testing for upgrades of the algorithms.

It is important to note that this procedure will not enable a valid comparison of a non-standard system to a standard system, since different data is presented to both systems (or algorithm/matching components). While it is possible to make comparisons by using multiple biometric systems in parallel (by having an individual present his/her finger to two different scanners), issues such as varying placement (bad alignment) or rotation are not resolved, and the procedure cannot be considered repeatable. An individual could use one scanner which could act as the input to multiple systems; however, this does not resolve resolution issues or issues of incompatibilities among scanner formats.

7.4.2.1 Step 1 – Preliminary Test – System Level

The purpose, result, and procedure for this test are the same as Step 1 in the Standard Testing Procedure described earlier.

7.4.2.2 Step 2 – Full Field Testing – System Level

Full field testing is conducted in order to 1) determine how a complete system will perform in the field, and 2) to obtain images to be stored in a database (relevant to that particular system) for algorithm testing. The setup for this test is the same as for field testing of standard systems, except that fingerprint data for every access attempt will be stored. The field testing is designed for valid user attempts only and not for imposter trials.

A log should be kept, documenting the following:

- the ID number of the individual attempting to gain access,
- the ID number of the impostor, if the person attempting to gain access is not the authorized user.
- whether the attempt was successful or not
- match scores
- comments/remarks about that attempt.

Like the standard system testing procedure, the field FRR can be obtained from these results. Once the field FRR is computed, the database must be pruned. This step removes all the garbage images from the database, thereby removing any effects that are not related to the algorithm. It should be noted again that all valid fingerprint images must be saved in the database, no matter what the print quality or skin condition is. Like Step 3 in the standard testing procedure, it is assumed that the system level FAR and the component level FAR are the same.

7.4.2.3 Step 3 – Database Testing (and Establishment of a System Specific Database)

Like Step 2 in the standard testing procedure, the algorithm is tested using a database to establish component level FAR and FRR. Once the database obtained in the previous step is pruned, the fingerprint images can be enrolled into the system as users, and the same procedure used as in Step 2 of the standard testing procedure.

The component level FRR will be obtained from these results; the difference in FRR between system level and component level testing is due to multiple factors such as the actual application, the hardware scanner, etc.

Fingerprint Class	class 1	class 2	class 3	...	class n
FAR	0.1	0.01	0.001		0.01
FRR	0.1	0.01	0.01		0.5

Figure 33. Example of how the performance of a fingerprint biometric system may vary depending on the characteristics of the input fingerprint. The performance of the overall system will also depend on the nature of the application.

Standard Database Test → $FAR_{C,A}$ / $FRR_{C,A}$ = Component Level (Algorithm) Error Rates

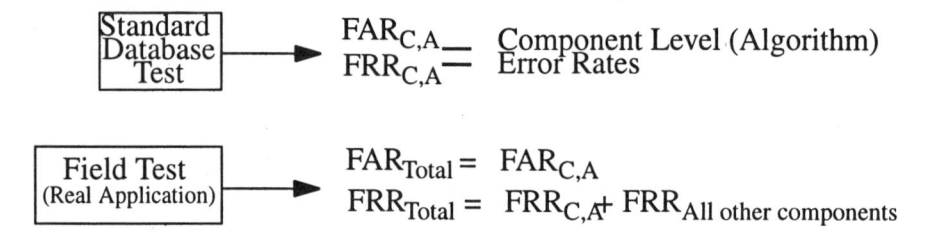

Field Test (Real Application) → $FAR_{Total} = FAR_{C,A}$
$FRR_{Total} = FRR_{C,A} + FRR_{All\ other\ components}$

Figure 34. How error rates relate to database testing (component level) and field testing (system level).

8 Chapter Summary

- The image processing steps used in fingerprint authentication can be augmented by fingerprint-quality analysis.

- A hierarchical approach to matching can enable better performance, especially in the presence of degradation.

- Reliability and uniqueness are the dominant parameters in determining conventional error rates. Parameters of system performance can be quantified and used to predict performance of a system.

- Overall error rates of a system do not provide a solid statistical measure for an *individual's* performance on that system; for this reason, a standard database containing prints representative of all possible users must be implemented in testing. The fingerprints in this database should be parsed in such a way that data describing the individual's personal characteristics (occupation, gender, age) and their particular characteristics (skin condition, print quality) are included.

- Testing should consist of three steps: preliminary testing, component level (database) testing, and field testing. The performance of a system once it is in the field will depend on factors such as the application environment; thus, an effort should be made to simulate the field conditions as much as possible during field testing.

References

[1] Stosz, J.D. and Alyea, L.A. (July 1994), "Automated System for Fingerprint Authentication Using Pores and Ridge Structure," *SPIE Proceedings, Automatic Systems for the Identification and Inspection of Humans*, Vol. 2277, pp. 210-223.

[2] Burt, P.J. and Adelson, E.H. (April 1983), "The Laplacian Pyramid as a Compact Image Code," *IEEE Transactions on Communications*, Vol. COM-31, No. 4, pp. 337-345.

[3] Stellafani, R. and Rosenfeld, A. (1971), "Some Parallel Thinning Algorithms for Digital Pictures," *JACM*, Vol. 18, No. 2, pp. 255-264.

[4] Roddy, A.R. and Stosz, J.D. (September 1997), "Fingerprint Features – Statistical Analysis and System Performance Estimates," *Proceedings of the IEEE*, Vol. 85, No. 9, pp. 1390-1421.

[5] Osterburg, J.W., Parthasarathy, T.W., Raghanvan, T.E.S., and Sclove, S.L. (December 1977), "Development of a Mathematical Formula for the Calculation of Fingerprint Probabilities Based on Individual Characteristics," *Journal of the American Statistical Association*, Vol. 72, No. 360., pp. 772-778.

Chapter 3:

Fingerprint Sub-Classification: A Neural Network Approach

FINGERPRINT SUB-CLASSIFICATION: A NEURAL NETWORK APPROACH

G.A. Drets
Centro de Cálculo, Facultad de Ingeniería
Universidad de la República
Julio Herrera y Reissig 565, C.P.11300, Montevideo
Uruguay

H.G. Liljenström
Theoretical Physics
Royal Institute of Technology
S-100 44, Stockholm
Sweden

Fingerprints provide the most widespread biometric unit of measure used for individual identification. In forensic applications, narrowing the search space is a crucial stage for successful and efficient identification. This chapter offers different aspects of fingerprint sub-classification. It shows the main components of automatic fingerprint identification systems and states motivations for including a sub-classification stage. FBI manual sub-classification procedure and automatic singular point detection methods found in the literature are reviewed. Finally, we present a multi-resolution neural network approach to the singular point detection problem, and a sub-classification procedure for loops which was tested on a subset of 955 NIST-14 fingerprints.

1 Introduction

All through the history of mankind, fingerprints and other dermatoglyphs, have been used for identification of individuals. Prehistoric drawings of the ridge patterns of a hand have been found in Nova Scotia. In the Babylonian civilization (about 2000 BC), fingerprints were used on clay tablets for business transactions. Also, there is text evidence that recaptured Babylonian army deserters were made to leave marks of their thumbs and fingers. In ancient China and Japan, more than 2000 years ago, thumb marks were used as a personal seal. There is also evidence of the use of fingerprints for identification purposes in the Roman Empire. In Persia, around the 14th century, fingerprints were used in various official government papers.

The modern era in fingerprint identification began at the end of last century, with Sir Francis Galton (1880) and Juan Vucetich (1891), and at the beginning of the present century with Sir Edward Henry (1901) [9]. Galton introduced, in his book *Finger Prints* [7], the notion of minutiae. Vucetich and Henry proposed two different classification systems. The Vucetich classification system is still used in most Spanish-speaking countries, while the Henry classification system is primarily used in English-speaking countries.

Human fingerprints have some desirable characteristics that make them the most widespread biometric mean of individual identification, before many others, such as tattoo, speech, face, signature or body measurements (also known as the Bertillon system). The desired characteristics include:

- The fingerprint pattern is kept constant during the lifetime of the individual.
- Fingerprints are formed in the very early stages of conception, determined by several genes from different chromosomes, and could be affected by variations in the environment [22]. This makes variations in fingerprint patterns almost unlimited, and each one of them unique. (As a matter of fact, to date, no two fingerprints have been found to be identical).
- Fingerprint identification is a non-invasive procedure.

Fingerprints are commonly associated with law enforcement applications, like criminal identification, but recently their application has been extended to more popular areas, such as access control, driver license applications, and bank transactions [14]. Law enforcement applications deal with large databases, where 10-fingerprint cards of thousands or millions of people are stored. Therefore, initial stages should partition the search space for a successful identification. Civil applications deal with small databases and should have instant response. In this chapter, we will be concerned with applications of the first kind.

With the explosive growth of computational power, the forensic task of fingerprint identification was aided by the development of Automated Fingerprint Identification Systems (AFIS). Contributors to fingerprint research come from academia, government agencies and private companies.

Different AFIS implementations have a structure similar to the one depicted in Figure 1. The first stage is image acquisition, where a fingerprint or a 10-fingerprint card is scanned. The next stage, pre-processing, strongly depends on the AFIS implementation. In general, pre-processing segments the image aiming to discard uninteresting areas in favor of the central part of the fingerprint. In some implementations the image is deskewed to nullify a possible rotation angle, and it is registered by placing the fingerprint in a standard position. Finally, the image is enhanced to emphasize the fingerprint ridge/valley structure and to eliminate noise. The output of the pre-processing stage may be a skeletonized binary image or a gray-scale enhanced image.

After preprocessing, two independent stages follow: classification and minutiae detection. The former reduces the search space, grouping the fingerprints in coarse classes according to global patterns. This, again, is implementation dependent. Usual classes that appear in the classification stage are, arch (tented or plain), loop (left and right[1]), whorl and scar (see Figure 2). The minutiae-detection stage localizes points such as ridge endings and bifurcations.

---

[1] In manual classification, loops are classified as radial or ulnar. Radial loops correspond to right loops in the left hand impression (and to left loops in the right hand impression). The reverse is valid for ulnar loop impressions. Manual classification depends on the orientation of the loop in the hand, whereas automatic classification normally corresponds to the orientation in the print.

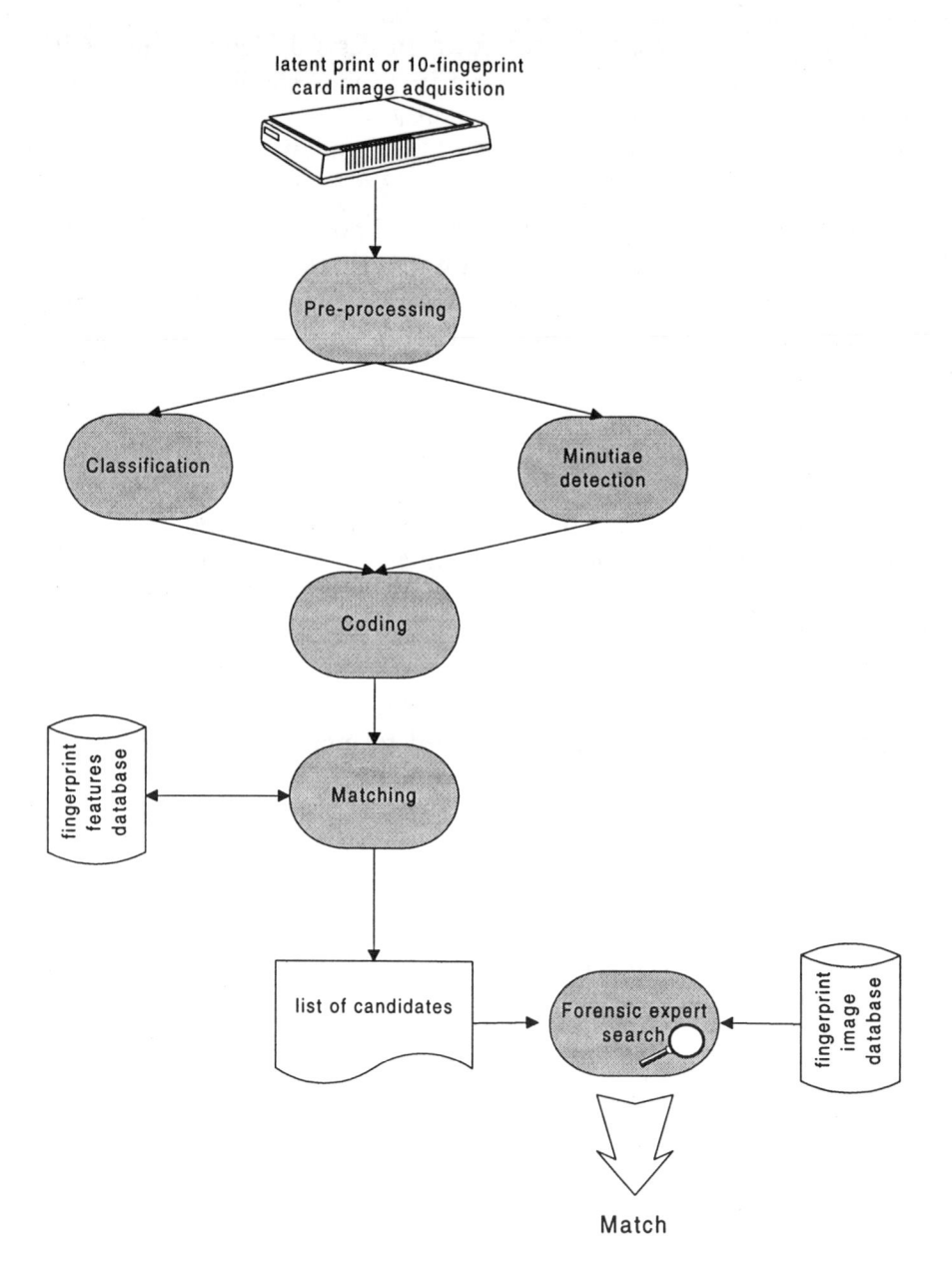

Figure 1. AFIS structure. In automatic fingerprint identification, the fingerprint goes through several stages in which features are extracted for comparison with others. The result is a list of similar fingerprints to the searched one, which are manually compared by a forensic expert.

Classification and minutiae are coded together into a feature vector. Matching compares the feature vector with the records of the features database. This stage could be considered as a fine-classification stage and its output is an ordered list of fingerprints similar to the input one.

Finally, a forensic expert manually compares the images of the matched fingerprints with the searched one.

Classification turns out to be essential to reduce the set of fingerprints that the matching stage will consider. Since fingerprints are unevenly classified (65% of the fingerprints are loops, 30% are whorls [6]) and the number of principal classes is small, the classification stage alone does not narrow the search enough in the features database. This motivates an intermediate stage known as *fingerprint sub-classification.*

2 Sub-Classification Techniques

In this section we define the singular (focal) points (SP), which are the base for fingerprint sub-classification procedures. We also present FBI manual sub-classification procedures for loops (known as ridge counting) and whorls (known as whorl tracing), as well as the FBI classification formula and searching procedures. Finally, we briefly review automatic SP detection methods found in the literature.

2.1 Singular Points

Fingerprints can show two different types of SPs, namely the core and the delta. According to Henry [9], the core point is defined as the topmost point on the innermost upward recurving ridge. The delta point is defined as the point on a ridge splitting into two branches, which extends to encompass the complete pattern area (see Figure 3).

The importance of SPs is threefold:

1. Some classification procedures are based on the number and position of SPs [11], [12], [16].
2. SPs are the basis for fingerprint sub-classification [6], [16].

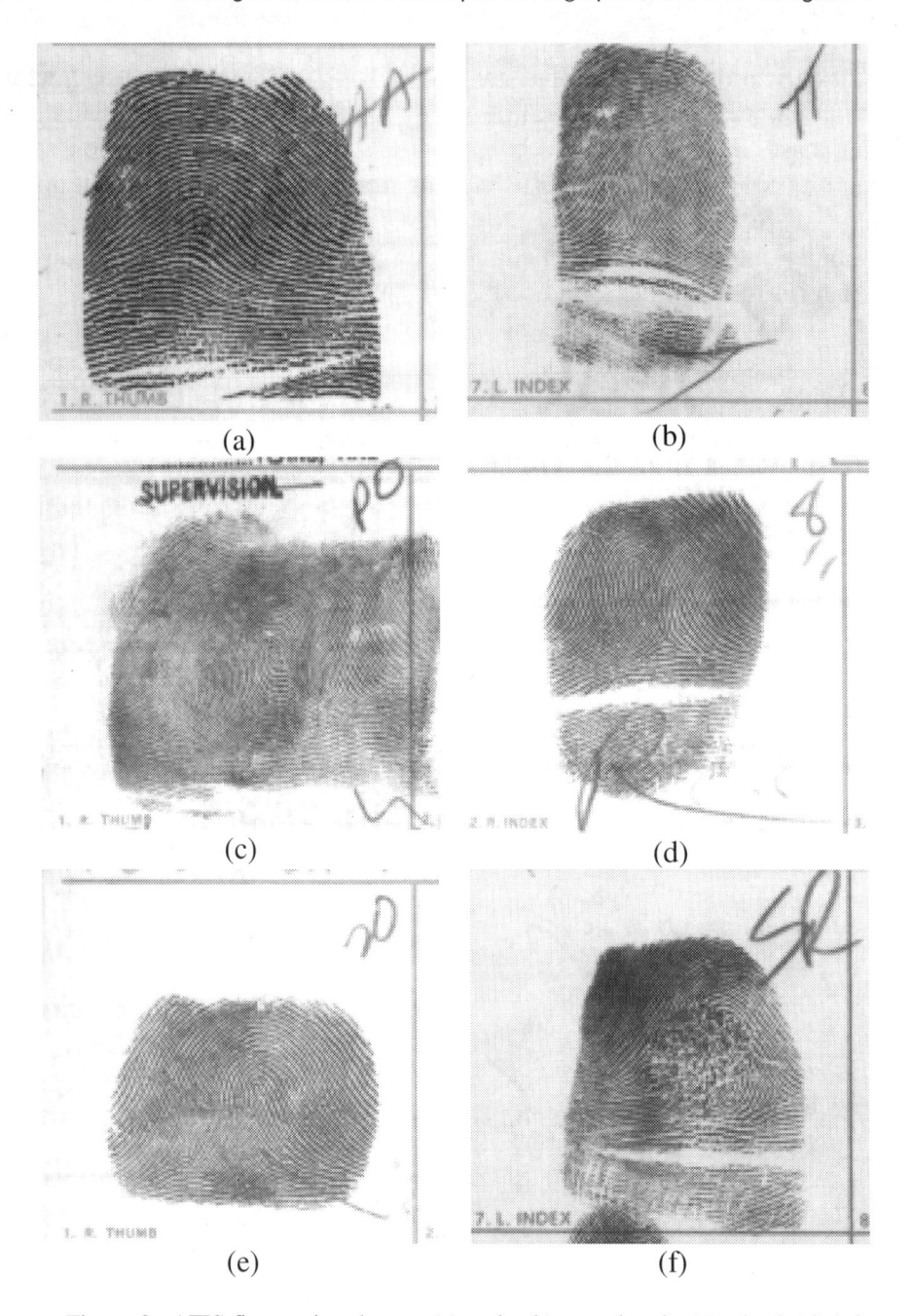

114   Intelligent Biometric Techniques in Fingerprint and Face Recognition

Figure 2. AFIS fingerprint classes. (a) arch, (b) tented arch, (c) whorl, (d) left loop, (e) right loop, (f) scar. (NIST-14 s0024501, s0024377, f0000002, f0000001, and f0026117, respectively).

3. SPs are commonly used as a registration mean, allowing scale, translation and rotation of the fingerprint image to standard positions. Registration is used in some implementations of classification and matching procedures [4], [8], [11], [19], [30].

2.2 Ridge Counting

As previously stated, loop sub-classification is also known as ridge counting, and it is defined as the number of ridges that intersect an imaginary line drawn between the core and the delta points. An example of ridge counting is depicted in Figure 3.

FBI procedures for ridge counting include some special rules:

- If the line crosses at the point of bifurcation or at an island, it is counted as two ridges.

- Fragments or dots are counted only in case their width is similar to other ridges.

As we will see later, the FBI uses the number of ridges to build the classification formula.

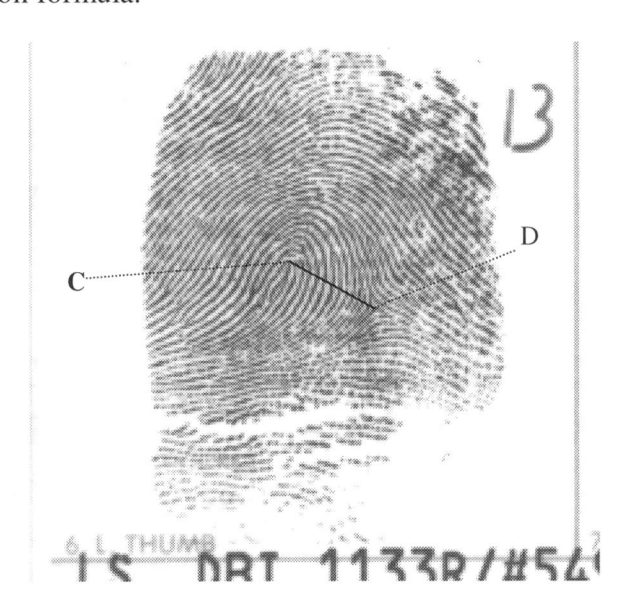

Figure 3. Ridge counting for fingerprint NIST-14 s0000066, number of lines between core (C) and delta (D) is 13.

2.3 Whorl Tracing

Whorls may have two or more deltas. For whorl sub-classification the ridge just below the extreme left delta is traced until the closest position to the extreme right delta is reached. In case the ridge traced ends, the tracing continues with the one below it. The number of ridges intervening between the rightmost delta and the point found in the traced ridge is counted. Figure 4 shows an example of whorl tracing.

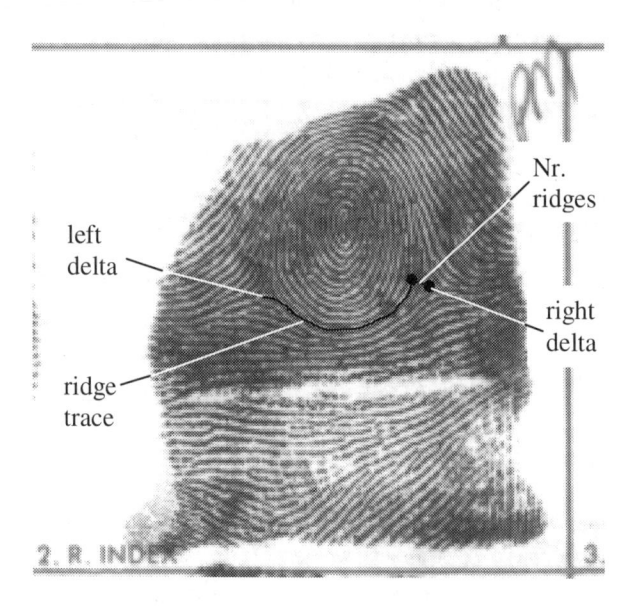

Figure 4. Whorl tracing for NIST-14 f0000022. The ridge traced passes over the right delta, but the number of ridges is not greater than 3 (meeting trace).

2.4 FBI Classification Formula

There are four elements that determine the FBI formula associated with a 10-fingerprint card: the fingerprint classification, the position within the card, the ridge count and the whorl trace [6]. The FBI classification formula comprises the following components which are briefly reviewed: (1) primary classification, (2) secondary classification, (3) sub-secondary classification, (4) the major division, (5) the final and (6) the key, and three extensions: (a) second-sub-secondary classification, (b) WCDX extension, and (c) a special loop extension.

1. **Primary classification:** This component is concerned with the position of whorls within the card. Each position in the fingerprint card is assigned a value as depicted in Table 1. The primary classification basically consists of summing for each hand the value corresponding to the whorls and adding one to the final value.

Table 1. Number assignment used for the primary classification.

16 Right Thumb	16 Index	8 Middle	8 Ring	4 Little
4 Left Thumb	2 Index	2 Middle	1 Ring	1 Little

2. **Secondary classification:** The fingerprints from the index fingers are classified as arch (A), tented arch (T), radial loop (R), ulnar loop (U) or whorl (W). Other fingerprints are classified as arch (a), tented arch (t) or radial loop (r).

3. **Sub-secondary classification:** This part of the formula is based on ridge count and whorl trace. Loops are grouped according to the ridge count value into I or O as shown in Table 2.

Table 2. Grouping of loops by ridge count for the sub-secondary classification.

Finger	Ridge count	Sub-secondary Classification
index	0-9	I
	10+	O
middle	0-10	I
	11+	O
ring	0-13	I
	14+	O

Whorls are grouped into inner (I), outer (O) or meeting (M) trace:
- I: the ridge traced passes over the rightmost delta and the number of ridges is larger than or equal to 3.

- **O:** the ridge traced passes bellow the rightmost delta and the number of ridges is larger than or equal to 3.
- **M:** otherwise.

4. Major division: Concerned with the whorl trace or ridge count of thumbs if present. The whorl trace is similar to the one described for the sub-secondary classification. The ridge count for thumbs is grouped into S (small), M (medium) or L (large) as depicted in Table 3.

Table 3. Grouping by ridge count for the major division.

Left Thumb (ridge count-major division)	Right Thumb (ridge count-major division)
01-11 S	01-11 S 12-16 M 17 + L
12-16 M	01-11 S 12-16 M 17 + L
17+ L	01-17 S 18-22 M 23+ L

5. The final: Ridge count of the right little finger —if this fingerprint is not a loop the left little finger or a ridge count in a whorl (not described here) is used.

6. The key: Ridge count of the first loop (except little fingers, which are considered in the final)

The extensions to the formula are:

  a. **Second sub-secondary classification:** changes the nomination of the sub-secondary classification from (I and O) to (S, M and L).
  b. **WCDX extension:** classifies the whorls into plain (W), central pocket (C), double loop (D) and accidental (X).
  c. **Special loop extension:** the all loop group is the most common within the 10-fingerprint card database [18]. Therefore, this extension allows to sub-divide it considering small groups of similar ridge count.

2.5 Search Procedure

As we have seen so far, counting ridges is one of the main components in fingerprint classification[2]. Since its value may be affected by various factors, such as noise and human subjectivity, the comparison of the number of ridges between two fingerprints should be relaxed.

When ±1-ridge changes the nomination (from I to O, S to M, M to L, etc.) in any of part of the formula, then both groups should be searched. In the case of the final and the key, groups within ±2-ridge count are searched.

2.6 Previous Automatic SP Detection Methods.

Fingerprint analysis is currently a field of active research. Particularly, the SP detection has been addressed by different approaches.

Early in the 1970s, Ausherman *et al.* [2] used the Fourier transform to locate core points. In the 1980s, Rao and Balck [20] made a syntactic description of the fingerprints and the singular points. In this approach, they have used strings to characterize patterns. The use of grammars for fingerprint identification was abandoned, due to the fact that the great diversity of patterns makes the approach unsuitable.

Wegstein [30] proposed a core detector known as the R92 registration procedure. The aim was to center incoming fingerprints, easing the matching stage. The R92 is based on groups of minutiae descriptors. It is still in use today by some classification procedures [4].

Srinivasan and Murthy [23] proposed the use of directional histograms for detecting singular points.

Kawagoe and Tojo [12] introduced the idea of using the "Poincaré index" for identifying core and delta points. When making a counter-clockwise turn around a certain point in the directional image, if the angle of the directions changes 180° then the point is labeled as core, and if the change is -180° it is labeled as delta. This is the most popular

---

[2] Counting ridges is also commonly used for representing relations between pair of minutiae. These relations are used in the matching stage [13][19].

approach today; it was adopted by Karu and Jain [11] and extended with fuzzy rules by Vizcaya and Gerhardt [26].

Finally, Lynch and Gaunt [16] and Trenkle [25] introduced the use of artificial neural networks for fingerprint SP detection. A sub-classification algorithm based on a multi-resolution neural SP detection method is presented in the next section.

3 A Neural Network Approach

Due to the great variety of fingerprint patterns, the application of Henry's definitions of core and delta, stated previously, is not straightforward. The FBI proposes more specialized rules for locating the SPs in different archetypes of fingerprints. The rules can be followed by a forensic expert, but are unfeasible to be implemented by a computer algorithm.

In addition to the intrinsic difficulties that fingerprints present, there are several sources of noise that may corrupt the fingerprint pattern:

- Differences in pressure used by the person taking the print.
- An uneven ink distribution in the fingerprint. While some parts may be over-inked leaving black spots, others may be under-inked, making it hard or impossible to count ridges, trace whorls or to distinguish useful points.
- The finger may slip or twist in the printing procedure, resulting in a smeared or blurred print.
- The appearance of foreign substances or perspiration in the fingerprint or paper, which corrupts the ridge-valley structure.
- The presence of extra elements, such as forensic dabs, writings or marks.

The problems discussed above provide a typical case in which artificial neural networks are known to manage better than conventional methods: when there are unknown rules relating the data (or too complex to write down in a computer code), and the presence of noise in the data.

The algorithm described in the following subsections implements a loop ridge-counting sub-system [5], that is part of an AFIS prototype [1]. It is formed by three major components (see Figure 5): (1) Preprocessing, (2) Singular Point Detection, and (3) Ridge Counting.

The preprocessing stage reads the input image and for every pixel computes the direction of the ridge at the point where the pixel is located. The directional field formed is the input to the singular point detection stage, which finds the position of the core and the delta points. The location of these points is used to draw a line between them and compute the intersection with the ridges. This process is described in more detail below.

3.1 Preprocessing

As shown in Figure 5, the preprocessing stage [4] consists of three steps:

Step 1: Segmentation. The goal of segmentation is to reduce the size of the input data, eliminating undesired background in favor of the central part of the print. A foreground mask is computed by binarizing the input image. This mask is eroded to eliminate noise, forensic dabs, and other marks. Then, the largest connected region is selected.

A tilted bounding box around the region found is computed, centered at the centroid of the mask and having the fingerprint orientation. Hence, the output of this step is a segmented and deskewed image, as the one shown in Figure 6 (a).

Step 2: Enhancement-I. In this step, square blocks of 32×32 pixels are extracted from the image with 8 pixels of overlapping. In each of these blocks, a band pass filter having the frequency of the ridges is applied, hence discarding undesired frequencies considered as noise.

The power spectrum of the remaining frequencies is amplified and the DC component is restored. Figure 6 (b) shows the result of enhancing NIST-14 fingerprint s0000068.

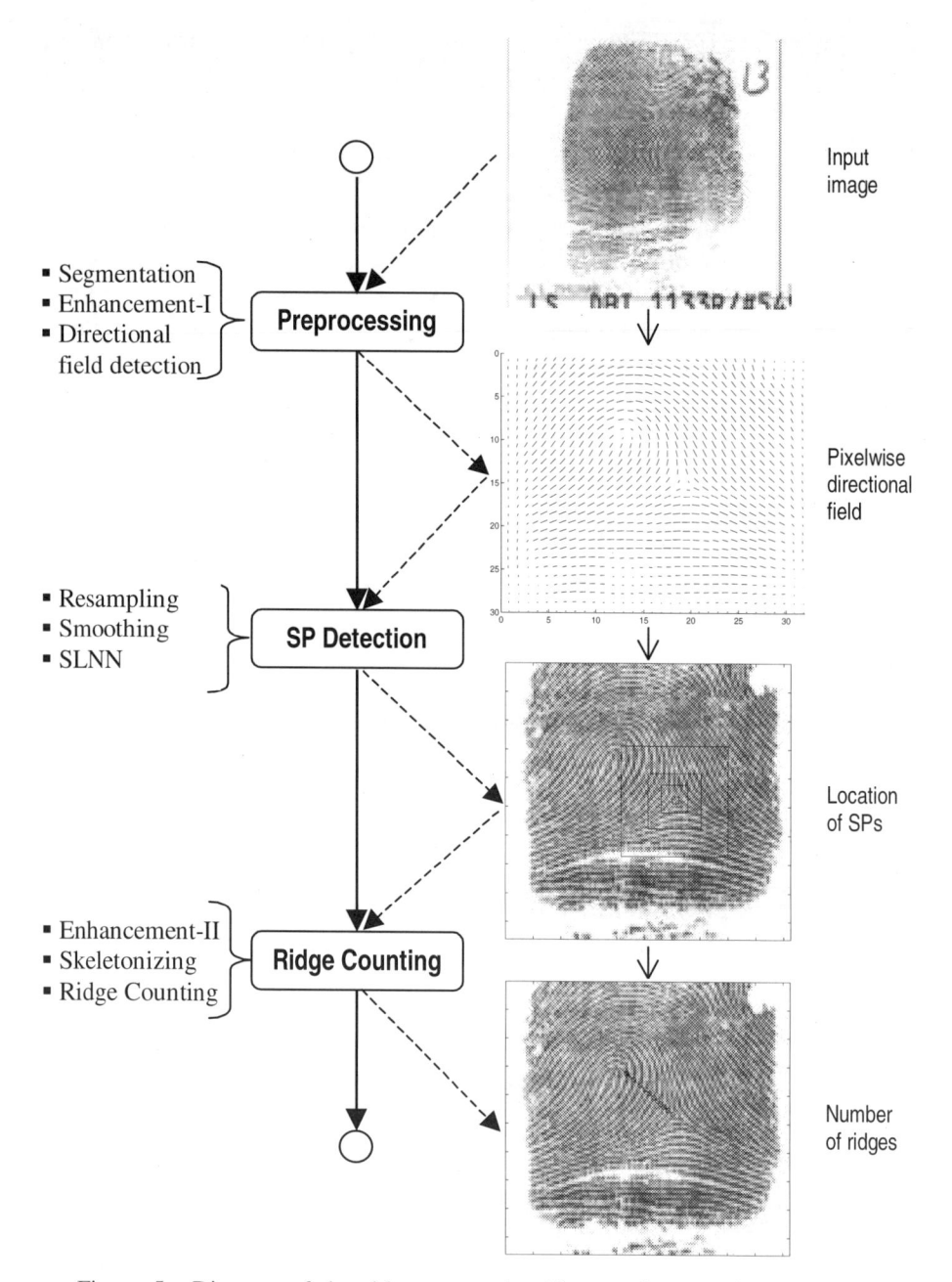

Figure 5. Diagram of the ridge count algorithm: main procedures (center), their components (left) and the image flow (right) (fingerprint NIST-14 s0000068). The Pre-processing stage finds the principal direction for each pixel of the image. Then, the SP Detection stage locates the core and the delta points. Finally, the Ridge Counting stage counts the number of ridges that intersect a line drawn between these two points.

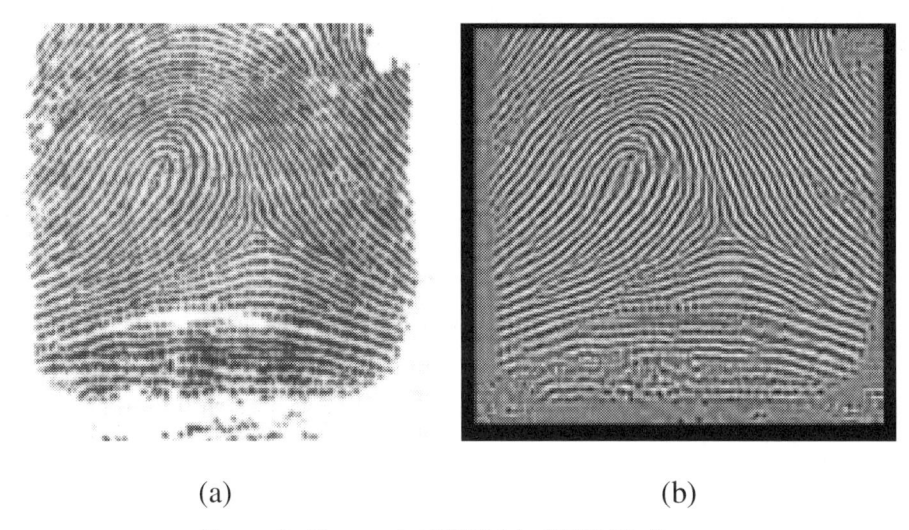

Fingerprint Sub-Classification: A Neural Network Approach 123

(a) (b)

Figure 6. Fingerprint NIST-14 s0000068 after segmentation (a) and after enhancement-I (b)

Step 3: Directional Field Detection. In this step, we compute, at each pixel location of the enhanced image (G), the local orientation of the ridge or valley. A common approach for finding ridge directions in fingerprints is the use of slits (see Figure 7), in which a discrete approximation of the ridge/valley direction is estimated. For each pixel (C), the values of the pixels in G corresponding to the 8 different slits (s_i, $i=1..8$) are summed. For example $s_2 = G(j-2,i+1) + G(j-4,i+2) + G(j+2,i-1) + G(j+4,i-2)$.

7	8		1		2	3
6	7	8	1	2	3	4
	6				4	
5	5		C		5	5
	4				6	
4	3	2	1	8	7	6
3	2		1		8	7

Figure 7. Eight slits are used to estimate the direction of the ridge/valley at pixel C.

The direction assigned to pixel C as suggested by Stock and Swonger [24] is computed in the following way:

$$dir(C) = \begin{cases} i \mid s_i = s_{max} & \text{if} \quad 4C + s_{min} + s_{max} > \frac{3}{8}\sum_{i=1}^{8} s_i \\ i \mid s_i = s_{min} & \text{otherwise} \end{cases}$$

Hence, the output of this step is a pixel-wise directional field of the same size as the segmented image.

3.2 Singular Point Detection

The SP detection is based on the fact that the directional field at the delta and core points is stable, i.e. the pattern repeats itself at different resolution levels. This idea is, in a sense, analogous to the self-similarity property of fractals [10] (see Figure 8). This fact is shown in Figure 9, where a fingerprint directional field is resampled at three different resolutions. Based on this hypothesis, we can use the same SP locator at different resolutions.

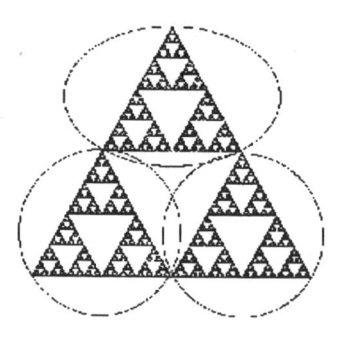

Figure 8. Fractal Sierpinski triangle: at different scales the same pattern repeats.

The multi-resolution approach has the advantage that, at low resolution the method considers the global pattern, and at high resolution, it searches for local features.

The algorithm is basically an iterative procedure at increasing resolution levels, composed of the following steps:

Step 1: Resampling. Since the pixel-wise directional field detected in the pre-processing stage is very noisy and coarsely quantized, it is necessary to resample it by averaging pixel values in local neighborhoods.

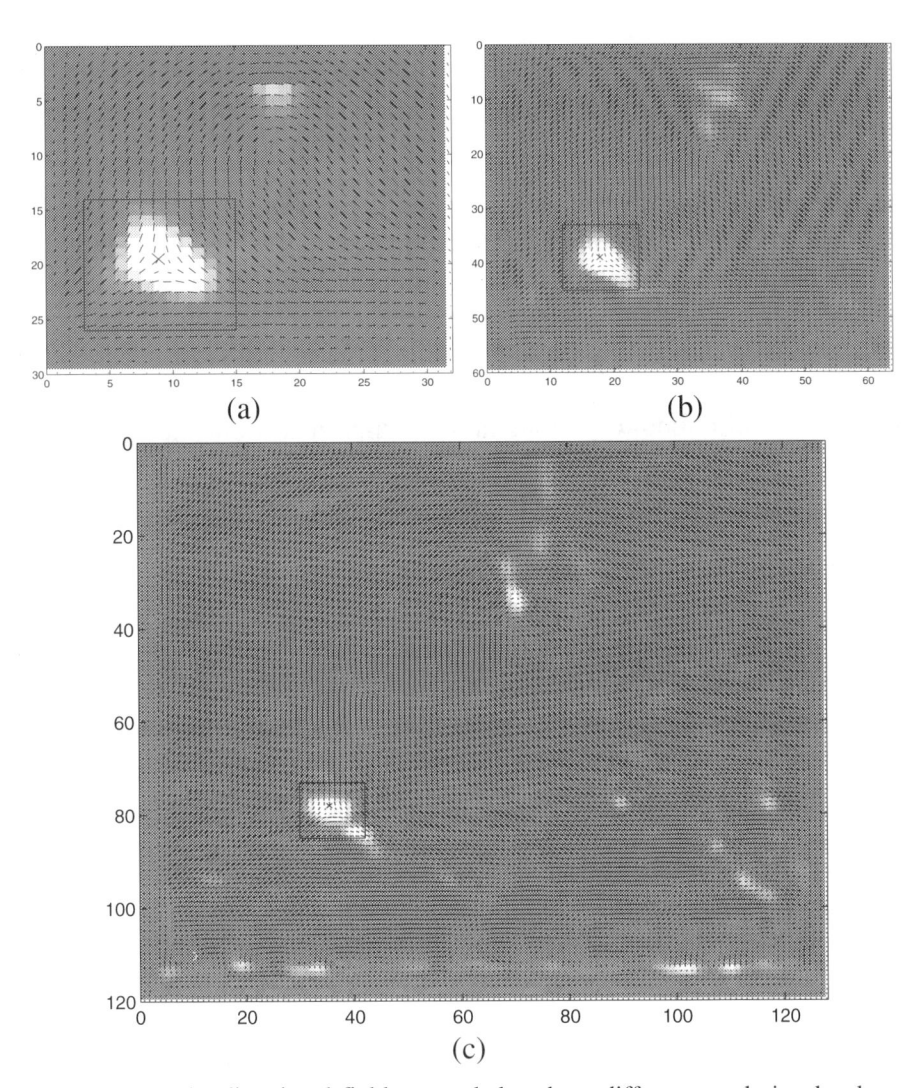

Fingerprint Sub-Classification: A Neural Network Approach

Figure 9. The directional field resampled at three different resolution levels: (a) 16, (b) 8, and (c) 4 pixels, with the neural network output overlapped and the square indicating the region of maximum response. Note how the delta and core patterns are preserved for these resolutions. (In the process only the directions within the plotted square are resampled.)

The discrete values of the pixel-wise directional field are represented as vectors of the form *(cos(2θ), sin(2θ))* [31]. The new representation of the directional field is divided into square blocks whose size depends on the level of the resolution for the iteration considered. In each block, the average value of the vectors is computed. Therefore, the output of this stage is a weighted directional field of the form $(\lambda cos(2\theta), \lambda sin(2\theta))$ as shown in Figure 9.

Step 2, Smoothing: This second smoothing step is done by convolving the directional field with a 3×3 average box filter. Figure 10 shows an example of the improvement in the SP location obtained by smoothing of the directional field. If the direction found comes from a good quality region, then the neighboring regions will very likely have the same direction and will not affect it. But if the region is noisy, then the direction will be corrected by the general trend. The smoothing of the directional field should be constrained, since it not only lessens noise but also corrupts points with large direction changes like the cores and deltas.

Step 3, Singular point detection by a sliding neural network: In this stage, a sliding neural network (SLNN) [25] moves across the directional field taking samples. At every step, the SLNN focuses a region of size bz ($bz = 8$ was selected) in the directional field which it uses as an input vector, and gives as output a value in the range [0,1], which could be interpreted as the confidence on the underlying pattern of being the searched SP.

The area of maximum response is located in the SLNN output map. If the normalized summed of SLNN output values for this area is greater than a certain threshold, the procedure continues from step 1, by resampling the directional field in this area at a higher resolution.

Figure 9 shows the SLNN response at three different resolution levels of the directional field (16, 8, and 4 pixels, respectively), and the area where the maximum was located.

It is possible to see the appearance of local maxima at high-resolution levels, which are not considered by the process since they are outside the area of maximum response of the previous resolution.

The SLNN is similar to the one proposed by Trenkle [25], but here the set of weighted vectors was used as input. The reason for this was to improve the response for noisy SP patterns.

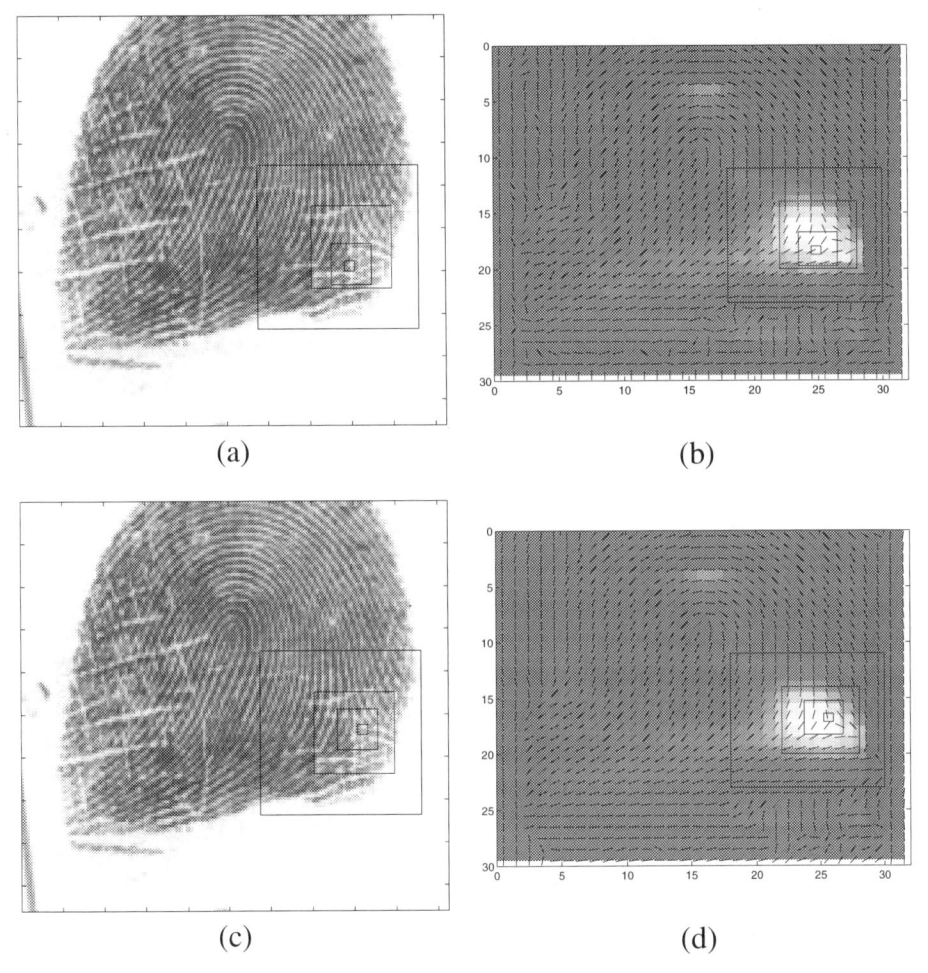

Fingerprint Sub-Classification: A Neural Network Approach

Figure 10. Smoothing of the directional field for NIST-14 s0024310: (a) delta point location and (b) directional field without smoothing. (c) Improved location of the delta point, due to the smoothing of the directional field (d).

Two multi-layer perceptron (MLP) [21] [32] networks were used (one for cores and another for deltas), with 128 input nodes (bz × bz × 2 components), 16 sigmoid nodes in the single hidden layer, and one sigmoid node in the output layer (Figure 11).

For training the networks, overlapping samples of SP and non-SP regions were extracted from the directional field of 40 randomly selected NIST-14 fingerprints. The training set was artificially extended to consider possible rotations of the patterns.

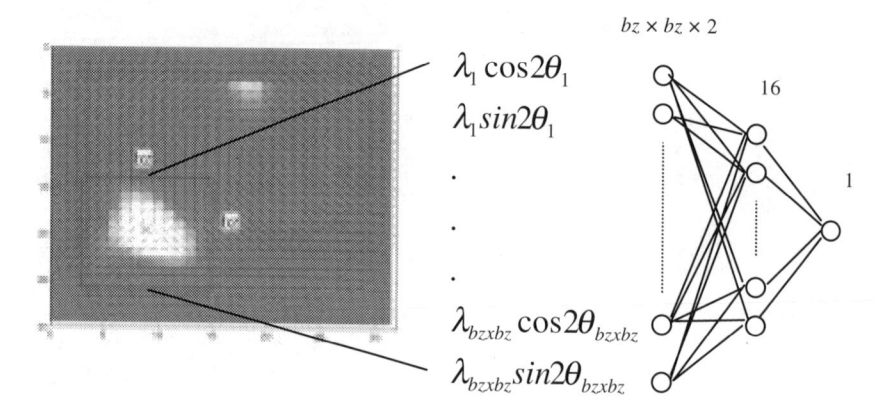

Figure 11. Architecture of the delta detector. Blocks of size bz × bz extracted from the directional field, feed a SLNN (each direction is represented as a vector, hence, the number of input nodes is bz × bz × 2). The network has 16 nodes in the hidden layer and 1 node in the output layer. A similar architecture is used for core detection.

3.3 Ridge Count

With core and delta located, the ridge counting stage follows. This process counts the number of intersections of the fingerprint skeleton with an imaginary line drawn between the two singular points.

Prior to the construction of the skeleton, a second and more powerful enhancement stage [3] is included for improving the ridge-valley structure and the local connectivity. This is done by applying, in local neighborhoods, a Gabor filter that depends on the ridge direction (θ), and frequency (ω):

$$f(x, y) = e^{-\frac{1}{2}\left(\frac{x^2}{\sigma_x^2} + \frac{y^2}{\sigma_y^2}\right)} \cos(2\pi\omega x)$$

$$f_\theta(\vec{x}) = f(R_{-\theta}\vec{x})$$

where $\vec{x}$ denotes the vector (x,y) and
$R_{-\theta}$ denotes the rotation by angle θ.

After enhancing the fingerprint, a simple binarization follows and the fingerprint skeleton is built. The intersections of the line drawn between the core and delta points with the ridges are found. Figure 12 shows examples of a ridge counting on NIST-14 fingerprints.

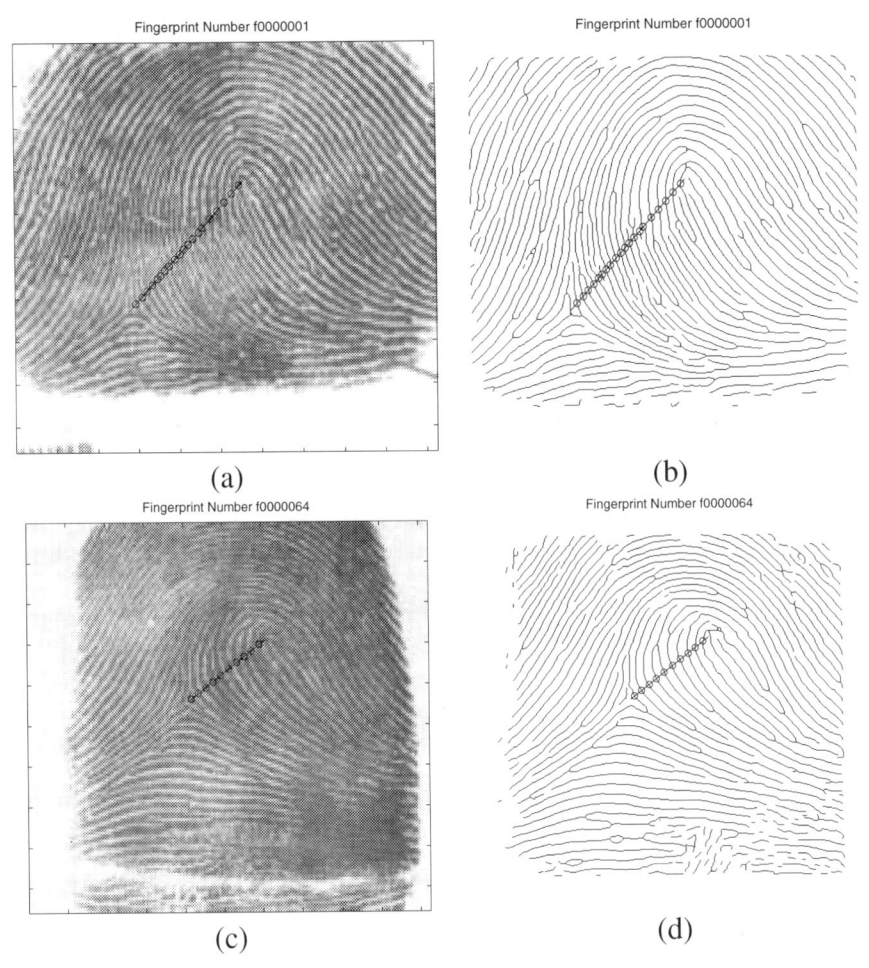

Figure 12. Ridge count resulting from the proposed procedure. Ridge count=20 for fingerprint NIST-14 f0000001 ((a) shows the intersections marked on the fingerprint and (b) on the skeleton). Ridge count = 10 for NIST-14 f0000064 (c) and (d).

3.4 Results

A subset of the NIST-14 fingerprint database was selected, since, together with NIST-4 and NIST-9 databases [27], [28], [29], these have become the *de facto* benchmarks for testing AFIS algorithms. The NIST-14 database consists of two different rolls for each fingerprint, called the *f* (file) and the *s* (search) roll respectively. The first 2000 NIST-14 were chosen, from which only those classified as non referenced loops were considered (443 f-roll and 512 s-roll fingerprints). There was no previous fingerprint rejection of low-quality fingerprints. This procedure was able to detect both core and delta for 367 f-roll prints (A) and 460 s-roll prints (B). The reason for the undetected SPs, in most of the cases, was due to the low quality of the images or due to an incomplete fingerprint roll. Hence, in these cases, it is not possible to detect the core, the delta or both points, even under visual inspection. Sets A and B had 339 pairs of fingerprints in common and results show that 81.4% of the s-rolls had a number of ridges that was within the two-ridge count band from its corresponding f-roll considered for the key and the final (see Table 4).

Table 4. The results shown represent the percentage of fingerprints (from a total of 339 pairs) and the ridge-count difference between the f and the s-roll.

Number of ridges	0	1	2	3	4	5	6	7	8	9	10	11	12
%	25.37	34.22	21.83	7.96	3.24	3.24	2.06	0.88	0	0.59	0	0.29	0.29

The occasionally large ridge-count difference was primarily due to corrupted skeletons, originating from low-quality fingerprints.

4 Conclusions and Discussion

Human fingerprints are the most widespread mean of person identification. The use of fingerprints in law enforcement applications deals with large databases, so classification aids to narrow the search space for further matching stage. The process of counting ridges between singular points – core and delta – is the basis for fingerprint sub-classification.

A sub-classification system was built based on a multi-resolution neural network singular point detector, and showed that it can be used to aid the forensic task of ridge counting, whorl-tracing, or finding relations between minutiae.

The algorithm is invariant to translations and scale changes, and tolerant to small rotations.

Although the results are quite satisfactory, we believe that the accuracy of the sub-classification procedure can be further improved. Some possible improvements include:

- The threshold used for determining if a region is a possible SP is the same for all resolutions. A threshold that depends on the resolution level is a better approximation since the stability hypothesis should be relaxed at increasing resolution levels.

- The localization of the maximum could be improved by using scale space theory, therefore, combining its location at different resolutions [15].

- Alternative and more robust methods for smoothing directions and counting ridges between points could be applied like the one proposed by Mardia *et al.* [17].

- Rejection of low-quality fingerprints.

To conclude, we believe that artificial neural networks can be a useful and powerful method to be considered for fingerprint identification.

Acknowledgments

This work was supported by grants from the National Council of Scientific and Technical Research (CONICYT-Uruguay) 96/94 (GD), from the Swedish International Development Cooperation Agency BITS URY0233 (GD and HL) and from The Swedish Research Council for Engineering Sciences, TFR 96-732 (HL).

We are also grateful to the Department of Numerical Analysis and Computer Science (NADA) at the Royal Institute of Technology in Sweden, where part of the work was done.

References

[1] Almansa, A., Curbelo, R., and Drets, G. (1996), "AFIS Prototype 0, Development of an Automatic Fingerprint Identification System," *Project BID/CONICYT 96/94 Report*, Universidad de la República, Uruguay.

[2] Aushermann, D.A., Fairchild, R.C., Moyers, R.E., Hall, W.D., and Mitchel, T.H. (1973), "A Proposed Method for the Analysis of Dermatoglyphics Patterns," *Proceedings of the International Society for Optical Engineering SPIE*, Vol. 40, pp. 171–180.

[3] Bergengruen, O. (1994), "Pre-processing of Poor Quality Fingerprint Images," *Proceedings of the XIV Int. Conf. of the Chilean Computer Science Society*, Concepción, pp. 43-54.

[4] Candela, G.T., Grother, P.J., Watson, C.I., Wilkinson, R.A., and Wilson, C.L. (1995), "PCASYS – A Pattern-level Classification Automation System for Fingerprints," *NISTIR 5647*, National Institute of Standards and Technology.

[5] Drets, G. and Liljenström, H.G. (1998), "Fingerprint Subclassification and Singular Point Detection," *International Journal of Pattern Recognition and Artificial Intelligence*, Vol. 12, No. 4, pp. 407–422.

[6] FBI (1984), *The Science of Fingerprints*, U.S. Department of Justice, Washington, D.C.

[7] Galton, F. (1892), *Finger Prints*, Macmillan, New York.

[8] Halici, U. and Ongun, G. (1996), "Fingerprint Classification Through Self-organizing Feature Maps Modified to Treat Uncertainties," *Proceedings of the IEEE*, Vol. 84, No. 10, pp. 1497-1512.

[9] Henry, E.R. (1900), *Classification and Uses of Fingerprints*, George Routledge and Sons, London.

[10] Hutchinson, J. (1981), "Fractals and Self-Similarity," *Journal of Mathematics*, Indiana University, Vol. 30, pp. 713-747.

[11] Karu, K. and Jain, A.K. (1996), "Fingerprint Classification," *Pattern Recognition,* Vol. 29, No. 3, pp. 389-404.

[12] Kawawoe, M. and Tojo, A. (1984), "Fingerprint Pattern Classification," *Pattern Recognition,* Vol. 17, No. 3, pp. 295-303.

[13] Kuji, K., Hoshino, Y., and Asai, K. (1990), "Automated Fingerprint Identification System (AFIS)," *NEC Research and Development Journal,* No. 96, pp. 143-146.

[14] Lee, H.C. and Gaensslen, R.E. (Eds.) (1991), *Advances in Fingerprint Technology,* Elsevier, New York.

[15] Lindeberg, T. (1994), *Scale Space Theory on Computer Vision,* Kluwer Academic Publishers, Dordrecht.

[16] Lynch, M.R. and Gaunt, R. (1995), "Application of Linear Weight Neural Networks to Fingerprint Recognition," *Proceedings of the 4th IEE International Conference,* No. 409, pp. 139-141.

[17] Mardia, K.V., Baczkowski, A.J., Feng, X., and Hainsworth, T.J. (1997), "Statistical Methods for Automatic Interpretation of Digitally Scanned Finger Prints," *Pattern Recognition Letters,* Vol. 18, pp. 1197-1203.

[18] McCabe, R.M., Wilson C.L., and Grubb, D. (1995), "Research Considerations Regarding FBI-IAFIS, Tasks and Requirements," *NISTIR 4892,* National Institute of Standards and Technology.

[19] Mehtre, B. (1993), "Fingerprint Image Analysis for Automatic Identification," *Machine Vision and Applications,* Vol. 6, pp. 124-139.

[20] Rao, C.V.K. and Black, K. (1980), "Type Classification of Fingerprints: A Syntactic Approach," *IEEE Trans. On Pattern Analysis and Machine Intelligence,* Vol. 2, pp. 223–231.

[21] Rumelhart, D.E. and McClelland, J.L. (Eds.) (1986), *Parallel Distributed Processing: Explorations in the Microstructure of Cognition,* Vol. 1, Cambridge, MA, MIT Press.

[22] Schaumann, B. and Alter, M. (Eds.) (1976), *Dermatoglyphics in Medical Disorders,* Springer-Verlag, Berlin.

[23] Srinivasan, V.S. and Murthy, N.N. (1992), "Detection of Singular Points in Fingerprint Images," *Pattern Recognition*, Vol. 25, No. 2, pp. 139–153.

[24] Stock, R.M. and Swonger, C.M. (1969), *Development of a Reader of Fingerprint Minutiae*, Cornell Aeronautical Laboratory, Technical Report CAL No. XM-2478-X-1:13-17.

[25] Trenkle, J.M. (1994), "Region of Interest Detection for Fingerprint Classification," *Proceedings of The International Society for Optical Engineering SPIE*, Vol. 2103, pp. 48–59.

[26] Vizcaya, P. and Gerhardt, L. (1997), "Multi-resolution Fuzzy Approach for Singularity Detection in Fingerprint Images," *Proceedings of The International Society for Optical Engineering SPIE*, Vol. 2935, pp. 46–56.

[27] Watson, C.I. (1993), *NIST Special Database 9. NIST 8-bit Gray Scale Images of Mated Fingerprint Card Pairs*, National Institute of Standards and Technology.

[28] Watson, C.I. (1993), *NIST Special Database 14. Mated Fingerprint Card Pairs 2*, National Institute of Standards and Technology.

[29] Watson, C.I. and Wilson, C.L. (1992), *NIST Special Database 4. Fingerprint Database*, National Institute of Standards and Technology.

[30] Wegstein, J.H. (1982), "An Automated Fingerprint Identification System," *NBS Special Publication 500-89*, National Bureau of Standards, U. S. Department of Commerce.

[31] Wilson, C.L., Candela, G.T., and Watson, C.I. (1993), "Neural Network Fingerprint Classification," *Journal of Artificial Neural Networks*, Vol. 1, No. 2, pp. 1–25.

[32] Werbos, P. (1974), *Beyond Regression: New Tools for Prediction and Analysis in the Behavioral Sciences*, PhD Thesis, Harvard, Cambridge, MA.

Chapter 4:

A Gabor Filter-Based Method for Fingerprint Identification

A GABOR FILTER-BASED METHOD FOR FINGERPRINT IDENTIFICATION

Y. Hamamoto
Faculty of Engineering, Yamaguchi University, Ube 755, Japan
hamamoto@csse.yamaguchi-u.ac.jp

A Gabor filter-based method for identifying fingerprints is proposed. A set of Gabor filters is applied to a fingerprint image. The Gabor feature is obtained by convolving the filter with the image. In pattern matching, a locally maximized similarity is adopted. The performance of the proposed method is demonstrated on the NIST database 14. From experimental results, we verified the effectiveness of combining Gabor features with the locally maximized similarity.

1 Introduction

The Gabor filter [1] is based on a multi-channel filtering theory for processing visual information in the early stages of the human visual systems. Gabor filters have been shown to be good fits to the receptive field profiles of simple cells in the striate cortex [2], [3]. An important property of Gabor filters is that they have optimal joint localization, or resolution, in both the spatial and the spatial-frequency domains [11]. For these reasons, the Gabor approach has been readily used for texture analysis [4], computer vision [5], and character recognition [6]. Some authors [8], [9] suggest that texture analysis can be devised to identify fingerprints. Motivated by these works, we propose a new Gabor filter-based method for identifying fingerprints. In our approach, the features describing line and edge segments around the core point can be extracted by using Gabor filters. An advantage of the proposed method is that it does not need true minutiae (ridge forks and ending), compared with other conventional methods [12], [13], [14]. Thus, the proposed method can be applied to those fingerprints where the image quality is low but still readable.

2 Gabor Features

A two-dimensional Gabor filter can be viewed as a sinusoidal plane of particular frequency and orientation, modulated by a Gaussian envelope. We will follow notations in [6]. The two-dimensional Gabor filter is defined by

$$f(x, y, \theta_k, \lambda) = \exp\left[-\frac{1}{2}\left\{\frac{(x\cos\theta_k + y\sin\theta_k)^2}{\sigma_x^2} + \frac{(-x\sin\theta_k + y\cos\theta_k)^2}{\sigma_y^2}\right\}\right]$$
$$\cdot \exp\left\{\frac{2\pi(x\cos\theta_k + y\sin\theta_k)}{\lambda}i\right\} \quad (1)$$

where σ_x and σ_y are the standard deviations of the Gaussian envelope along the x- and y-dimensions, respectively, and λ and θ_k are the wavelength and orientation, respectively. The Gabor filter is shown in Figure 1. The spread of the two-dimensional Gaussian is given by standard deviations σ_x and σ_y. Here, σ_x and σ_y are defined by using the value of λ. That is, σ_x and σ_y are a function of λ, respectively.

A rotation of the $x - y$ plane by an angle θ_k will result in a Gabor filter at orientation θ_k. The θ_k is defined by

$$\theta_k = \frac{\pi}{n}(k-1), \quad k = 1, 2, \cdots, n \quad (2)$$

where n denotes the number of orientations.

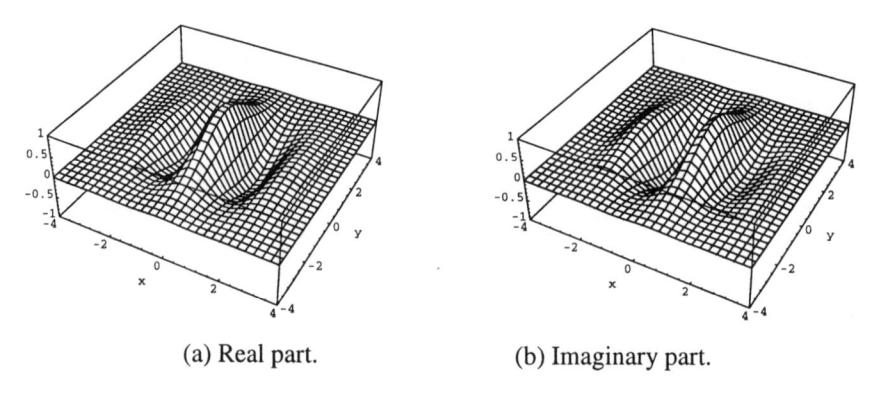

(a) Real part. (b) Imaginary part.

Figure 1. Gabor filter with $\theta_k = 0$, $\lambda = 2\sqrt{2}$, $\sigma_x = 0.5\lambda$, and $\sigma_y = 0.5\lambda$

A Gabor Filter-Based Method for Fingerprint Identification

For instance, when $n = 4$, four values of orientation θ_k are used: $0, \pi/4, \pi/2, 3\pi/4$. The Gabor feature can be viewed as the response of the Gabor filter located at a sampling point. The response is obtained by convolving the filter with an image. The Gabor feature corresponding the filter with particular orientation θ_k and wavelength λ at a sampling point (X, Y), denoted as $g(\cdot)$, is defined as:

$$g(X, Y, \theta_k, \lambda) = \left| \sum_{x=-X}^{N-X-1} \sum_{y=-Y}^{N-Y-1} I(X+x, Y+y) f(x, y, \theta_k, \lambda) \right| \quad (3)$$

where $I(x, y)$ denotes an $N \times N$ gray-scale image and $|z|$ denotes the absolute value of a complex number z.

Figure 2 shows an example of extracting the Gabor feature. Gabor filters tend to detect line and edge segments; in fingerprint images, combinations of such features especially around the core point seem to be good discriminating features.

The selection of sampling points is crucial in our method. In order to determine sampling points, one needs to detect a core point of the fin-

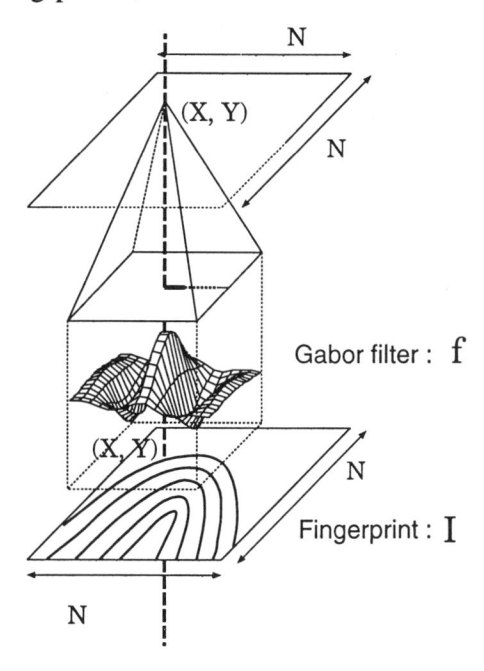

Figure 2. Filtering at a sampling point (X, Y).

gerprint, because minutiae like endings and bifurcations around the core point have significant discriminatory information. In order to detect the core point, we used Shimizu's algorithm [10] because of its simplicity. The detected core point is itself one sampling point. In addition to the core point, 8 points on the circle centered at the core point were also used as sampling points. That is, 9 sampling points were used. Figure 3 shows an example of sampling points. For each sampling point, the Gabor features can be obtained by varying orientations, with the value of λ fixed (shown in Figure 4). The use of n orientations leads to n Gabor features for each sampling point. After all, $9n$ features are extracted for each fingerprint. The Gabor vector consists of $9n$ features. Figure 5 shows an example when $n = 8$.

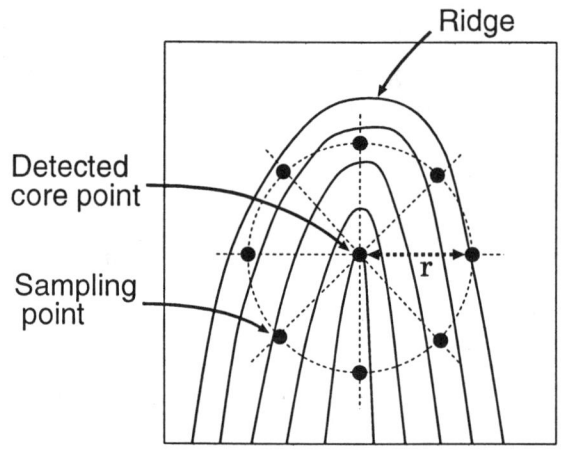

Figure 3. Sampling points.

3 Identification

In pattern matching between two fingerprints, the accuracy of the detected core point is significantly crucial. The inexact core point results in a degradation of the identification performance. In order to overcome this problem, we adopted a locally maximized similarity. The basic idea comes from Yamada et al. [7]. Now, consider pattern matching between fingerprints A and B shown in Figure 6. First, the detected core point is shifted by $(\Delta u, \Delta v)$ and consequently 8 sampling points on the circle are also shifted by $(\Delta u, \Delta v)$. Then, Gabor features are extracted on the shifted sampling points. The resulting Gabor vector is called a shifted Gabor vector. A locally maximized similarity S is now defined by:

A Gabor Filter-Based Method for Fingerprint Identification 141

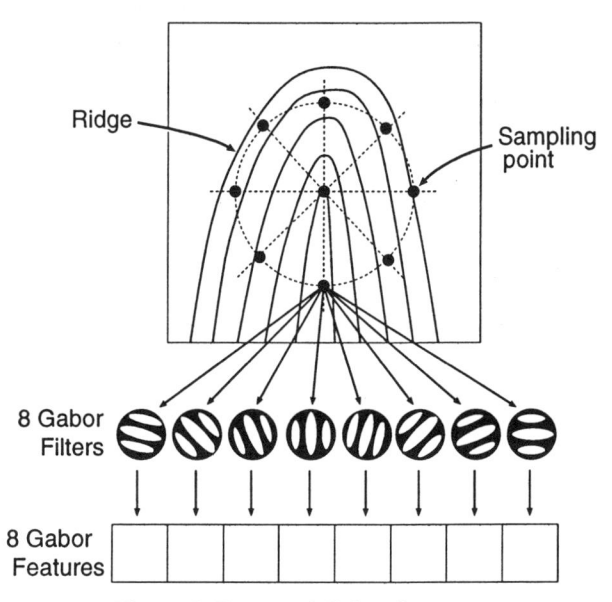

Figure 4. Extracted Gabor features.

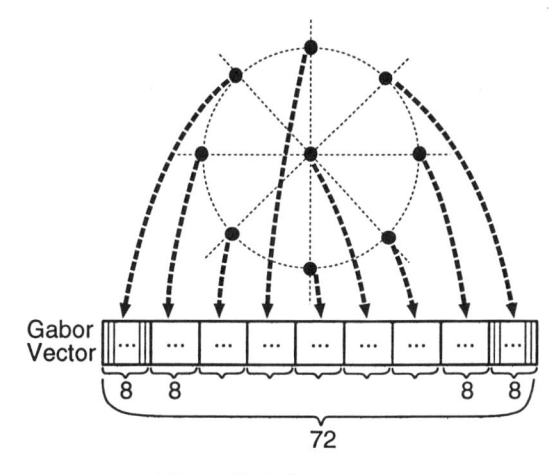

Figure 5. Gabor vector.

$$S = \max_{\substack{-R \leq \Delta u \leq R \\ -R \leq \Delta v \leq R}} \frac{x_{(\Delta u, \Delta v)}^T y}{\|x_{(\Delta u, \Delta v)}\| \|y\|} \quad (4)$$

where $x_{(\Delta u, \Delta v)}$ is a shifted Gabor vector, y is a non-shifted Gabor vector and R is a parameter concerning the shift operation. Note that when $R = 0$, the locally maximized similarity becomes the commonly used simple similarity. As R increases, the computational time increases. If

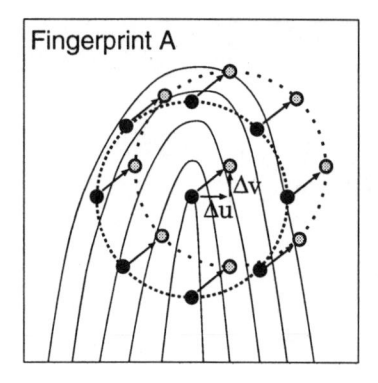

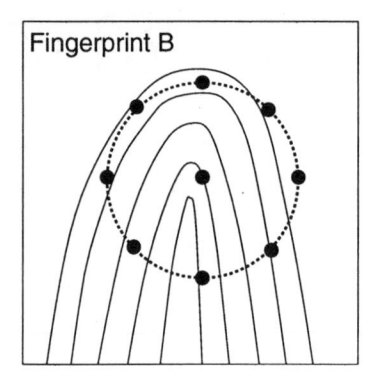

Figure 6. Shift of sampling points.

the value of R is properly selected, the locally maximized similarity can provide flexible pattern matching.

In identification, the following decision rule is done: if the similarity between two fingerprints is more than a threshold t, they come from the same person, otherwise they come from different persons. Then, there are two types of error: one results from deciding $x \in w_i$ and $y \in w_i$ when $x \in w_i$ and $y \in w_j$, denoted by ε_1, and the other results from deciding $x \in w_i$ and $y \in w_j$ when $x \in w_i$ and $y \in w_i$, denoted by ε_2. The total error ε is a weighted sum of these errors:

$$\varepsilon = (1-w)\varepsilon_1 + w\varepsilon_2 \qquad (5)$$

where w is a weight. The total error is a function of t. Next, we discuss the optimization of t. In order to minimize the total error ε, one must find an optimal value of threshold, t^*. This can be automatically done by using a jack-knife procedure. For each run, 8 candidate ts were tried: $0.6, 0.65, 0.7, 0.75, 0.8, 0.85, 0.9, 0.95$.

Step 1. Select a pair of fingerprints.

Step 2. For each t, estimate the total error on the remaining fingerprints.

Step 3. Find the optimal value of t^* which minimizes the total error.

Step 4. Identify the pair by using the selected threshold, t^*.

Step 5. Repeat steps 1-4 L times where L is the number of all possible pairs.

A Gabor Filter-Based Method for Fingerprint Identification 143

The pairs were chosen so that no pair was selected more than once. Note that in the jack-knife procedure, the training samples are completely separated from the test samples.

Figure 7. Examples of fingerprints.

4 Experimental Results

We studied the effectiveness of combining Gabor features with the locally maximized similarity on the NIST database 14 [17]. The used data subset contains 80 fingerprints from 40 persons. From each person, 2 fingerprints are obtained. Each fingerprint is presented as a 352×352 image with 256 gray levels. Figure 7 shows four samples in the data subset. The performance of the proposed method depends significantly on the parameters, i.e., the radius r, wavelength λ, standard deviations σ_x

and σ_y, and shift parameter R. First, we studied the influence of these parameters on the identification performance. From experimental results, the optimal values of r, λ, σ_x, σ_y and R are as follows:

$$\begin{aligned} \text{Radius } r &: 60 \\ \text{Wavelength } \lambda &: 7\sqrt{2} \\ \text{Standard deviation } \sigma_x &: 4.9\sqrt{2} \\ \sigma_y &: 4.9\sqrt{2} \\ \text{Shift parameter } R &: 2 \end{aligned}$$

With regard to orientation, the value of n was set to 8. This means that 72 Gabor features are extracted for each fingerprint.

4.1 Experiment 1

The purpose of this experiment is to study the effectiveness of the proposed method. Figures 8 and 9 show results. Experimental results show that the proposed method works well, compared with those obtained by the simple similarity. Moreover, the results seem to be much less sensitive to the accuracy of the core point. This advantage comes from the use of the locally maximized similarity.

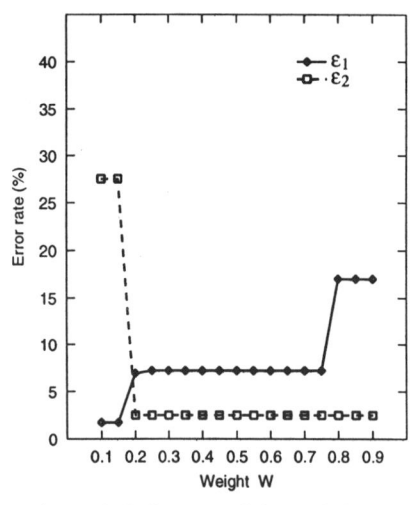

Figure 8. Influence of the weight w on the error rate ($R = 0$)

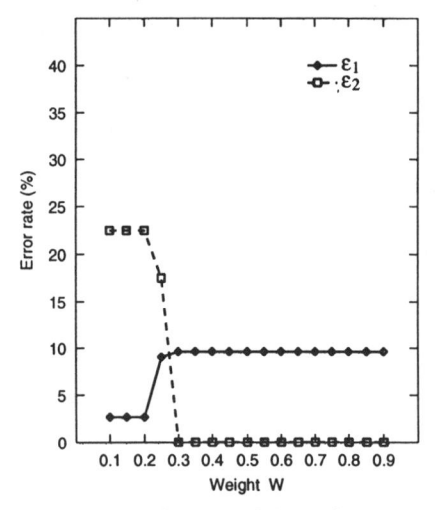

Figure 9. Influence of the weight w on the error rate ($R = 2$)

It should be pointed out that the selection of λ was crucial in this approach. From experimental results, the effectiveness of combining the locally maximized similarity and Gabor features is verified.

4.2 Experiment 2

In Experiment 1, a gray-scale fingerprint image is used. The purpose of this experiment is to study the effect of using a binary fingerprint image. We used the binarization method presented by Mehtre [15]. In this method, contextual filters are used to improve the quality of fingerprint images. The design of a contextual filter consists of two steps. In the first stage, the direction image from an input gray-scale fingerprint image is generated. In the second stage, filters corresponding to these directions are generated. The direction image represents the local orientation of the ridges. The direction which occurs maximum number of times is chosen

gray-scale image (a)　　　　　　binary image (a)

gray-scale image (b)　　　　　　binary image (b)

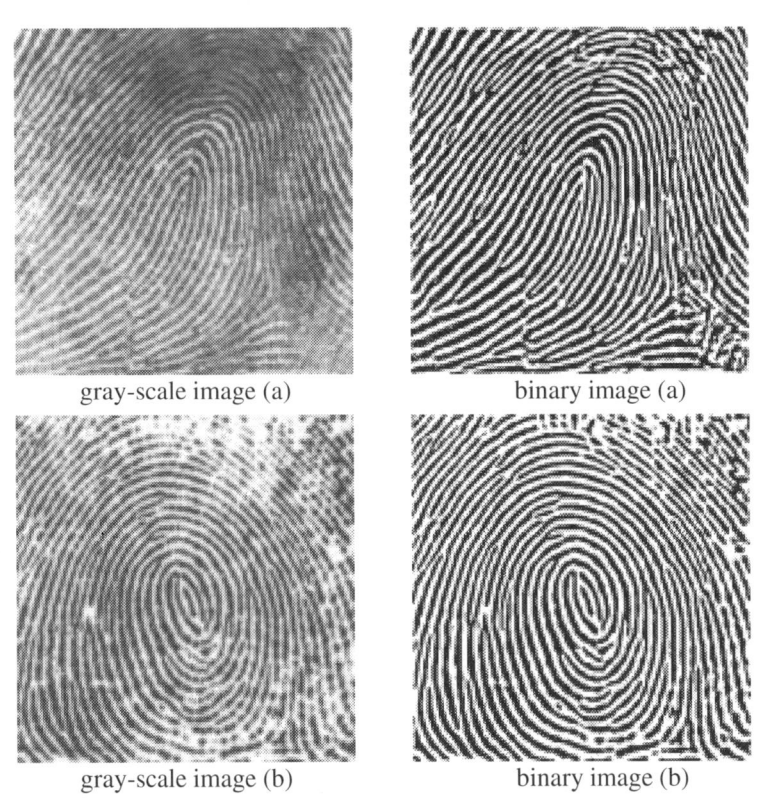

Figure 10. Binarization by Mehtre's method.

as the direction of a given block (of 16 × 16 pixels). The value of directions is from 1 to 8. Figure 10 shows two pairs of a gray-scale image and the resulting binary image. Results are shown in Figures 11 and 12. From results, we see that the use of the binary image is very effective in identifying fingerprints. This is mainly due to the enhancement of the ridges.

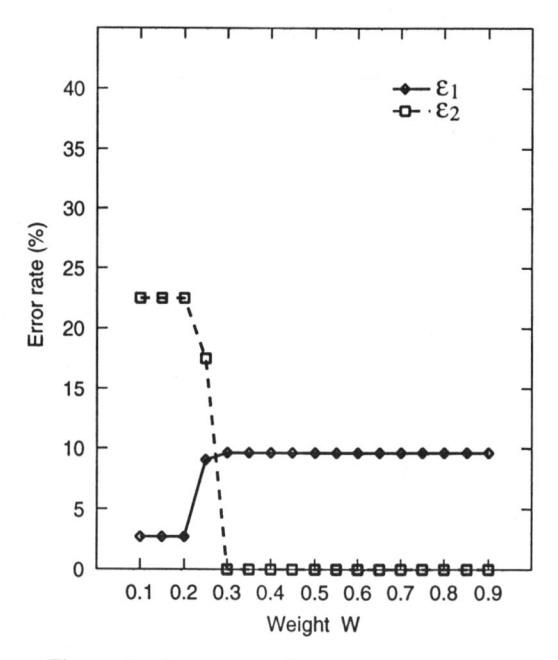

Figure 11. Results for the gray-scale image.

4.3 Experiment 3

The purpose of this experiment is to study the performance of a Gabor filter-based method for pre-classifying fingerprints into 5 classes. Here, 5 classes consists of left loop, right loop, whorl, arch, and others. Each class label is given by human experts. Examples from each class are shown in Figure 13. The used algorithm is as follows:

Step 1. Compute the binary image for each fingerprint image by the Mehtre's method.

Step 2. Extract 72 Gabor features for each binary fingerprint image.

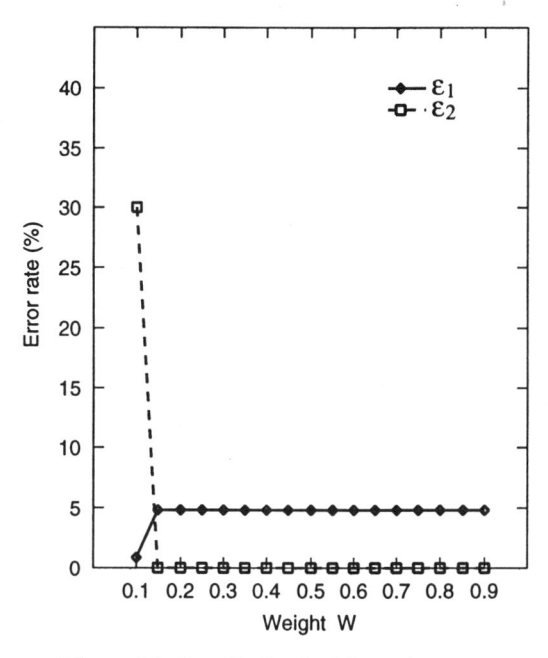

Figure 12. Results for the binary image.

Step 3. Transform the 72-dimensional feature vector into the two-dimensional feature vector by the Sammon's method [16].

Results are shown in Figure 14. Classes A and B are perfectly separated. Classes B and C are slightly overlapped. We believe that by optimizing parameter values, the pre-classification performance can be improved.

5 Conclusion

We have proposed a Gabor filter-based method for identifying fingerprints. The performance of the proposed method was demonstrated on the NIST database 14. From experimental results, we conclude that the method developed here should be considered in the design of a fingerprint identification system.

Acknowledgment

We would like to thank S. Nakai and M. Miki for their help.

Left Loop   Right Loop

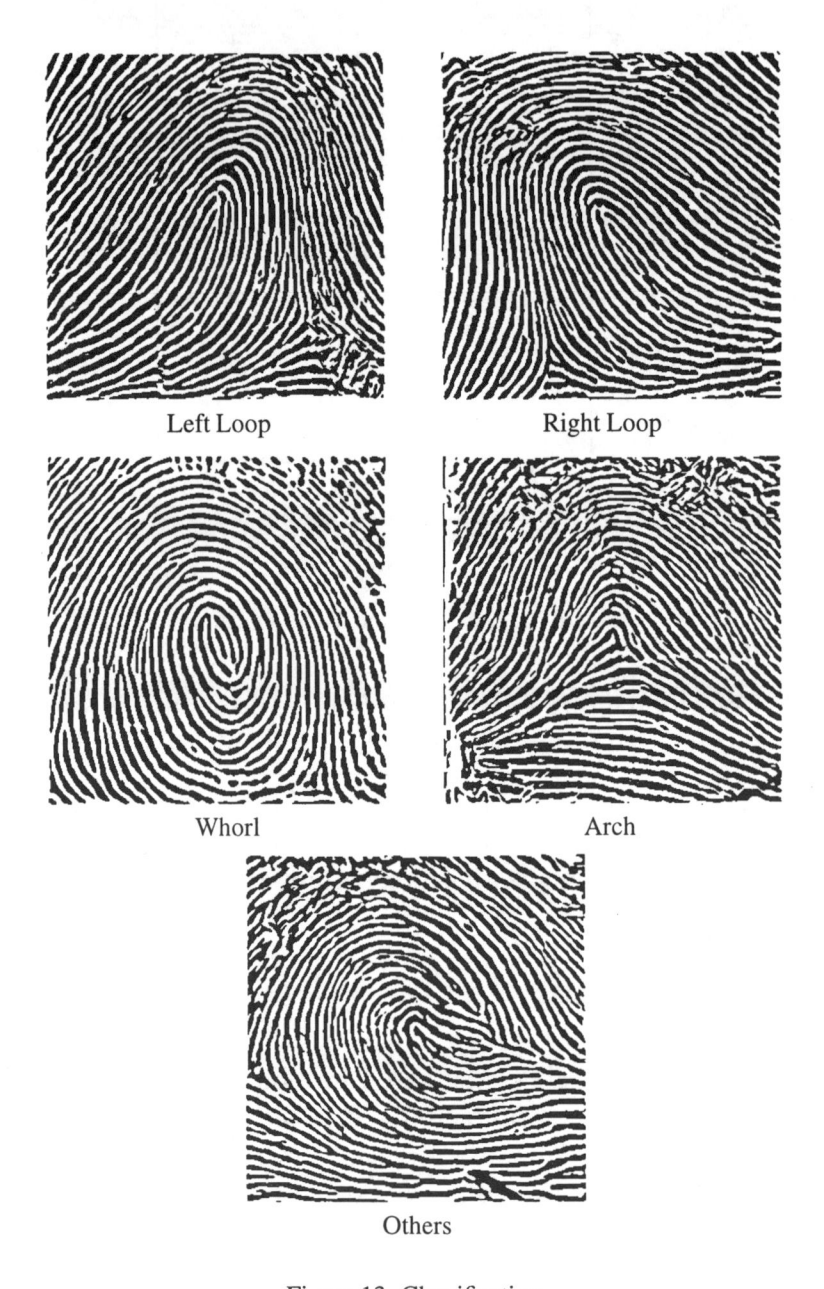

Whorl   Arch

Others

Figure 13. Classification.

A Gabor Filter-Based Method for Fingerprint Identification 149

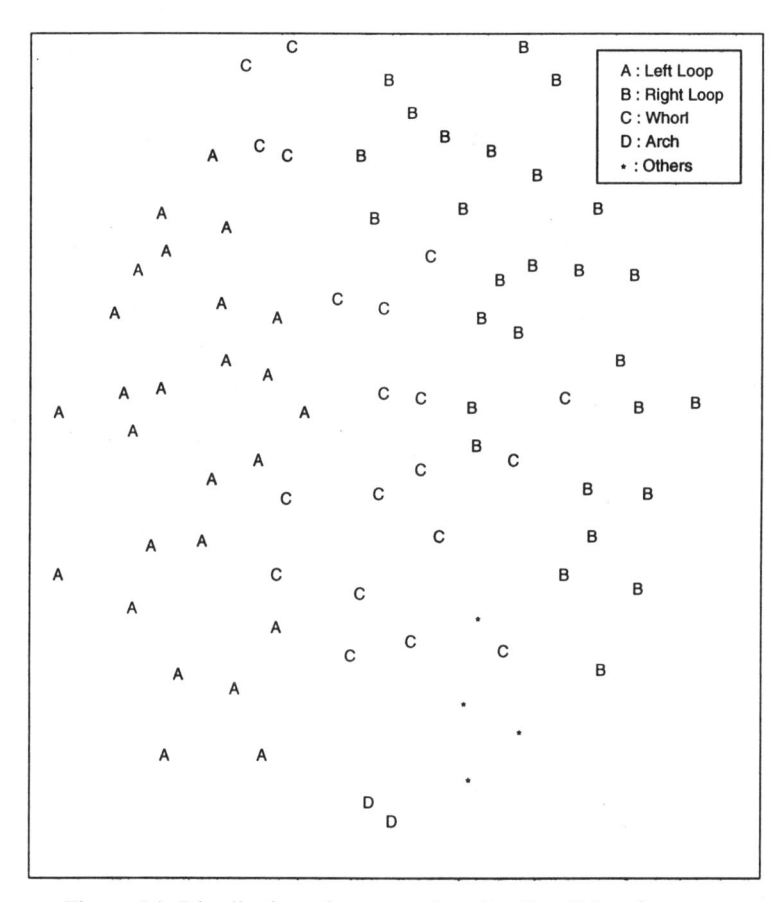

Figure 14. Distribution of patterns described by Gabor features.

References

[1] Gabor, D. (1946), "Theory of communication," *J. Inst. Elect. Engr.*, Vol. 93, pp. 429-457.

[2] Marcelja, S. (1980), "Mathematical description of the responses of simple cortical cells," *Journal of the Optimal Society of America*, Vol. 70, pp. 1297-1300.

[3] Daugman, J.G. (1980), "Two-dimensional spectral analysis of cortical receptive field profiles," *Vision Res.*, Vol. 20, pp. 847-856.

[4] Turner, M.R. (1986), "Texture discrimination by Gabor functions," *Biological Cybernetics*, Vol. 55, pp. 71-82.

[5] Porat, M. and Zeevi, Y.Y, (1988), "The generalized Gabor scheme of image representation in biological and machine vision," *IEEE Transactions on Pattern Analysis and Machine Intelligence*, Vol. 10, pp. 452-468.

[6] Hamamoto, Y., Uchimura, S., Masamizu, K. and Tomita, S. (1995), "Recognition of handprinted Chinese characters using Gabor features," *Proc. of the 3rd Int. Conf. Document Analysis and Recognition*, pp. 819-823, Montreal.

[7] Yamada, H., Saito, T. and Mori, S. (1981), "An improvement of correlation method – Locally maximized correlation –," *IECE Transactions of the Institute of Electronics and Communication Engineers of Japan*, Vol. J64-D(10), pp. 970-976 (in Japanese).

[8] Chang, T.L. (1980), "Texture analysis of digitized fingerprints for singularity detection," *Proc. Int. Conf. Pattern Recognition*, pp. 478-480, Miami.

[9] Weatherall, D. (1986), "Automated inspection of surface texture," *Sensor Review*, pp. 27-28, January.

[10] Shimizu, A. and Hase, M. (1984), "Detection method of the core point in fingerprint," *IECE Transactions of the Institute of Electronics and Communication Engineers of Japan*, Vol. J67-D(3), pp. 383-384 (in Japanese).

[11] Daugman, J.G. (1985), "Uncertainty relation for resolution in space, spatial-frequency, and orientation optimized by two-dimensional visual cortical filters," *Journal of the Optimal Society of America*, Vol. 2, pp. 1160-1169.

[12] Wegstein, J.H. (1970), "Automated fingerprint identification," NBS Tech. Note, 538.

[13] Hoshino, Y., Asai, K., Kato, Y. and Kiji, K. (1980), "Automatic reading and matching for single-fingerprint identification," *65th International Association for Identification Conference*, pp. 1-7, Ottawa, Canada.

[14] Hrechak, A.K. and Mchugh, J.A. (1990), "Automated fingerprint recognition using structural matching," *Pattern Recognition*, Vol. 23, pp. 893-904.

[15] Mehtre, B.M. (1993), "Fingerprint image analysis for automatic identification," *Machine Vision and Applications*, Vol. 6, pp. 124-139, Springer-Verlag.

[16] Sammon Jr, J.W. (1969), "A nonlinear mapping for data structure analysis," *IEEE Trans. Computers*, Vol. C-18, 5, pp. 401-409.

[17] Watson, C.I. (1993), "NIST Special Database 14-Mated Fingerprint Card Paris 2," NIST.

Chapter 5:

Minutiae Extraction and Filtering from Gray-Scale Images

MINUTIAE EXTRACTION AND FILTERING FROM GRAY-SCALE IMAGES

Dario Maio and Davide Maltoni
DEIS, CSITE-CNR
University of Bologna, Bologna
ITALY

This chapter surveys the principal minutiae extraction and minutiae filtering techniques proposed in the literature, and describes in detail two direct gray-scale methods introduced by the authors. The first method, dedicated to the minutiae extraction problem, adopts a ridge-line tracking strategy which does not require preprocessing, binarization nor thinning as *a priori* steps. The second one allows the accuracy of the automatic extraction algorithm to be improved; for this purpose a neural network classifier is trained to analyze the minutiae neighborhoods and to decide whether they are valid or not.

1 Introduction

Most automatic systems for fingerprint comparison are based on minutiae matching [12], [24], [29] and [46]. *Minutiae*, or Galton's characteristics [8], are local discontinuities in the fingerprint pattern. Although several types of minutiae can be considered (the most common are shown in Figure 1a), usually only a coarse classification is adopted to deal with the practical difficulty in discerning the different types with sufficient certainty. The American National Standards Institute proposed a minutiae taxonomy based on four classes: *terminations*, *bifurcations*, *trifurcations* (or crossovers) and *undetermined* [1]. In this chapter the FBI identification model [46], which uses only termination and bifurcation, is adopted; each minutia entry contains the minutiae class, the coordinates and the angle that the tangent to the minutia forms with the horizontal direction (Figures 1b and 1c). Figure 2a shows a pre-processed fingerprint portion where the ridge lines appear as dark traces on a light background; two terminations (1,2) and one bifurcation (3) can be easily detected. It should be noted that on the complementary image (Figure 2b) three corresponding minutiae take the same positions, but their type is exchanged: terminations appear as bifurcations and vice versa (this property is known as termination/bifurcation duality). The problem of automatic minutiae extraction has been thoroughly studied but never completely solved. The main difficulty is that fingerprint quality is often too low[1] due to the critical nature of the acquisition process; noise and contrast deficiency can produce false minutiae and hide valid minutiae.

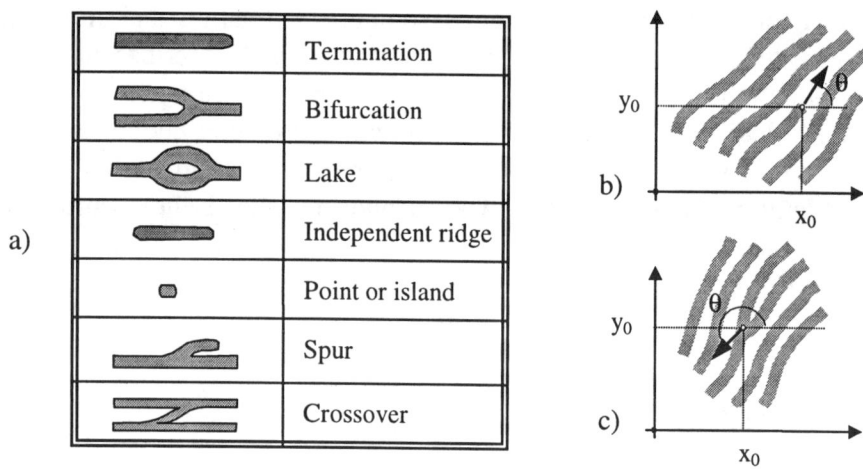

Figure 1. In Figure 1a the most common minutiae types are reported. Figure 1b shows a termination minutia: (x0, y0) are the minutia coordinates; θ is the angle that the minutia tangent forms with the horizontal direction. Figure 1c shows a bifurcation minutia: θ is now defined by means of the termination minutia existing in the complementary image in correspondence with the original bifurcation.

---

[1] The fingerprint acquisition process is rather critical. The most famous technique, known as the "ink technique," often produces images including regions which miss some information due to excessive inkiness or to ink deficiency. The techniques which use optical prisms [9] and holograms [15] require a high degree of accuracy during the acquisition process, that is, the finger pressure on the optical surface must be adequate. Furthermore, in some subjects, especially manual workers and elderly people, the prominence of the ridge lines can be considerably lower and the fingerprint pattern can be unreadable.

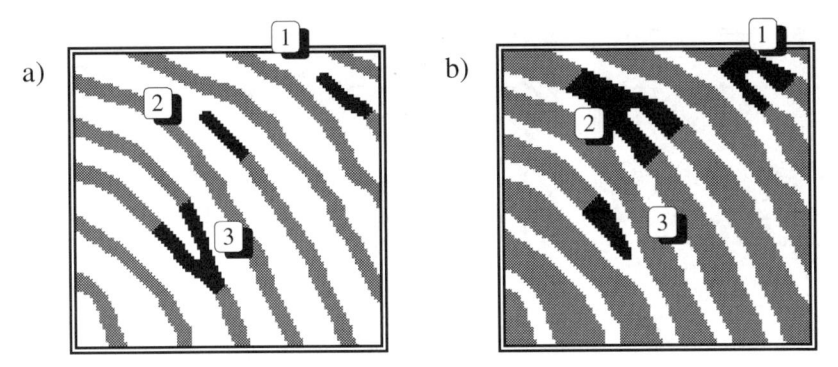

Figure 2. The termination/bifurcation duality on a binary image and its complementary image.

In particular two types of degradation usually affect fingerprint images [25]:

1. the ridge lines are not strictly continuous since they sometimes include small breaks (*gaps*).
2. parallel ridge lines are not always well separated due to the presence of cluttering noise.

Several approaches to automatic minutiae extraction have been proposed: although rather different from one other, most of these methods transform fingerprint images into binary images through an ad hoc algorithm. The images obtained are submitted to a thinning process which allows for the ridge-line thickness to be reduced to one pixel (Figure 3). Finally, a simple image scan allows to detect the pixels which correspond to minutiae: in fact it is easy to prove that the pixels corresponding to minutiae are characterized by a *crossing number* [2] different by 2.

Few other attempts, generally based on neural networks, are known in the literature which extract minutiae without requiring a binarization. In [20] the authors proposed a new detection approach, where the features are extracted directly from the gray-scale image without binarization and thinning; this choice is motivated by the following considerations:

- A lot of information may be lost during the binarization process.
- Binarization and thinning are time-consuming.
- The binarization techniques which were experimented proved to be unsatisfactory when applied to low-quality images.

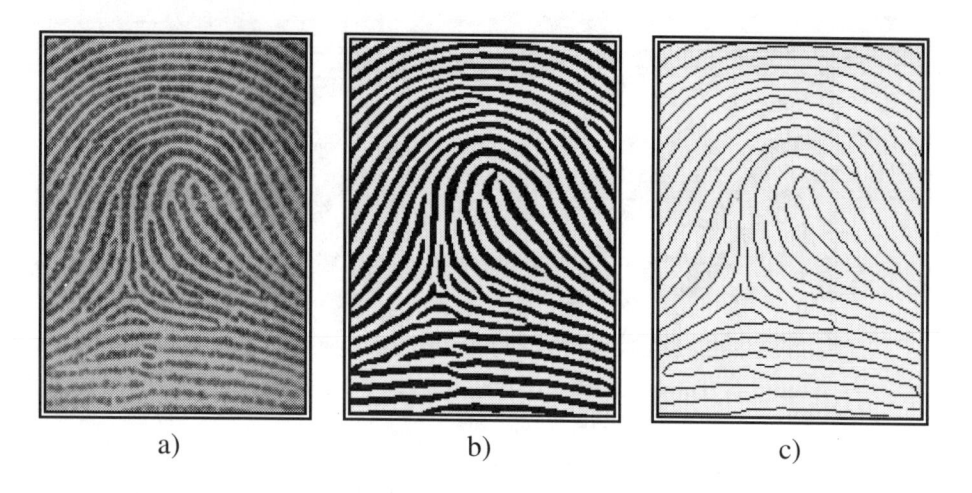

Figure 3. Figure 3a shows a fingerprint gray-scale image; Figure 3b shows the image obtained after a binarization of the image 3a; Figure 3c shows the image obtained after a thinning of the image 3b.

The basic idea of the above mentioned method is to track the ridge lines on the gray-scale image, by "sailing" according to the local orientation of the ridge pattern. A set of starting points is determined by superimposing a grid on the gray-scale image; for each starting point, the algorithm keeps following the ridge lines until they terminate or intersect other ridge lines (minutiae detection). The comparison with other binarization-based schemes proved the efficiency and accuracy of this method. Some experimental results (as reported in Section 3), over a database of medium-quality fingerprints at 500 dpi resolution, exhibited: dropped = 4.51%, false = 8.52% and exchanged = 13.03% (where dropped refers to the percentage of ground true minutiae not located by the algorithm, false refers to the percentage of false minutiae and exchanged indicates the percentage of minutiae whose type, termination or bifurcation, is confused).

Nowadays, several emerging fingerprint applications, such as access control, electronic commerce, electronic voting, etc., require real-time performance and very low-cost acquisition sensors which usually provide images at low resolutions. On hardware Pentium 200 MHz the extraction method described in [20] allows minutiae extraction to be performed in less than 0.3 seconds on 256×256 images, but, obviously, its accuracy decreases with the quality of the sensor used (which is mainly determined by the resolution). For example, by using a prototypal sensor, which works at about 350 dpi, the authors measured: dropped = 11.8%, false = 24.2% and exchanged = 13.1%.

To improve these performances, a new filtering technique has been proposed in [21]. Each minutia, as detected by the algorithm [20], is reclassified through a Neural Network (NN) whose output determines if it is a termination, a bifurcation, or a false minutia. Obviously, this kind of post-processing can reduce only false and exchanged minutiae, but not dropped minutiae. Anyway, significant improvements have been obtained in terms of cumulative error (dropped + false + exchanged).

Section 2 of this chapter surveys the different approaches presented in the literature by separately considering binarization-thinning-based techniques, other extraction methods, and minutiae filtering. In Section 3 the algorithm presented in [20] is described in more details and its performance is compared with those of other techniques. Finally, Section 4 is dedicated to the discussion of the Neural-Network filtering approach reported in [21].

2 Minutiae Extraction and Filtering: Literature Review

2.1 Binarization-Thinning Based Techniques

The general problem of image binarization, sometimes referred to as segmentation, has been widely studied in the fields of image processing and pattern recognition [28]. The easiest approach uses a *global threshold t* and works by setting to 0 the pixels whose gray level is lower than *t* and to 1 the remaining. In general, different portions of an image can be characterized by different contrast and illumination, and consequently a single threshold is not sufficient for a correct segmentation; for this reason the *local threshold* techniques change *t* locally, by adapting its value to the average local intensity. In the specific case of fingerprint images, which are usually of poor quality, a local threshold method cannot always guarantee acceptable results; hence, ad-hoc solutions exploiting specific knowledge of this application domain (e.g. striped local-oriented pattern, smooth orientation changes, ridge-line duality, etc.) have been proposed. In the following the main contributions are briefly summarized:

- The FBI "minutiae reader" [37] binarizes the image through a composite approach based on a local threshold and a "slit comparison" formula which compares pixel alignment along 8 discrete directions; in fact, it can be observed that for each pixel belonging to a ridge line there exists an orientation (the ridge-line orientation) whose average local intensity is lower than those of the remaining orientations (which intersect one or more ridge lines).

- Moayer and Fu [26] proposed a binarization technique based on the iterative application of a Laplacian operator and a pair of dynamic thresholds. At each iteration, the image is convolved through a Laplacian operator and the pixels whose intensity is external with respect to the range bounded by the two thresholds are set to 0 and 1. The thresholds are progressively moved toward a unique value so that a secure convergency is obtained. A similar approach has been proposed by Xiao and Raafat in [41] and [42] where, after the convolution step, a local threshold is employed to deal with regions differently contrasted.

- A fuzzy approach to image enhancement and the use of an adaptive threshold, aimed to preserve the same number of 1 and 0 pixels for each neighborhood, form the basis of the binarization technique proposed by Verma *et al.* in [40]. The image is initially partitioned in small regions which are processed separately. Each region is submitted to the following step: smoothing, fuzzy coding of the pixel intensities, contrast enhancement, binarization, 1's and 0's counting, fuzzy decoding, parameters adjusting. The sequence is repeated until the number of 1's approximately equals the number of 0's.

- Coetzee and Botha [6] proposed a binarization technique based on the use of the edges in conjunction with the gray-scale image. Edge extraction is performed through Marr-Hildreth algorithm [23]. A sophisticated technique, which operates on small local window through a blob-coloring routine, allows the edge and the intensity information to be fused.

- O'Gorman and Nickerson presented in [27] a technique for enhancement and binarization based on the convolution of the image with some filters oriented according to the *directional image*. The directional image may be conceived as a matrix whose elements represent the tangent direction to the ridge lines of the original image. The filters are computed parametrically with respect to the ridge-line characteristic at the resolution used: min. and max. ridge-line width, min. and max. ridge-line inter-distance, max. bending. This kind of filtering performs a local regularization which is very robust with respect to noise, since its relies on directional information which can be reliably extracted even on poor images; on the other hand, this method requires the convolution of each point with a large mask, resulting in a very time-consuming process. Similar methods have been developed by other authors [7], [25].

- Ratha *et al.* proposed in [30] an interesting approach to fingerprint binarization which is based on maxima searching along ridge-line orthogonal sections. The basic idea of this method is very similar to that independently developed by the authors in [20], where it is employed for the ridge-line following algorithm (see Section 3).

- Sherlock *et al.* [34] [35] defined a technique for fingerprint enhancement and binarization, which performs a frequency-domain filtering through position-dependent filters. The filters, which are constituted by directional bandpass, enhance the images according to the local ridge-line orientation and, at the same time, remove the noise associated to both the low and the high frequencies. Watson *et al.* proposed a similar approach in [44], where each 32×32 image block is independently processed by enhancing the dominant frequency; the blocks are partially overlapped to reduce the border effects.

- Sherstinsky and Picard reported in [33] a complex method for fingerprint binarization which employs a dynamic non-linear system called "M-lattice," which is based on the reaction-diffusion model first proposed by Turing in 1952 to explain the formation of animals patterns such as zebra stripes.

Among the techniques reported, some provide good results when applied to high-quality fingerprints, but usually the most time ones are sufficiently robust in the presence of noise. Figure 4 shows the results obtained by binarizing a good fingerprint portion through some of the methods described.

With the aim of improving the quality of the binary images, some researchers introduced regularization techniques which usually work by filling holes and by removing small gaps and other artifacts produced by noise. To this purpose, Coetzee and Botha [6] tracked the ridge-lines edges through an adaptive window technique, whereas Ratha *et al.* in [30] used a morphological "open" operator [11] whose structuring elements is a 3×3 box oriented according to the ridge-lines local direction.

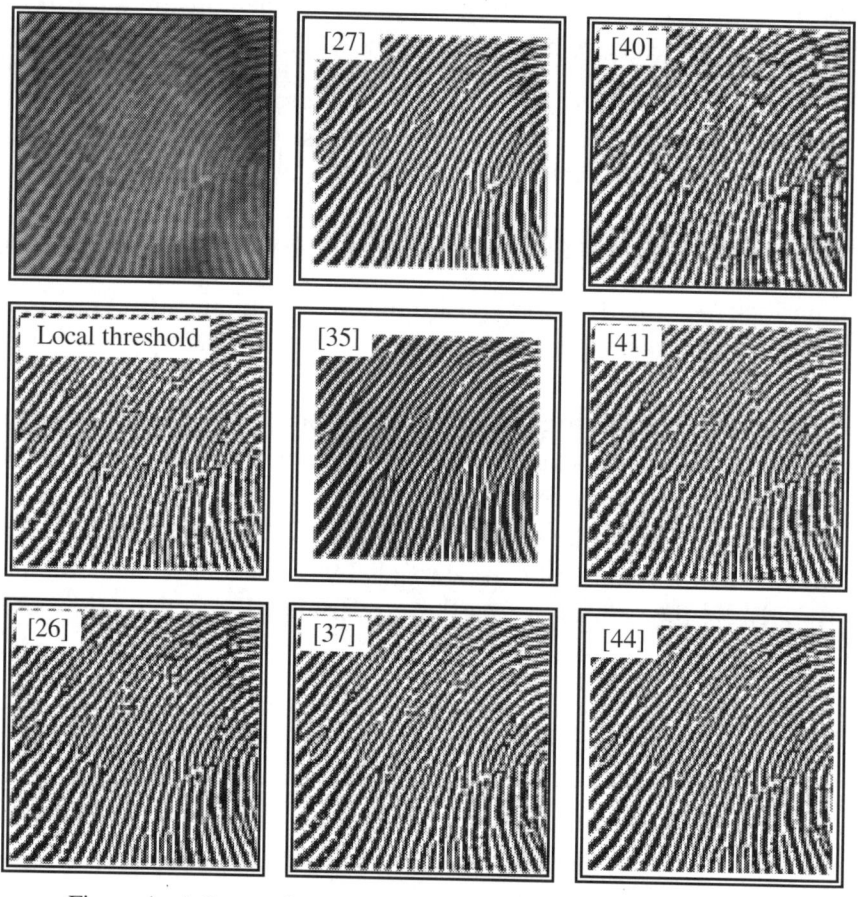

Figure 4. A fingerprint portion of good quality and its binarization through some of the methods discussed in this section.

As far as thinning techniques are concerned [17], a large number of approaches are available in the literature due to the central role of this processing step in many applications: character recognition, document analysis, map and draws vectorization, etc. Hung [14] used the algorithm [2] by Arcelli and Baja; Ratha *et al.* [30] adopted a technique included by HIPS library, Mehtre [25] employed the parallel algorithm described in [39]. Finally, Coetze and Botha [6] used the Baruch's method [3].

2.2 Other Techniques

Few works have been proposed for automatic minutiae extraction which do not conform to the scheme binarization-thinning:

- In the work [45] by Weber, the gray-scale fingerprints are enhanced by a bandpass filtering in the frequency domain and binarized via a local threshold; instead of using a conventional thinning the author proposed an algorithm which detects the minutiae starting from the thick ridges in the binary image.
- E. Székely and V. Székely [38], starting from binary images, developed a minutiae-detection technique based on the computation of the directional image divergence. The foundation of this method consists in using a divergence operator in order to discern fingerprint pattern discontinuities which correspond to minutiae. Unfortunately, acceptable results are produced only on very good-quality fingerprints.
- Leung *et al.* introduced in [18] a neural network-based approach to the minutiae detection, where a multilayer perceptron analyzes the output of a rank of Gabor's filters applied to the gray-scale image. The image is initially transformed into the frequency domain where the filtering takes place; the resultant magnitude and phase signals constitute the input of a neural network composed by 6 sub-networks, each of which is responsible for detecting minutiae at a specific orientation; a final classifier is employed to combine the intermediate responses. Another neural network schema is presented in [19] where a tree-layer perceptron is trained to extract the minutiae starting from thinned binary images.
- Maio and Maltoni introduced in [20] a direct gray-scale minutiae extraction technique, described in detail in Section 3.

2.3 Minutiae Filtering

Post-processing techniques, based on simple structural considerations, are widely used to discard many of the false minutiae which usually affect thinned binary fingerprint images. In [14], [41] the most common false minutiae structures are reported and some ad-hoc rules to remove them are proposed (Figure 5). For example, it is extremely unlikely that two or more minutiae are very close to each other, or that two termination minutiae face each other at short distance.

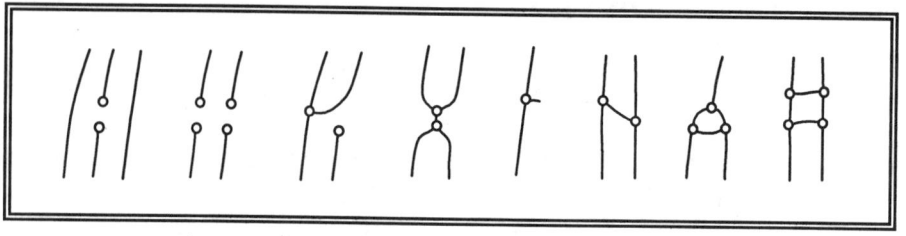

Figure 5. The most common false minutiae structures which affect thinned binary image.

A different filtering technique, which operates directly on the gray-scale images, has been proposed in [21]; the extraction algorithm [20] is employed to detect the minutiae and each minutia neighborhood, after a normalization step, is independently verified through a neural network classifier. Section 4 is dedicated to the description of this approach.

3 A Direct Gray-Scale Minutiae Extraction Approach

The basic idea of this method is to track the ridge lines on the gray-scale image, by "sailing" according to the fingerprint directional image. A set of starting points is determined by superimposing a square-meshed grid on the gray-scale image. For each starting point, the algorithm keeps following the ridge lines until they terminate or intersect other ridge lines (termination or bifurcation detection). A labeling strategy is adopted to examine each ridge line only once and locate the intersections between ridge lines.

Minutiae Extraction and Filtering from Gray-Scale Images 165

3.1 Ridge-Line Following Algorithm

Let **I** be an $a \times b$ gray-scale image with g gray levels, and $gray(i,j)$ be the gray level of pixel (i,j) of **I**, $i = 1,...a$, $j = 1,...b$. Let $z = S(i,j)$ be the discrete surface corresponding to the image **I**: $S(i,j) = gray(i,j)$, $i = 1,...a$, $j = 1,...b$. By associating bright pixels with gray levels near to 0 and dark pixels with gray levels near to $g-1$, the fingerprint ridge lines (appearing dark in **I**) correspond to surface ridges, and the spaces between the ridge lines (appearing bright in **I**) correspond to surface ravines (Figure 6).

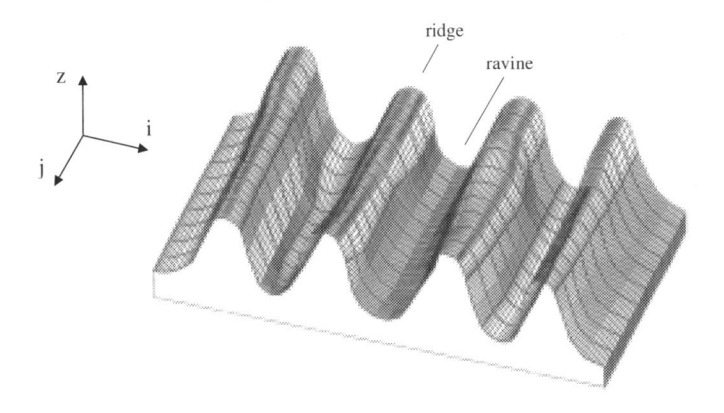

Figure 6. A surface S corresponding to a small area of a fingerprint is shown (the surface is depicted as continuous due to representation problems).

From a mathematical point of view, a ridge line is defined as a set of points which are local maxima along one direction. The ridge-line extraction algorithm attempts to locate, at each step, a local maximum relative to a section orthogonal to the ridge direction. By connecting the consecutive maxima, a polygonal approximation of the ridge line can be obtained.

Let (i_s, j_s) be a local maximum of a ridge line of **I**, and φ_0 be the direction of the tangent to the ridge-line in (i_s, j_s); a pseudo-code version of the ridge-line following algorithm is:

ridge-line following(i_s, j_s, φ_0)
 {   end := false ;
     $(i_c, j_c) := (i_s, j_s)$;

$\varphi_c := \varphi_0$;
while ($\neg$ end)
{ $(i_t, j_t) := (i_c, j_c) + \mu$ *pixel along direction* φ_c;
  $\Omega :=$ *section set centered in* (i_t, j_t) *with direction* $\varphi_c + \pi/2$ *and length* $2\sigma + 1$;
  $(i_n, j_n) :=$ *local maximum over* Ω;
  store (i_n, j_n);
  end := *check stop criteria on* $(i_c, j_c), (i_t, j_t), (i_n, j_n)$;
  $(i_c, j_c) := (i_n, j_n)$;
  $\varphi_c :=$ *tangent direction in* (i_c, j_c);
}
}

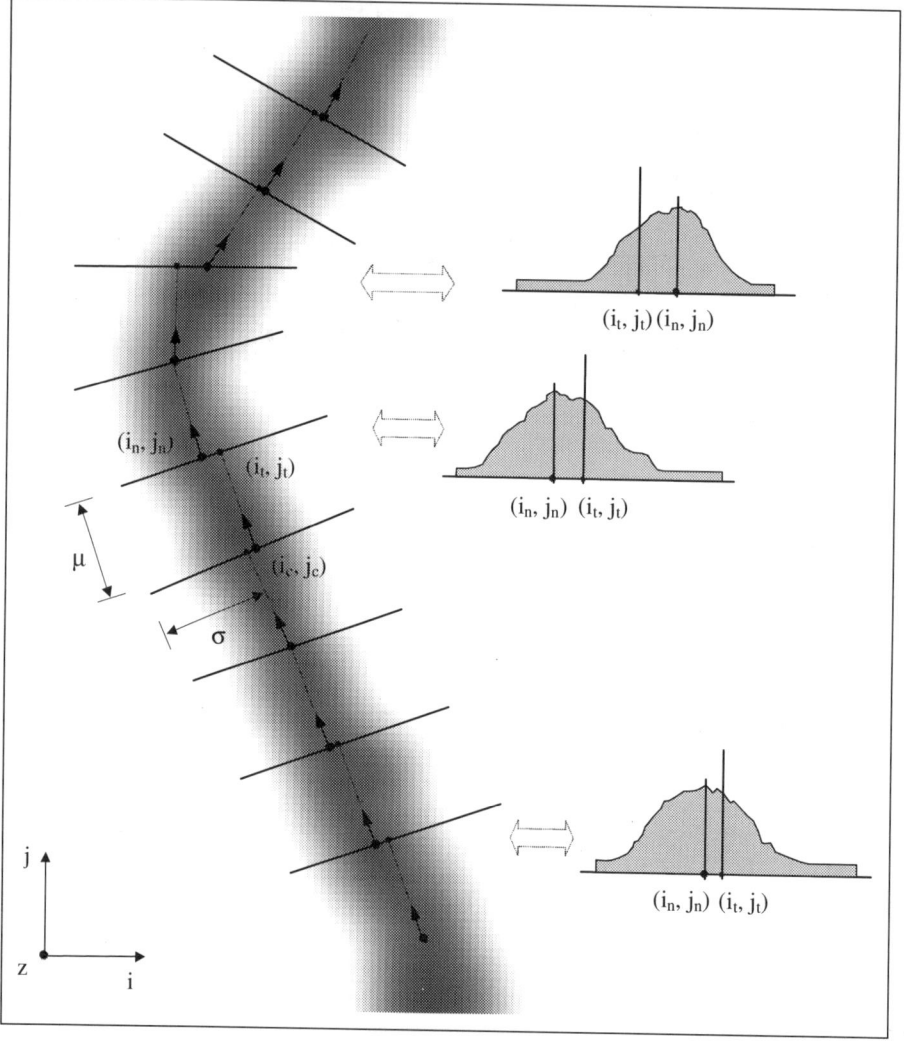

Figure 7. Some ridge-line following steps; on the right, some sections are shown.

Minutiae Extraction and Filtering from Gray-Scale Images 167

The algorithm runs until a stop criterion becomes true. At each step, it computes a point (i_t, j_t), moving μ pixels from (i_c, j_c) along direction φ_c. Then, it computes the *section set* Ω as the set of points belonging to the section segment lying on the ij-plane and having median point (i_t, j_t), direction orthogonal to φ_c and length $2\sigma+1$. A new point (i_n, j_n), belonging to the ridge line, is chosen among the local maxima of the set Ω. The point (i_n, j_n) becomes the current point (i_c, j_c) and a new direction φ_c is computed (Figure 7). μ and σ are parameters whose optimal value can be determined according to the average thickness of the image ridge lines. The main algorithm steps, namely, sectioning and maximum determination, computation of the direction φ_c and testing of the stop criteria, are discussed in the following sub-sections.

3.1.1 Sectioning and Maximum Determination

Determining a local maximum of the section set Ω is a very important step. In principle, the maximum can be computed simply by comparing the gray levels of the points belonging to Ω. Noise and contrast deficiency make this technique unsuitable, except for excellent-quality images. Figures 8a, 8b show two sections which intersect 5 and 6 ridges respectively. In both of these sections it is easy to locate the ridges, but detecting the corresponding maxima is not straightforward.

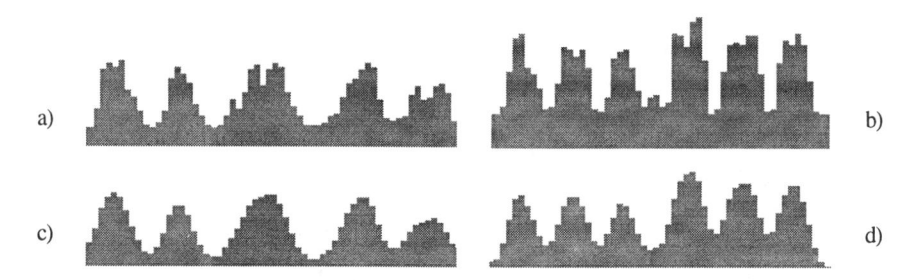

Figure 8. Figures 8a, 8b show two sections belonging to regions with different ridge-line density. Figure 8c, 8d report the same sections after the regularization.

Hence, a two-step processing aimed at regularizing the section silhouette has been introduced which makes the determination of the local maxima more reliable: the first step is based on a local average of the gray levels of the pixels belonging to $2h+1$ (h integer) parallel adjacent sections. The second step is based on a convolution with a small 1-d mask **d** resembling a Gaussian silhouette (the mask size is

$2p+1$, p integer). Actually, the whole processing can be conceived as a convolution with a 2-dimensional $2h+1 \times 2p+1$ mask obtained by shifting **d** by $2h+1$ pixels along the direction orthogonal to the ridge-line. O'Gorman and Nickerson in [27] and Mehtre in [25] performed the image enhancement in a similar way, but they carried out the enhancement throughout the image whereas the method here described regularizes only a subset of points which are determined during the ridge-line following. Figure 8c, 8d show the regularized profiles of Sections 8a and 8b respectively.

3.1.2 Tangent Direction Computation

At each step, the algorithm computes a new point (i_t, j_t) by moving μ pixels from the current point (i_c, j_c) along direction φ_c. The direction φ_c represents the ridge-line local direction and can be computed as the tangent to the ridge in the point (i_c, j_c). The method used in [20], as proposed by Donahue and Rokhlin [7], employs a gradient-type operator to extract a directional estimate from each 2×2 pixel neighborhood, which is then averaged over a local window by least-squares minimization to control noise. This method allows for an *unoriented* direction to be computed. The computation of an *oriented* direction is subordinate to the choice of an orientation. For each step of the ridge-line following, the orientation is chosen in such a way that φ_c comes closest to the direction computed at the previous step. The technique used to compute the tangent directions, although rather efficient and robust, can become computationally expensive if the local windows used are large (if their side is 19 or more pixels) and the number of directions to be computed is very high. A more efficient implementation scheme can be obtained by pre-computing the directional image over a discrete grid (Figure 9) and then determining the direction φ_c through Lagrangian interpolation.

3.1.3 Stop Criteria

The stop criteria (i.e., the events which stop the ridge-line following) are:
1. *Exit from interest area.* The new point (i_t, j_t) is external to a rectangular window **W** which represents the sub-image whose minutiae have to be detected.

2. *Termination.* No local maxima, such that the segments having extremes (i_c, j_c) (i_n, j_n) form angles less than β (threshold value) with the direction $φ_c$, could be found in Ω: a termination minutia has been detected. According to this criterion the ridge-line following stops independently on the gray level of the current region, and the algorithm can work both on saturated regions and on contrast-deficient regions with no need for a particular tuning.

3. *Intersection.* The point (i_n, j_n) has been previously labelled (see Section 3.2) as belonging to another ridge line: a bifurcation minutia has been detected.

4. *Excessive bending.* The segment delimited by (i_c, j_c) (i_n, j_n) forms with the *ridge-line local direction* an angle greater than ψ (threshold value). The ridge-line local direction is defined as the average of the directions of the segments (i_c, j_c) (i_n, j_n) relative to the last k steps (k = 2,...4). This criterion allows for the ridge-line following to be stopped when the ridge line direction changes suddenly. In fact, due to the ridge-line continuity, excessive bending always denotes an error in the ridge-line following.

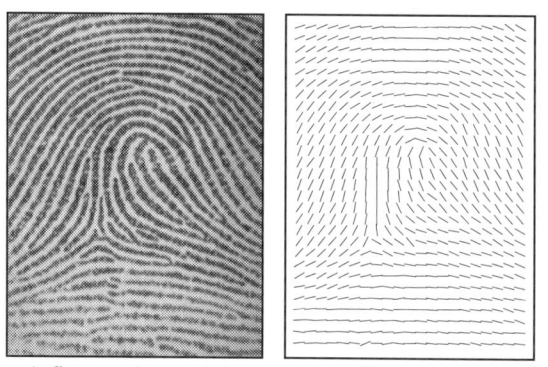

Figure 9. A fingerprint and the corresponding directional image computed over a grid whose granularity is 8 pixels.

3.2 Minutiae Detection

In the previous section an algorithm capable of extracting a ridge line given a starting point and an oriented direction has been presented. When a ridge line terminates or intersects another ridge line (originating a minutia) the algorithm stops and returns the minutia characteristics (type, coordinates and direction) of the minutia found.

It is now necessary to define a schema for extracting all the ridge lines in the image and, consequently, detecting all the minutiae. The main problems arise from the difficulty of examining each ridge line only once and locating the intersections with the ridge lines already extracted. To this purpose an auxiliary image **T** of the same dimension as **I** is used. **T** is initialized by setting its pixel values to 0; every time a new ridge line is extracted from **I**, the pixels of **T** *corresponding* to the ridge line are labeled by assigning them an identifier. The pixels of **T** corresponding to a ridge line are the pixels belonging to the polygonal, ε-pixels thick, which links the consecutive maximum points (i_n, j_n) located by the ridge-line following algorithm on the ridge line (Figure 10).

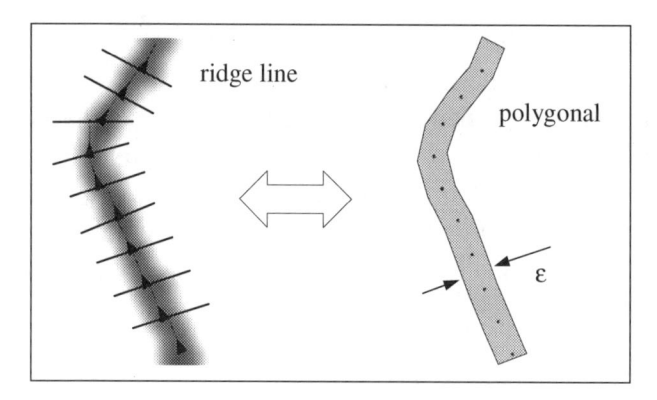

Figure 10. A ridge line and the corresponding polygonal (ε-pixels thick).

Let G be a regular square-meshed grid, with granularity ν pixels, superimposed on the window **W**; for each node of G the minutiae detection algorithm searches the nearest ridge line by using a sectioning technique very similar to that already described in Section 3.1.1 and tracks it by means of the ridge-line following routine. Since the initial point can be everywhere in the middle of a ridge line the tracking is executed alternately in both the directions. The auxiliary image **T,** which is updated after each ridge-line following, provides a simple and effective way to discover ridge-line intersections and to avoid multiple tracking. Figure 11 shows the results obtained by applying the minutiae detection algorithm to a sample fingerprint.

3.3 Performance Evaluation

In this section, the direct gray-scale minutiae-detection approach [20] (A) is compared with other four different schemes based on binarization and thinning, which have been derived from [37] (B), [26] (C), [40] (D), and [27] (E), respectively. It is worth remarking that the primary goal of papers [37] and [27] is minutiae detection, while in [26] and [40] binarization and thinning are steps for the classification and/or recognition of fingerprint patterns.

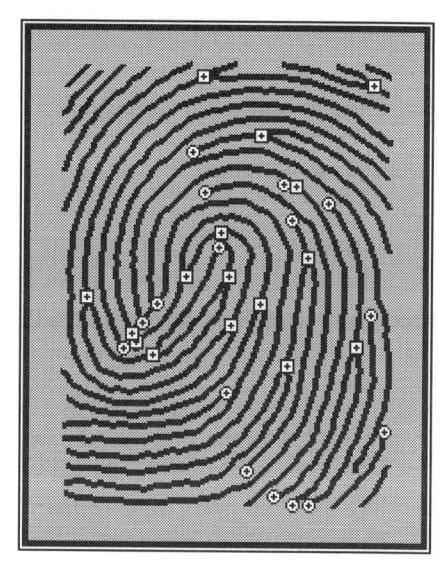

Figure 11. Minutiae detection on a sample fingerprint. The ridge lines are represented through the corresponding polylines of **T**. The termination minutiae are denoted by circles while the bifurcation minutiae are denoted by squares.

In order to perform the comparison, a sample set of fingerprint images belonging to different sources and exhibiting a different degree of image quality has been assembled. The sample set has the following composition: 7 fingerprints taken from the NIST fingerprint database [43] (500 dpi), 4 fingerprints from an FBI sample set (500 dpi), and 3 fingerprints acquired through an opto-electronic device based on a prism (about 350 dpi). By using the contrast and consistency index proposed in [7], the fingerprints have been coarsely classified according to their quality into *good* and *poor*. On each fingerprint a human expert has marked the *certain* minutiae, neglecting the minutiae located in regions with poor contrast, where minutiae detection cannot even be performed manually. Automatic minutiae detection has been achieved through the technique A and through the binarization-based techniques B, C, D and E, respectively. The parameter values used for A are: section length → σ = 7, sectioning planes → h = 1, threshold angles → $\beta = \psi = 30°$, polygonal thickness → ε = 3, grid granularity → v = 2 for all the fingerprints in the sample set; the step μ of the ridge-line following is set to 5 for the 500 dpi fingerprints and to 3 for the three 350 dpi fingerprints. Two different values for the parameter μ are needed to process images taken at a different resolution. In approaches B and C the binarization process has been preceded by a smoothing operation based on the convolution with a Gaussian 5×5 pixels mask. This operation regularizes the starting image, so that the approaches B and C give better results. In approaches B, C, D and E, the thinning process has been carried out through Baruch's algorithm [3], which provides good results on fingerprints. In B, C, D and E, the minutiae detection on the binary skeleton has been performed by labelling as minutiae those pixels whose crossing number is different from 2. In all approaches, A, B, C, D and E, the minutiae detected have been filtered by removing:

- the minutiae belonging to regions where the image contrast (computed as in [7]) is less than half of the average image contrast.
- the pairs of termination minutiae which are less than k pixels (k = 6) distant from each other.
- the sets of bifurcation minutiae (except one minutia for each set) belonging to a neighborhood with diameter k pixels (k = 6).

Figure 12 shows the *certain* minutiae manually detected on fingerprints no. 1 and no. 13 of the sample set. Figures 13, 14 and 15 show the automatic extraction through approaches A, B, C, D, and E on the same fingerprints. Table 1 reports the results in terms of undetected minutiae (*dropped*), non-existent minutiae (*false*) and type-exchanged minutiae (*exchanged*). Tables 2, 3 and 4 report the average error percentage relative to the classes *good*, *poor* and to the whole sample set, respectively. Table 5 reports the average computational times spent in automatic minutiae extraction measured on a PC 80486-DX 50 MHz. The graphics in Figure 16 compare the average error percentage and the average computational times obtained with the five approaches.

Minutiae Extraction and Filtering from Gray-Scale Images 173

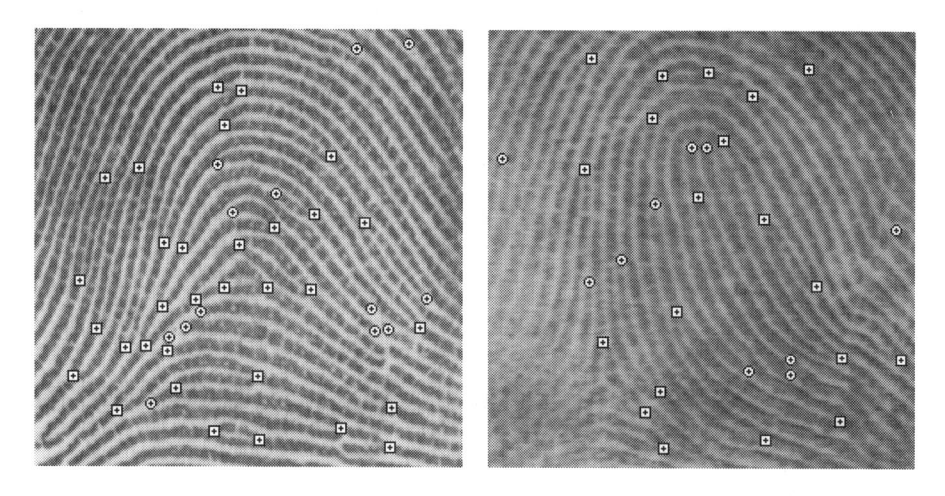

Figure 12. The figure shows the *certain* minutiae manually detected by a human expert on fingerprints no. 1 (on the left) and no. 13 (on the right) of the sample set.

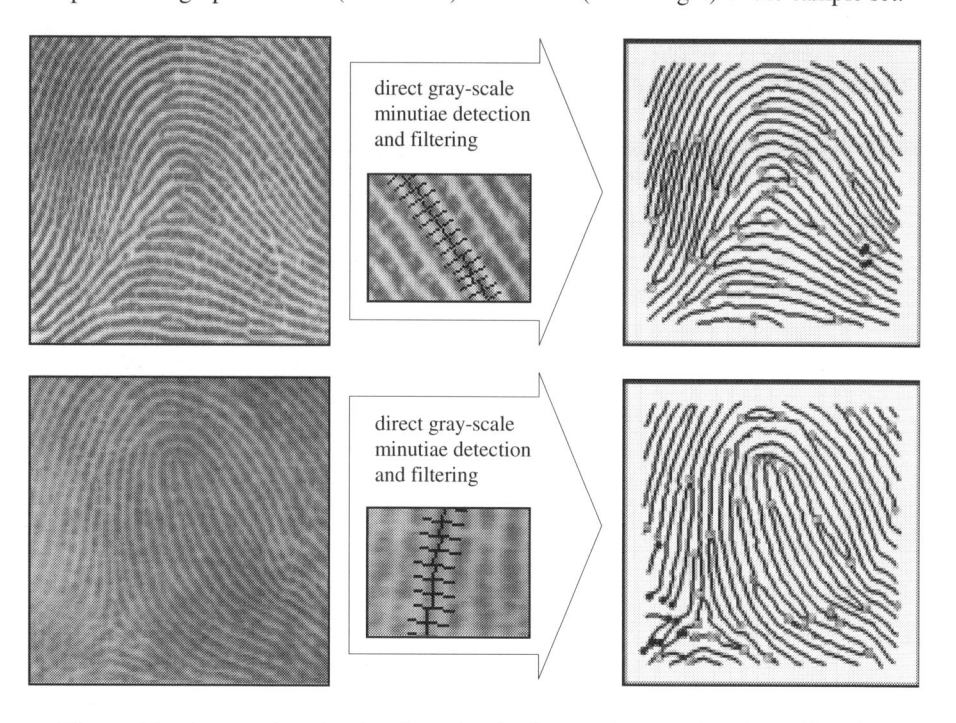

Figure 13. Automatic minutiae detection in fingerprints no. 1 and no. 13 using approach (A). Inside the arrow-box a snapshot of a direct gray-scale ridge-line following is shown. In the output image the ridge lines are represented through the corresponding polylines of **T**. The minutiae are denoted by small white circles (termination minutiae) and squares (bifurcation minutiae). Both black squares and black circles denote filtered minutiae.

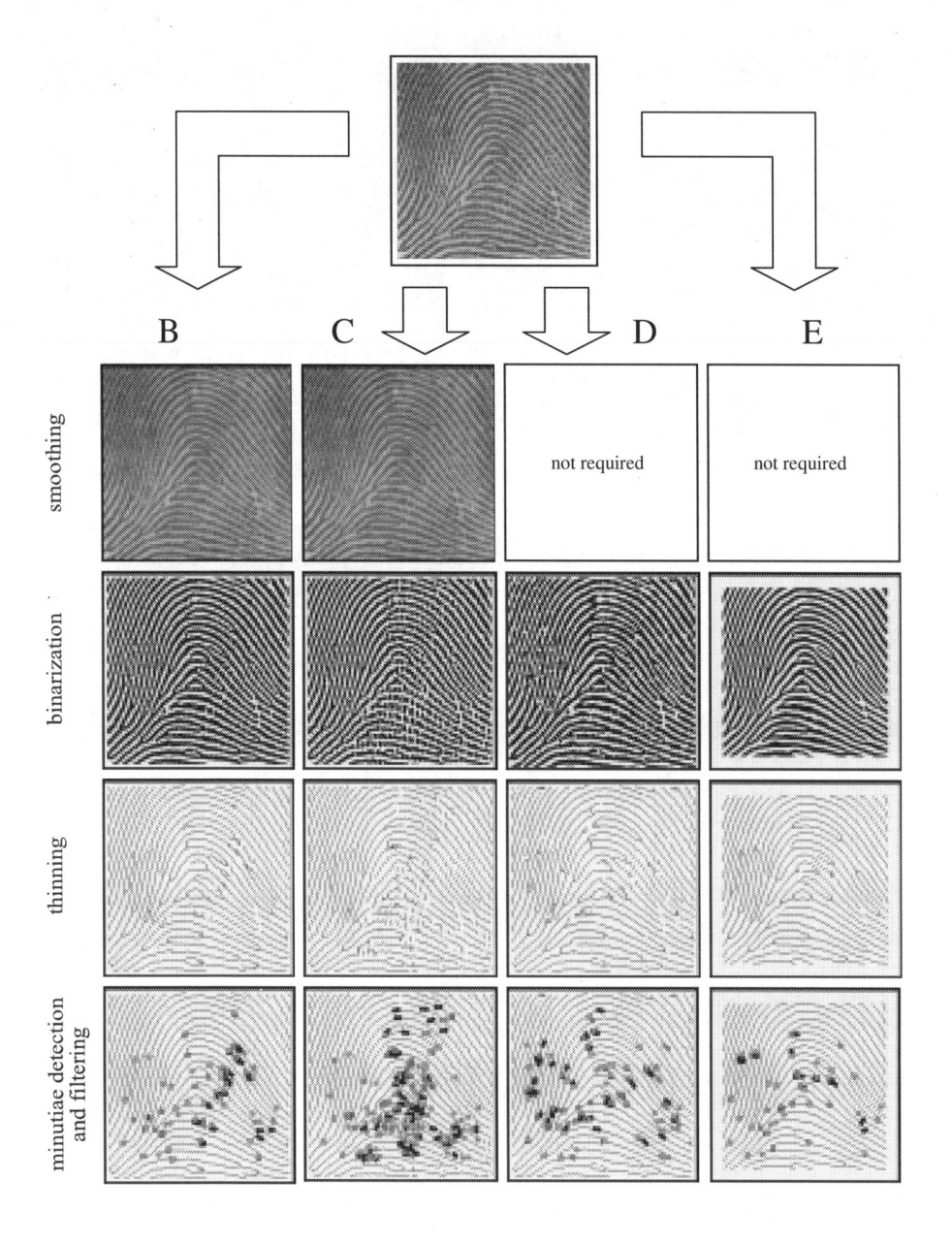

174   Intelligent Biometric Techniques in Fingerprint and Face Recognition

Figure 14. Automatic minutiae detection in fingerprint no.1 (belonging to the class *good*) using approaches B, C, D and E. Each column shows the results of the processing steps of the corresponding approach. The minutiae are denoted by small white circles (termination minutiae) and squares (bifurcation minutiae). Both black squares and black circles denote filtered minutiae.

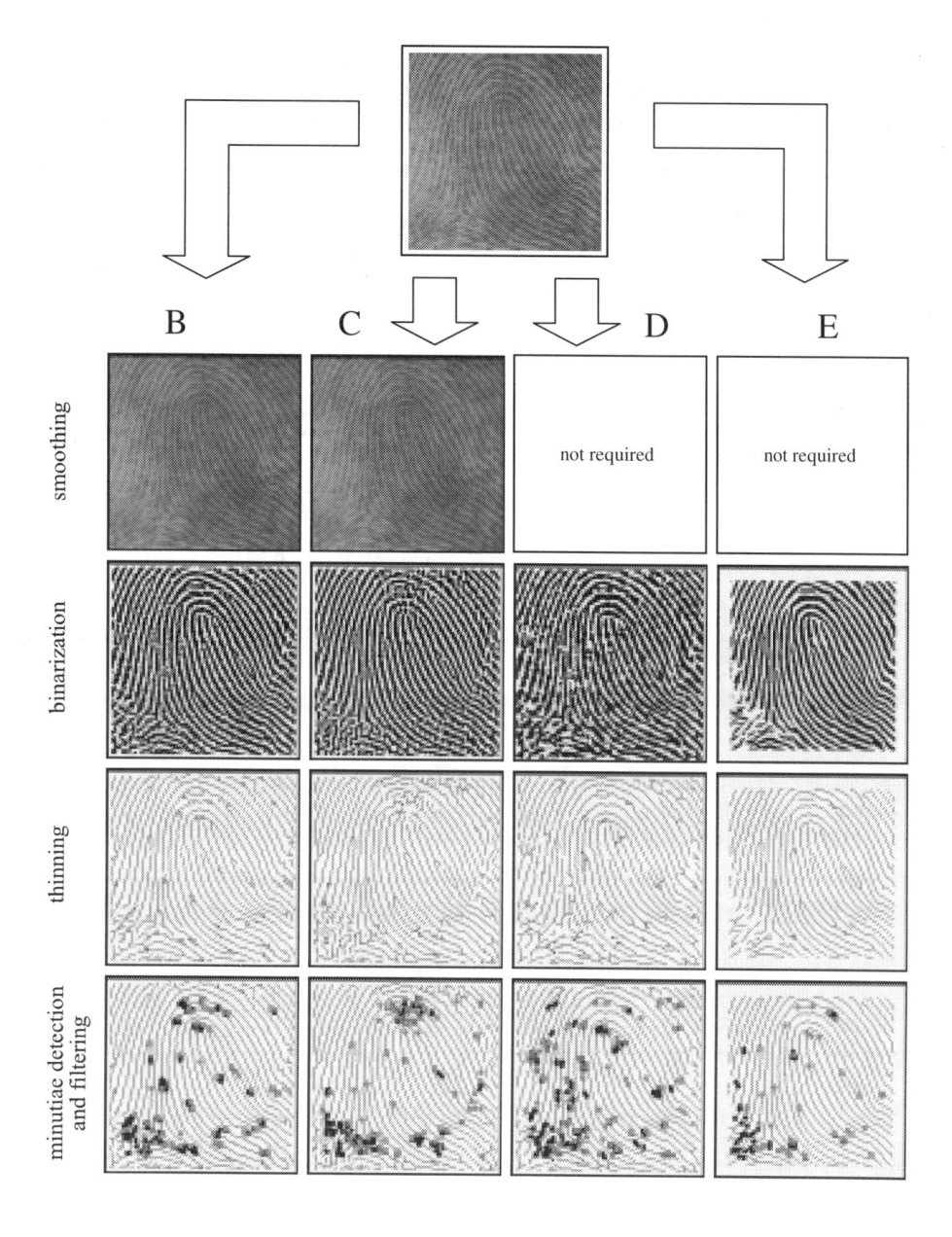

Minutiae Extraction and Filtering from Gray-Scale Images 175

Figure 15. Automatic minutiae detection in fingerprint no.13 (belonging to the class *poor*) using approaches B, C, D and E. Each column shows the results of the processing steps of the corresponding approach. The minutiae are denoted by small white circles (termination minutiae) and squares (bifurcation minutiae). Both black squares and black circles denote filtered minutiae.

Table 1. Automatic minutiae detection. The second column indicates the number of *certain* minutiae manually detected. d, f and x denote the number of *dropped* minutiae, *false* minutiae and *exchanged* minutiae, respectively.

		A			B			C			D			E		
Fing no.	Minutiae	d	f	x	d	f	x	d	f	x	d	f	x	d	f	x
1	33	0	2	7	0	25	2	1	102	7	0	53	6	0	5	4
2	29	3	1	4	0	20	2	2	24	1	1	34	2	2	4	0
3	28	1	2	4	0	24	0	1	35	1	1	28	2	0	4	4
4	37	3	0	4	0	15	3	4	32	0	1	42	3	2	2	2
5	22	0	0	3	0	38	1	0	80	2	1	18	0	0	8	2
6	23	0	0	4	0	3	2	0	25	3	0	19	2	0	2	1
7	31	2	1	2	3	13	2	2	27	2	0	40	3	2	0	2
8	31	1	0	3	0	2	0	1	10	3	0	5	4	0	0	3
9	21	1	10	1	1	115	1	0	180	1	0	153	1	0	24	2
10	22	1	0	4	0	53	1	0	104	3	0	100	1	0	8	4
11	32	3	5	4	1	22	4	2	22	3	3	73	4	1	5	2
12	33	3	8	2	0	23	3	4	45	1	1	79	3	0	10	5
13	20	0	0	4	0	48	2	0	57	3	0	81	2	0	7	5
14	37	0	5	6	1	43	5	3	67	5	1	57	4	0	11	2

Table 2. Average error percentage relative to the fingerprints of the class *good*.

Good	A	B	C	D	E
Dropped minutiae	4.27%	1.28%	4.70%	1.71%	2.56%
False minutiae	2.56%	59.83%	143.16%	102.14%	10.68%
Exchanged minutiae	13.25%	5.13%	8.12%	9.40%	7.69%
Total error	20.09%	66.24%	155.98%	113.25%	20.94%

Table 3. Average error percentage relative to the fingerprints of the class *poor*.

Poor	A	B	C	D	E
Dropped minutiae	4.85%	1.82%	5.45%	3.03%	0.61%
False minutiae	16.97%	184.24%	287.88%	329.09%	39.39%
Exchanged minutiae	12.73%	9.70%	9.70%	9.09%	12.12%
Total error	34.55%	195.76%	303.03%	341.21%	52.12%

Minutiae Extraction and Filtering from Gray-Scale Images

Table 4. Average error percentage relative to the fingerprints of the whole sample set.

Whole set	A	B	C	D	E
Dropped minutiae	4.51%	1.50%	5.01%	2.26%	1.75%
False minutiae	8.52%	111.28%	203.01%	195.99%	22.56%
Exchanged minutiae	13.03%	7.02%	8.77%	9.27%	9.52%
Total error	26.07%	119.80%	216.79%	207.52%	33.83%

Table 5. Average computational time (in seconds) taken for the minutiae detection on PC 80486-DX 50 MHz architecture.

Average computational time (sec.)	A	B	C	D	E
Directional image	0.51	-	-	-	0.51
Smoothing	-	3.90	3.90	-	-
Binarization	-	2.25	3.08	2.64	15.73
Thinning	-	3.11	3.11	3.15	2.45
min. detection and filtering	2.22	0.51	0.67	0.66	0.33
Total time	2.73	9.77	10.76	6.45	19.03

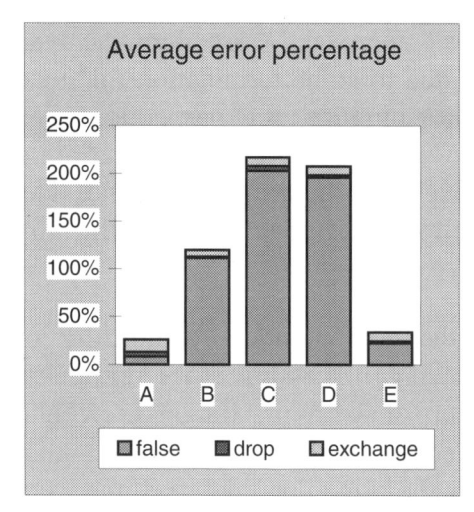

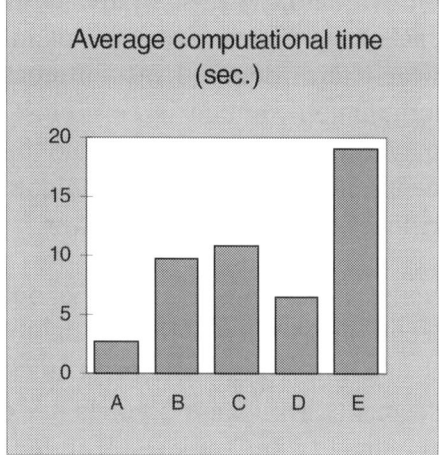

Figure 16. A comparison between the average error percentage and the average computational times of the five different approaches.

The results achieved by technique A on a real sample of 150 fingerprints, acquired through an opto-electronic device based on a prism, are very similar to those obtained for the class *good* of the sample set considered here (Table 2, approach A).

The following conclusions can be drawn:

- the average error percentage, in terms of *dropped* and *exchanged* minutiae, as produced by approach A is comparable to the errors produced by the other approaches, although slightly larger.
- the average error percentage, in terms of *false* minutiae, as produced by approach A is considerably lower than the errors produced by the other approaches.
- the average computational time of approach A is considerably lower than the time of the other approaches.
- approach E, whose performance in terms of total error is comparable with that of A, is one order of magnitude slower than A.

The large number of false minutiae determined by approaches B, C and D (especially on class *poor*) is due to the irregularity of the binary traces produced by the binarization process. Regularization techniques, similar to that presented in [6], can substantially reduce the number of false minutiae. Most of the errors of approach A are minutiae type exchanges. These errors are mainly due to some termination minutiae which are detected as bifurcation minutiae. In particular, if a termination minutia is very close to another ridge line the algorithm may skip the termination and intersect the adjacent ridge line. The neural network filtering technique introduced in Section 4 significantly reduces the amount of this error.

Some final considerations about the computational complexity of technique A are provided in the following. It is assumed, for simplicity, that a fingerprint pattern is made up of a set of straight horizontal segments, which are ξ-pixels thick and ξ-pixels distant from each other (Figure 17). In fact, even if in reality the ridge line thickness may vary from about 4-5 pixel to 12-15 pixel (at the resolution used), ξ can be assumed to be the mean value. Under this assumption, the number of ridge lines in a $n \times n$ image is $n/2\xi$. If μ is the step of the ridge-line following algorithm, n/μ steps are necessary to extract a whole ridge

Minutiae Extraction and Filtering from Gray-Scale Images

line. Therefore, the algorithm performs $n^2/(2\xi\mu)$ steps in order to extract all the ridge lines and then all the minutiae.

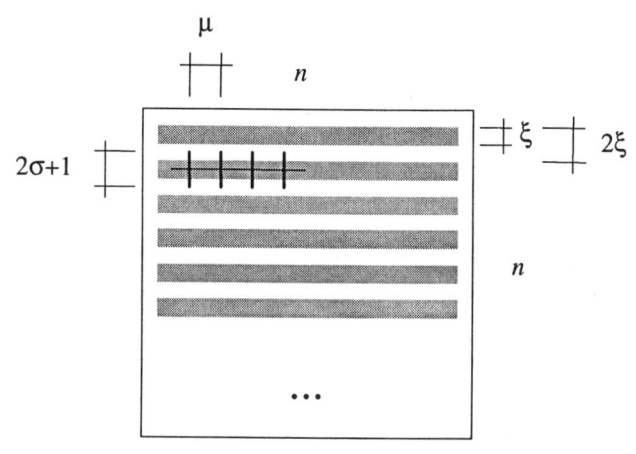

Figure 17. A simplified fingerprint pattern

At each step the algorithm performs the following operations:

1. computation of the new point (i_t, j_t)
2. construction of the section segment Ω
3. convolution with the Gaussian mask **d**
4. maxima searching
5. checking the stop criteria
6. update of **T**
7. computation of the new direction φ_c

By using Bresenham's algorithm [4] to compute Ω and to update **T**, and by representing the angles through discrete values, all the computations can be performed by means of integer arithmetic. Table 6 summarizes the elementary operations carried out at each ridge-line following step. All the parameter values can be approximately estimated according to the ridge thickness ξ; in the following it is assumed: $\sigma = 2p+1 = \xi$ and $\mu = \varepsilon = 2h+1 = \frac{1}{2}\xi$. The computational complexity of the technique A (except for the computation of the directional image) is then:

$$n^2 \left(\frac{3/2\, \xi^2 + 8\,\xi + 30}{\xi^2} \right) \text{sum}, \quad (\approx 3\, n^2 \quad \text{if } \xi = 8)$$

$$n^2\left(\frac{2\xi^2+2\xi+18}{\xi^2}\right) \text{ multiplication}, \quad (\approx 2.5\,n^2 \quad \text{if } \xi=8)$$

$$n^2\left(\frac{11/2\,\xi+8}{\xi^2}\right) \text{ test}. \quad (\approx 0.8\,n^2 \quad \text{if } \xi=8)$$

Hence, the total number of operations executed for the direct gray-scale minutiae detection is only a few times the number of pixels of the whole image. It is important to note that, by using techniques based on binarization and thinning, the binarization step alone requires n^2 pixel-neighborhoods to be processed, so that the time taken for the whole detection is undoubtedly longer than in approach A.

Table 6. Elementary operations carried out at each following step. $(2\sigma+1)$ is the length of the section segment, $(2h+1)$ is the number of planes used by the first step of the regularization, $(2p+1)$ is the dimension of the mask **d** and ε is the thickness of the polygonal traces used to update **T**.

	sum	multiplication	test
1	2	2	-
2	$11+2(2\sigma+1)$	4	3
3	$(2h+1)(2\sigma+1)$	$(2p+1)(2\sigma+1)$	-
4	-	-	$2(2\sigma-1)$
5	3	1	7
6	$\varepsilon(2\mu+7)$	2ε	3ε
7	12	11	-
Total	$28+(2\sigma+1)(2h+3)$ $+\varepsilon(2\mu+7)$	$18+(2p+1)(2\sigma+1)$ $+2\varepsilon$	$10+2(2\sigma-1)+3\varepsilon$

4 Neural Network-Based Minutiae Filtering

In [21], minutiae filtering is conceived as a post-processing step which can be used to refine the output of an automatic minutiae extraction algorithm. Each minutia, as detected by the algorithm [20], is normalized and analyzed through a Neural Network (NN) classifier. It is obvious that this kind of post-processing can reduce only false and exchanged minutiae, but not dropped minutiae. As reported in Section 2.2, some approaches to minutiae extraction with NN have been presented in the literature [18], [19]. It is worth noting that there is a fundamental difference between a NN approach for minutiae location and for minutiae filtering. In fact, in the above-mentioned methods the networks are directly used to locate the minutiae in the images, and to this purpose each possible position must be processed. On the other hand, minutiae filtering usually requires only a small set of candidates to be verified. Actually, some attempts have been carried out by using NNs for minutiae location on gray-scale images, but the following problems were encountered:

- Too many false errors were produced; in fact, even if, after an appropriate training, the false percentage error can be reduced to very low values, to compute the average number of false minutiae per fingerprint the above value must be multiplied by the huge number of possible positions.
- It is computationally too heavy for real-time applications.

The results obtained by other researchers using NNs for direct localization exhibit the same problems: see for example [31], [36] where NNs have been employed for the detection of human faces in complex backgrounds.

The rest of this section is organized as follows: Subsection 4.1 describes the normalization and the dimensionality reduction performed on each minutia neighborhood before passing it to the neural classifier; Subsection 4.2 introduces the neural network topology and the training phase; finally, in Subsection 4.3 some results and concluding remarks are reported.

4.1 Neighborhood Normalization and Dimensionality Reduction

One of the most recurrent dilemmas in pattern recognition applications is at which level the invariance must be dealt with: that is, whether the extracted features should be normalized or the classifier itself should be tolerant to feature variations [47]. In the field of NNs several attempts have been made for developing invariant classifiers: weight-sharing

[32] and high-order NNs [10] are examples of networks which allow certain degrees of invariance to be obtained. However, in these cases the network topology is much more complex (especially when the rotation invariance is a strong requirement) and consequently the learning time can become prohibitive. To solve the invariance problem, in [18] Leung used six different NNs, each trained to detect minutiae within a specific range of directions.

In [21] the minutiae neighborhoods are normalized before passing them to the classifier in order to reduce the classification efforts. As shown in Figure 18a, minutiae neighborhoods of the same type (termination or bifurcation) can be very different from one other due to the minutia orientation (*rotation*), to the local ridge-line density (*scale*) and to other perturbations (local image intensity, intra-class minutiae variation and noise). The normalization step takes into account both rotation and scale:

- The orientation normalization consists of rotating the minutiae neighborhoods according to their local ridge-line orientation. For local orientation several techniques have been proposed. In [20] the authors used the method [7] which is very effective and robust for the computation of a fingerprint directional image (Figure 19b).
- The scale normalization consists of re-scaling the minutiae neighborhoods according to their local ridge-line density. As far as ridge line density is concerned, it should be noted that this value may noticeably vary from fingerprint to fingerprint, and also among different minutiae of the same fingerprint. In [22] a new approach to the computation of the local ridge-line density, based on the partial derivatives of sinusoidal signals, is presented (Figure 19c).

Figure 18b shows the same minutiae neighborhoods of Figure 18a after the normalization. Figure 20 shows the images obtained by averaging over 1000 normalized minutiae neighborhoods; from these images it appears evident the well known termination/bifurcation duality. As can be noted in Figure 18b, the upper and the lower part of the circular neighborhoods usually contain information which are relevant for the classification; therefore elliptical neighborhoods, whose horizontal axis coincide with the original width and whose vertical axis length is 0.5 times the original height, are cropped from the normalized neighborhoods. At 350 dpi resolution this corresponds to dealing with 23×11 elliptical neighborhoods containing exactly 217 pixels. To take advantage from the termination/bifurcation duality, both the original neighborhood and its negative version should constitute the input of the neural classifier, so that 434 neurons would be necessary for the first layer. To avoid the problems related to training of large networks, the dimensionality of the normalized neighborhoods is reduced through the KL transform [16]. The authors found experimentally that the first 26 principal components carried 90% of the neighborhood information.

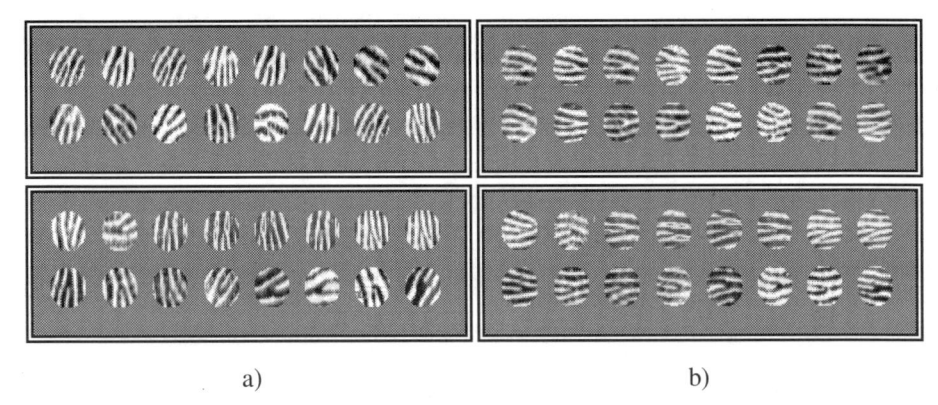

Figure 18. (a) Shows non-normalized minutiae neighborhoods (termination minutiae at the top, bifurcation minutiae at the bottom); in (b) the same neighborhoods have been normalized with respect to rotation and scale.

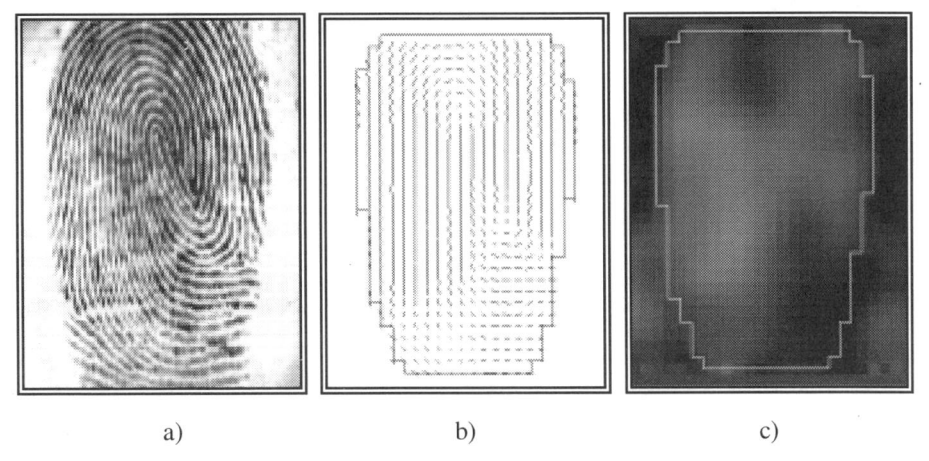

Figure 19. Figure 19b reports the directional image of the fingerprint 19a, whereas Figure 19c shows its local density map (the white blocks correspond to higher density regions).

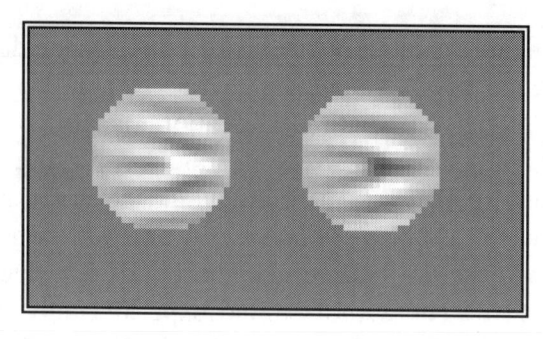

Figure 20. The mean termination and bifurcation neighborhoods obtained by averaging over 1000 normalized minutiae neighborhoods.

4.2 Neural Network Classifier

Several types and topologies of NNs have been used in the literature for classification purposes. In [5] the performances of 9 different classifiers (4 of which are NN classifiers) are compared in the context of fingerprint classification. In general, multi-layer perceptrons, radial basis functions and probabilistic NNs were proven to be very effective for classification purposes in different applications.

In [21] a typical three-layer architecture has been adopted, where a partial weight-sharing allows the termination/bifurcation duality to be exploited (Figure 21).

The input layer contains two groups of 26 neurons; the first group receives the 26 principal components extracted by the normalized minutia neighborhoods, and the second group the 26 principal components extracted by the corresponding negative images. Each group is fully connected to 10 neurons in the second layer, but the connection weights of the two groups are shared. This constraint requires the same type of processing to be performed both on the positive and the negative neighborhood. Finally, the output layer contains 2 units which are fully connected to the hidden layer. The output configurations [1,0], [0,1], [0,0] denote, respectively, a termination minutia, a bifurcation minutia and the absence of any minutia. This network has more degrees of freedom with respect to a three-layer (26-10-2) perceptron trained both on positive and negative version of the same neighborhood[2], and used twice for each classification; in fact, in this case both the positive and negative information convey contemporary to the third layer which acts as a final decider.

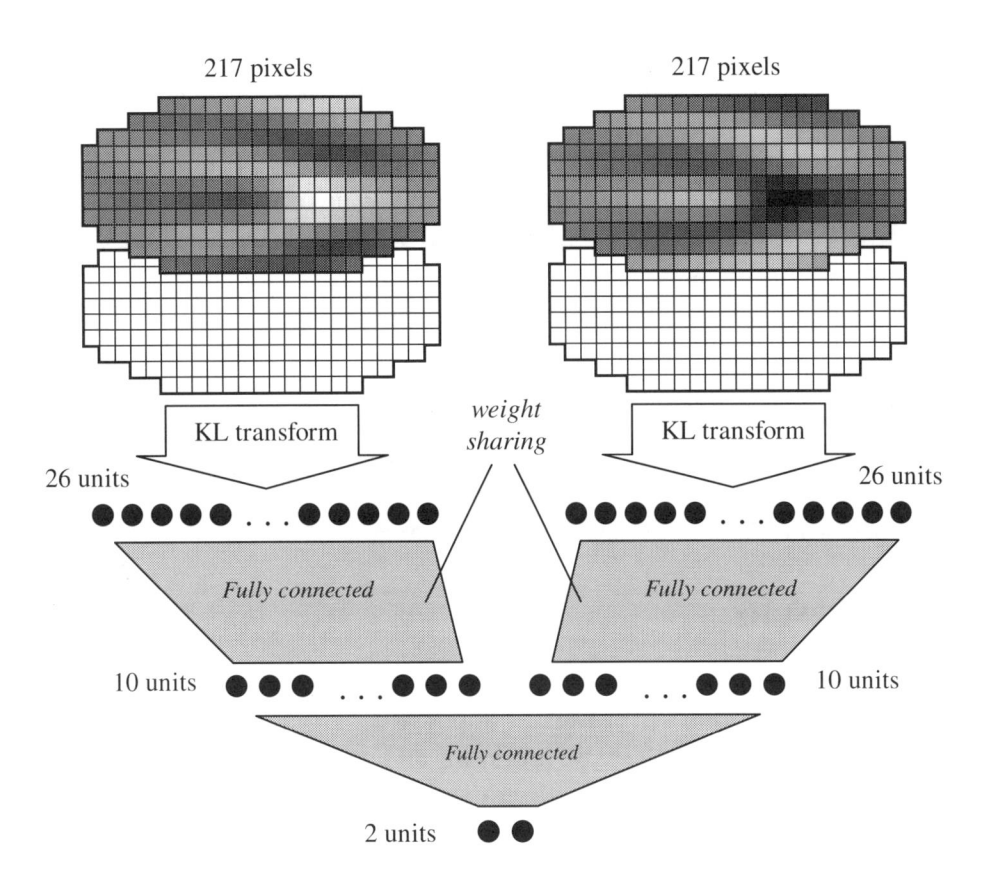

Figure 21. The Neural Network classifier architecture.

A database of 62 fingerprint images, acquired through a prototypal optical sensor which works at about 350 dpi, was used for training and testing the classifier. From the first 31 images the following sets were extracted:

TML: 100 termination minutiae manually selected.
BML: 100 bifurcation minutiae manually selected.

---

[2] Training with negative neighborhoods requires provision of inverted outputs in the case of true minutiae and the same output for false minutiae.

FML: 100 positions manually selected where no minutiae were present.
TAL: 300 true termination minutiae automatically detected by the algorithm [20].
BAL: 300 true bifurcation minutiae automatically detected by the algorithm [20].
FAL: 300 false minutiae detected by the algorithm [20].

1/3 TAL + 1/3 BAL + 1/3 FAL = 300 examples were employed for calculating the covariance matrix for the KL transform; TML + BML + FML + 2/3 TAL + 2/3 BAL + 2/3 FAL = 900 examples constituted the training set for the neural classifier.

When applied to images 32-62, the extraction algorithm detected 954 minutiae. A subset of 739 minutiae was used as a test set. This subset is constituted by those minutiae which could be reliably manually labeled in *true termination* (303), *true bifurcation* (240) and *false minutiae* (196). The corresponding extraction error (without filtering) are: dropped = 11.8%, false = 24.2% and exchanged = 13.1%.

4.3 Results

For the implementation and training of the NN, the authors used the SPR/ANNLIB library by Delft Pattern Recognition Group [13], a powerful and flexible tool for defining and training neural networks. 6000 iterations allowed a 98.2% convergence to be achieved over the training set by using the standard back-propagation algorithm. Table 7 reports the results obtained on the test set in terms of confusion matrix. The three types of classification errors can be directly extracted by the confusion matrix (dropped and exchanged classification errors are computed with respect to the total number of terminations and bifurcations which constitute the test set, whereas the false classification errors refer to the number of errors in the test set): D_C (Dropped) =17.5%, E_C (Exchanged) = 2.21% and F_C (False) = 27.0%. Let D_A, F_A, E_A be the percentage errors produced by the extraction algorithm (as referred to the number of ground true minutiae) then, if the final decision is taken according to the output of the classifier (disregarding of the minutiae-extraction algorithm response), the overall errors D, F, E become:

Minutiae Extraction and Filtering from Gray-Scale Images 187

$D = D_A + (1-D_A) \cdot D_C = 11.8\% + 88.2\% \cdot 17.5\% = 27.2\%$
$F = F_A \cdot F_C = 24.2\% \cdot 27.0\% = 6.5\%$
$E = (1-D_A) \cdot E_C = 1.9\%$

where, $(1-D_A)$ and F_A denote respectively the percentage of true and false minutiae considered by the classifier.

Table 7. Confusion matrix of the classification results.

True classification of the extraction algorithm output

NN classification		Termination	Bifurcation	Error
	Termination	267 (88.2%)	8 (3.3%)	20 (10.4%)
	Bifurcation	4 (1.3%)	169 (70.5%)	32 (16.6%)

Dropped = 17.5%   Exchanged = 2.21%   False = 27.0%

Summarizing, the filtering method described in this section, in spite of a certain increase in dropped errors, determines significant reductions in false and exchange errors, respectively. The total extraction error (11.8% + 24.2% + 13.1% = 49.1%) is reduced to (27.2% + 6.5% + 1.9% = 35.6%). Finally, since the average number of minutiae per fingerprint is about 50, a small computational effort is required.

Bibliography

[1] American National Standards Institute (1986), *Fingerprint Identification − Data Format for Information Interchange*, New York.

[2] Arcelli, C. and Baja, G.S.D. (1984), "A Width Independent Fast Thinning Algorithm," *IEEE tPAMI*, Vol. 4, No. 7, pp. 463-474.

[3] Baruch, O. (1988), "Line thinning by line following," *Pattern Recognition Letters*, Vol. 8, No. 4, pp. 271-276.

[4] Bresenham, J. (1965), *IBM System Journal*, Vol. 4, No. 1, pp. 25-30.

[5] Candela, G.T. and Chellappa, R. (1993), "Comparative Performance of Classification Methods for Fingerprints," *Tech. Report TR 5163*, National Institute of Standards and Technology.

[6] Coetzee, L. and Botha, E.C. (1993), "Fingerprint recognition in low quality images," *Pattern Recognition*, Vol. 26, No. 10, pp. 1441-1460.

[7] Donahue, M.J. and Rokhlin, S.I. (1993), "On the Use of Level Curves in Image Analysis," *Image Understanding*, Vol. 57, No. 2, pp. 185-203.

[8] Galton, F. (1892), *Finger Prints*, Macmillan, London.

[9] Gamble, F.T., Frye, L.M., and Grieser, D.R. (1992), "Real-time fingerprint verification system," *Applied Optics*, Vol. 31, No. 5, pp. 652-655.

[10] Giles, C.L. and Maxwell, T. (1987), "Learning, Invariance and Generalization in High-order Neural Networks," *Applied Optics*, Vol. 26, pp. 4972-4978.

[11] Gonzalez, R.C. and Woods, R.E. (1992), *Digital Imaging Processing*, Addison-Wesley, Reading, Massachusetts, USA.

[12] Hollingum, J. (1992), "Automated Fingerprint Analysis Offers Fast Verification," *Sensor Review*, Vol. 12, No. 3, pp. 12-15.

[13] Hoekstra, A., Kraaijveld, M.A., Ridder, D., and Schmidt, W.F. (1996), *The Complete SPRLIB & ANNLIB*, Pattern Recognition Group, Delft University of Technology, the Netherlands.

[14] Hung, D.C.D. (1993), "Enhancement and feature purification of fingerprint images," *Pattern Recognition*, Vol. 26, 1661-1671.

[15] Igaki, S., Eguchi, S., Yamagishi, F., Ikeda, H., and Inagaki, T. (1992), "Real-time fingerprint sensor using a hologram," *Applied Optics*, Vol. 31 No. 11, pp. 1794-1802.

[16] Jolliffe, I.T. (1986), *Principle Component Analysis*, Springer Verlag, New York.

[17] Lam, L., Lee, S.-W., and Suen, C.Y. (1992), "Thinning methodologies: A comprehensive survey," *IEEE tPAMI*, Vol. 14, No. 9, pp. 869-885.

[18] Leung, M.T., Engeler, W.E., and Frank, P. (1990), "Fingerprint Image Processing Using Neural Network," *Proc. 10th conf. on Computer and Communication Systems*, pp. 582-586, Hong Kong.

[19] Leung, W.F., Leung, S.H., Lau, W.H., and Luk, A. (1991), "Fingerprint Recognition Using Neural Network," *Proc. of the IEEE workshop Neural Network for Signal Processing*, pp. 226-235.

[20] Maio, D. and Maltoni, D. (1997), "Direct Gray-Scale Minutiae Detection in Fingerprints," *IEEE tPAMI*, Vol. 19, No. 1.

[21] Maio, D. and Maltoni, D. (1998), "Neural Network based Minutiae Filtering in Fingerprints," *Proc. of 14th ICPR*, Brisbane, Australia.

[22] Maio, D. and Maltoni, D. (1998), "Ridge-line Density Estimation in Digital Images," *Proc. of 14th ICPR*, Brisbane, Australia.

[23] Marr, D. and Hildreth, E.C. (1980), "Theory of Edge Detection," *Proc. R. Soc. London*, B 207, 187-217.

[24] Mehtre, B.M. and Murthy, N.N. (1986), "A minutia based fingerprint identification system," *Proc. 2nd Int. Conf. Advances in Pattern Recognition and Digital Techniques*, Calcutta, India.

[25] Mehtre, B.M. (1993), "Fingerprint Image Analysis for Automatic Identification," *Machine Vision and Applications*, Vol. 6, No. 2-3, pp. 124-139.

[26] Moayer, B. and Fu, K. (1986), "A tree system approach for fingerprint pattern recognition," *IEEE tPAMI*, Vol. 8, No. 3, pp. 376-388.

[27] O'Gorman, L. and Nickerson, J.V. (1989), "An approach to fingerprint filter design," *Pattern Recognition*, Vol. 22, No. 1, pp. 29-38.

[28] Pal, N.R. and Pal, S.K. (1993), "A review on image segmentation techniques," *Pattern Recognition*, vol. 26, No. 9, pp. 1277-1294.

[29] Pernus, F., Kovacic, S., and Gyergyek, L. (1980), "Minutiae-based fingerprint recognition," *Proc. 5th ICPR*, pp. 1380-1382.

[30] Ratha, N.K., Chen, S., and Jain, A.K. (1995), "Adaptive flow Orientation-Based Feature Extraction in Fingerprint Images," *Pattern Recognition*, Vol. 28, No. 11, pp. 1657-1672.

[31] Rowley, H.A., Baluja, S., and Kanade, T. (1995), "Human Face Detection in Visual Scenes," *Tech. Report CMU-CS-95-158R*, Carnegie Mellon University.

[32] Rumelhart, D.E., Hinton, G.E., and Williams, R.J. (1986), "Learning Internal Representation by Error Propagation," in D.E. Rumelhart and J.L. McClelland (Eds.), *Parallel Distributed Processing: Explorations in the Microstructures of Cognition*, Vol. 1, pp. 318-362. MIT Press, Cambridge.

[33] Sherstinsky, A. and Picard, R. (1994), "Restoration and Enhancement of Fingerprint Images Using M-Lattice – A Novel Non-Linear Dynamical System," *Tech. report #264*, MIT Media Laboratory Perceptual Computing Group.

[34] Sherlock, B.G., Monro, D.M., and Millard, K. (1992), "Algorithm for enhancing fingerprint images," *Electronics letters*, Vol. 28, No. 18, pp. 1720-1721.

[35] Sherlock, B.G., Monro, D.M., and Millard, K. (1994), "Fingerprint Enhancement by Directional Fourier filtering," *IEEE Proceedings Vision, Image and Signal Processing*, No. 141, pp. 87-94.

[36] Soulie, F.F. (1997), "Connectionist Methods for Human Face Processing," *Proc. of NATO FACES 97*, Stirling UK.

[37] Stock, R.M. and Swonger, C.W. (1969), "Development and evalutation of a reader of fingerprint minutiae," *Tech. report CAL No. XM-2478-X-1:13-17*, Cornell Aeronautical Labaratory.

[38] Székely, E.N. and Székely, V. (1993), "Image recognition problems of fingerprint identification," *Microprocessors and Microsystems*, Vol. 17, No. 4, pp 215.

[39] Tamura, H. (1978), "A Comparison of Line Thinning Algorithms from Digital Topology Viewpoint," *Proc. 4th ICPR, pp. 715-719, Japan.*

[40] Verma, M.R., Majumdar, A.K., and Chatterjee, B. (1987), "Edge detection in fingerprints," *Pattern Recognition*, Vol. 20, No. 5, pp. 513-523.

[41] Xiao, Q. and Raafat, H. (1991), "Fingerprint image postprocessing: a combined statistical and structural approach," *Pattern Recognition*, Vol. 24, No. 10, pp. 985-992.

[42] Xiao, Q. and Raafat, H. (1991), "Combining statistical and structural information for fingerprint image processing, classification and identification," in *Pattern Recognition: Architectures, Algorithms and Applications*, World Scientific Series in Computer Science, pp. 335-354.

[43] Watson, C.I. and Wilson, C.L. (1992), *Fingerprint Database. National Institute of Standards and Technology, Special Database 4*, April 18.

[44] Watson, C.I., Candela, G.T., and Grother, P.J. (1994), "Comparison of FFT Fingerprint Filtering Methods for Neural Network Classification," Tech rep. U.S. Departement of Commerce NIST.

[45] Weber, D.M. (1992), "A Cost Effective Fingerprint Verification Algorithm for Commercial Applications," *Proc. of the 1992 South African Symposium on Communication and Signal Processing*, pp. 99-104.

[46] Wegstein, J.H. (1982), *An automated fingerprint identification system*, U.S. Government Publication, Washington, D.C.

[47] Wood, J. (1996), "Invariant Pattern Recognition: A Review," *Pattern Recognition*, Vol. 29, No. 1, pp. 1-17.

Chapter 6:

Feature Selective Filtering for Ridge Extraction

FEATURE SELECTIVE FILTERING FOR RIDGE EXTRACTION

A. Erol
Halici Software House, METU Technopark
Middle East Technical University, 06531, Ankara, Turkey
ali@halici.com.tr

U. Halici and **G. Ongun**
Computer Vision and Artificial Neural Networks Research Lab.
Dept. of Electrical and Electronics Eng.
Middle East Technical University, 06531, Ankara, Turkey
ugur-halici@metu.edu.tr

Low-quality images of fingerprints form the main source of errors in fingerprint recognition systems. Most of the algorithms in the literature perform well with high-quality fingerprint images but result in an excessive number of false minutiae with low-quality images. Enhancement of fingerprint images plays an important role in fingerprint recognition systems. In this chapter, a module which performs both enhancement and ridge extraction is presented. The algorithm is based on processing of an input image with linear filters tuned to extract ridges at a specific orientation and frequency.

1 Introduction

Fingerprints are the oldest and most extensively studied biometric technique of identification. They are usually associated with criminology but recently they have also been utilized in many civilian applications such as access control and financial security [10].

In a manual fingerprint matching process, experts use a set of minute features, which were established 100 years ago by Galton [4]. These features are called minutiae. A fingerprint image consists of regularly spaced ridge curves. The spacing between two ridges is called valley (See Figure 1). There are two basic types of minutiae called ridge endings and ridge bifurcations. Ridge endings are the points where the ridge curves end and the bifurcations are the intersection points of ridge curves. There are also other types of minutiae, which can be expressed in terms of these two basic minutiae types [3]. The locations of minutiae have been proven to be unique for each person. This fact is also utilized in most of the automated fingerprint recognition systems. Minutiae detection on a fingerprint image forms the core of such systems [1], [6], [7], [9], [11], [16], [21].

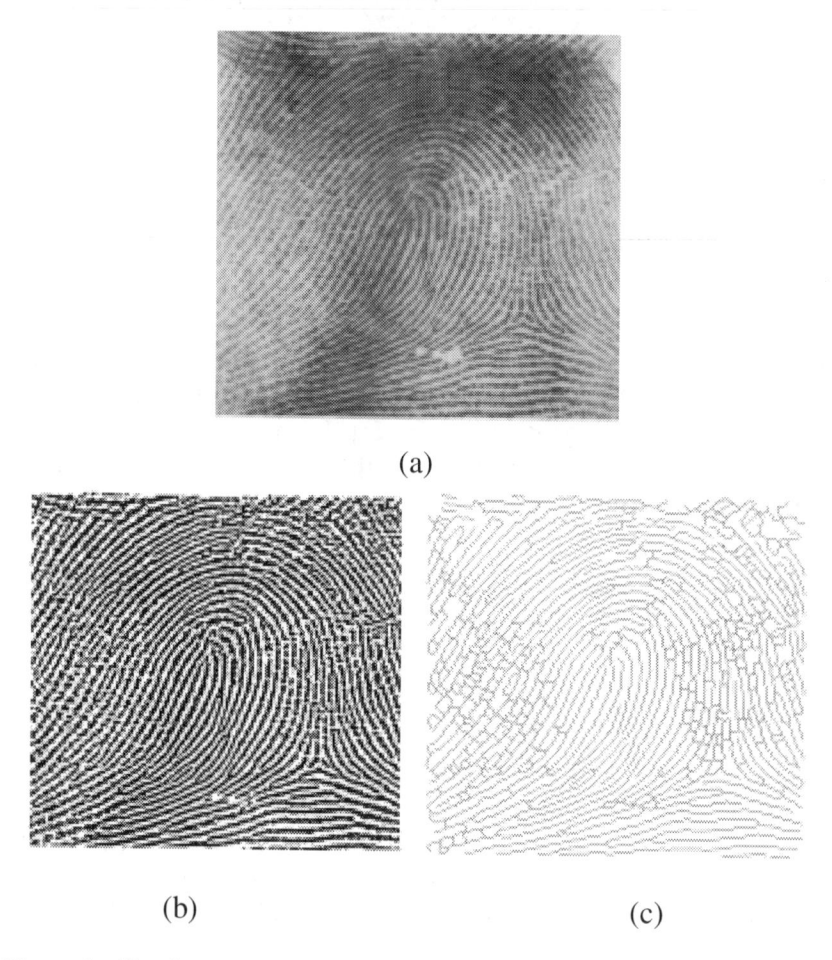

(a)

(b) (c)

Figure 1. Usual processing of fingerprint image a) A gray scale fingerprint image. b) Binarized version of the image shown in (a). c) Thinned version of the image shown in (b).

Extraction of minutiae is a difficult problem. A common approach to minutiae extraction is to obtain a binary image (Figure 1-b) from the

Feature Selective Filtering for Ridge Extraction

gray scale input image (Figure 1-a) and thin this binary image (Figure 1-c). On the thinned image minutiae detection is a simple process. If the ridge curves are represented by white pixels, counting the number of black to white transitions in the 3×3 neighborhood of a white pixel is enough to classify the pixel as a ridge pixel (2 transitions), an end pixel (1 transition) and a bifurcation pixel (3 or more transitions). The problem with this approach is that it mostly ends up with false minutiae. So, heuristic postprocessing is applied to the set of detected minutiae and some of these are eliminated. This process is called minutiae reduction and it is sometimes designed as a very complicated process [7], [21]. False minutiae are inevitable because of distortions like scars, sweat, dryness, etc., but usually the algorithm itself creates false minutiae in the binarization process. The thinned image in Figure 1-c contains many false minutiae-like bridges (lines connecting two parallel ridges) and breaks (discontinuities in the ridge curve). The main source of false minutiae is low-quality regions on the fingerprint image. An approach to cope with this problem is to design a good fingerprint enhancement module that supports the binarization process [12], [17], [18].

The aim of the fingerprint enhancement is to obtain an image where the ridges and valleys are smoother and easily distinguishable, which is an intermediate step between a full binary image and the unprocessed raw image. The resulting image is then processed by simple binarization methods like local thresholding or ridge valley binarizer [20].

In this chapter, a method proposed in [2] that performs enhancement and binarization, which can also be called ridge extraction, is presented. The enhancement algorithm is designed as a filtering process that tries to extract ridges. When there is high evidence for the existence of a ridge, the binarization is performed directly otherwise just an enhancement operation is performed. Filtering is performed by oriented feature selective Gabor-like band pass filters. The parameters of these filters are position dependent and determined by the local textural features of the fingerprint image [2].

In the following section, the design of the feature selective filter is explained in detail. In Section 3, implementation of a binarization system is given, and in Section 4, the experimental results are reported. Section 5 concludes the chapter.

2 The Filter Design

The enhancement algorithm given in this chapter can be viewed as processing of the image by a space (time) varying system that changes its filter coefficients using the local properties of the input signal. The filter is a restricted version of Gabor filters that is designed to give maximal response to ridges at a specific orientation and spacing on the fingerprint image [2]. The impulse response of the filter is given by the product of a Gaussian and a cosine plane wave,

$$f(x,y) = e^{\frac{-(x^2+y^2)}{2\sigma^2}} \cos(k_x x + k_y y) \quad (1)$$

$$F(w_x, w_y) = \sqrt{2\pi}\sigma \left(e^{\frac{-((w_x-k_x)^2+(w_y-k_y)^2)\sigma^2}{2}} + e^{\frac{-((w_x+k_x)^2+(w_y+k_y)^2)\sigma^2}{2}} \right) \quad (2)$$

where σ is the variance of the Gaussian, $\mathbf{k}=[k_x, k_y]^T$ is the wave vector of the plane wave and $F(w_x,w_y)$ is the Fourier Transform. A 3D plot of the filter in spatial and frequency domain are given in Figure 2. In the spatial domain, when variance is high enough, the filter responds maximally to ridges with spacing frequency equal to the magnitude of the wave vector and oriented orthogonal to its direction. When $\mathbf{k}$ and σ are appropriate a filter like the one shown in Figure 2-a is obtained. In this figure, the main lobe in the middle matches to the ridge and the side lobes with smaller amplitude correspond to the neighboring ridges.

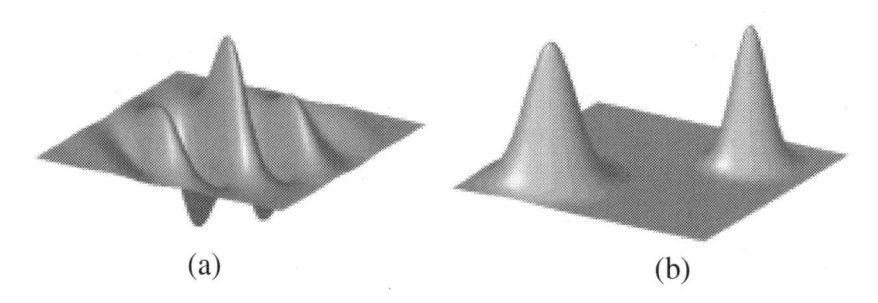

(a) (b)

Figure 2. 3-D plots of ridge extraction filters:
(a) spatial domain; (b) frequency domain.

The region with negative amplitude between the side lobes and the main lobe corresponds to the valleys. In medium-variance levels, the two side lobes disappear and the filter functions as a directional smoothing filter. In very-low-variance levels it just performs smoothing. In the frequency domain the filter can be viewed as an orientation selective bandpass filter whose bandwidth is inversely proportional to σ.

The main features used in filter construction are local ridge spacing and local ridge orientation. The ridge spacing determines both the magnitude of the wave vector and the variance of the Gaussian. The magnitude of the wave vector corresponds to the frequency of the cosine along the wave vector. Given the orientation angle θ and the ridge spacing S, **k** is determined by the equations below:

$$S = \frac{2\pi}{\sqrt{k_x^2 + k_y^2}} \qquad \tan(\theta) = \frac{-k_x}{k_y} \qquad (3)$$

See Figure 3 for a physical interpretation of these parameters.

Spatial Domain   Frequency Domain

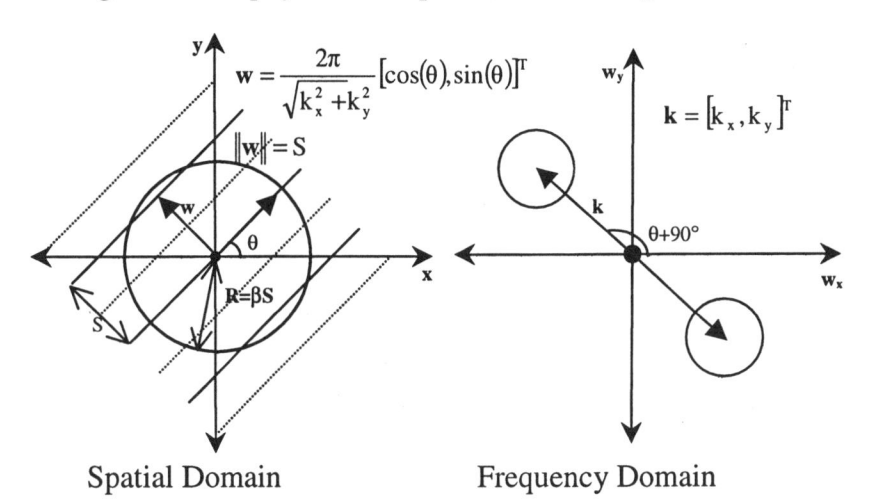

Figure 3. Parameters of ridge extraction filters in spatial and frequency domain. Left: spatial domain. Right: frequency domain. In both figures the circles represent the effective region of the Gaussian. In the figure on the left the solid lines and broken lines correspond to maxima and minima of the plane wave, respectively.

The variance of the Gaussian determines the bandwidth of the filter. A reasonable criterion for choosing a variance value is keeping the radius of effective region of the Gaussian close to S. If the radius is much larger than S, existence of many regularly spaced ridges and valleys is conditioned. If the radius is too small, a filter with no selectivity but just smoothing capability is obtained (See Figure 3). Also in frequency domain it is reasonable to have high bandwidth at high frequencies and low bandwidth in low frequencies as in multiresolution approaches like wavelets.

Our experiments have shown that the filter construction rules stated above works very well when the orientation and ridge spacing are estimated correctly. However errors in the estimation of orientation distorts the image and results in loss of true minutiae or creation of spurious minutiae. It is usually not possible to estimate orientation correctly neither in the low quality nor in the high-curvature regions of the image. As a solution to the problem, the orientation angle is complemented with a measure of orientation certainty. This orientation certainty factor is used to adjust the variance of the Gaussian. The regions with low orientation certainty are processed using a filter with variance less than those used in high-certainty regions.

After a number of experiments the following formula has been constructed for determining the variance parameter of the filter:

$$e^{\frac{-(R)^2}{2\sigma^2}} = \alpha \tag{4}$$

$$R = \beta S \tag{5}$$

$$\beta = (1 + 0.75\log(C)) \qquad C \in [0.1, 1] \tag{6}$$

In these equations R denotes the radius of the effective region of the filter and α is a small constant used for adjusting the desired bandwidth. Outside the region defined by R the Gaussian should have values less than α (see (4)). As stated before R should be proportional to ridge spacing S. They are related linearly by a constant β as in (5) (See also Figure 3). The value of β is calculated by (6) where C denotes the orientation certainty value and is clipped to the range [0.1,1]. In Figure 4, filters with different ridge spacing and orientation certainty parameters are shown.

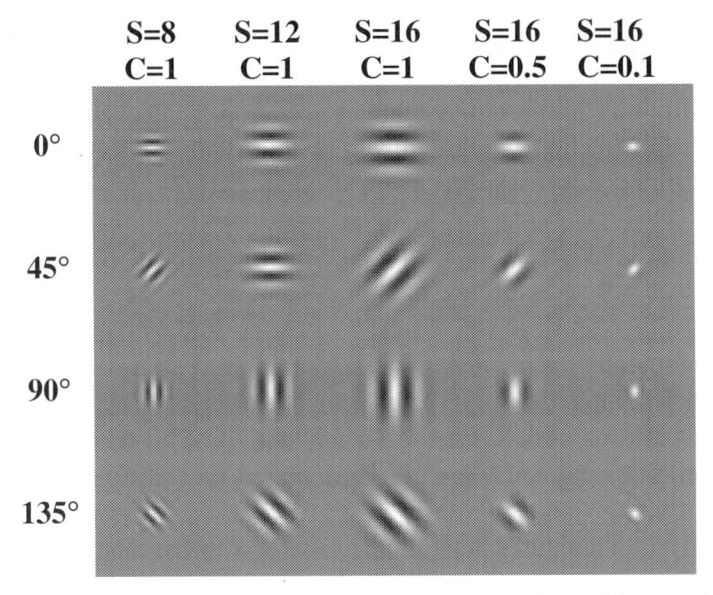

Figure 4. Ridge extraction filters at different orientations, ridge spacing and orientation certainty levels. Up to the middle column of the table the ridge spacing S is increased which means a decrease in frequency. After the middle column the frequency is kept constant but certainty C is decreased which means that bandwidth of the filter is increased.

The filters are actually IIR (infinite impulse response) filters but the desired value for α allows one to approximate it with a small FIR (finite impulse response) filter. The FIR filter size is chosen to be a 4S × 4S grid for $\alpha = 10^{-1}$. The coefficients outside this region are very close to zero. It is observed that ridge spacing varies from 6 pixels to 16 pixels and the average is about 9 pixels which makes a filter size of 36×36 on the average.

Since different regions of the image are processed with different filters, the coefficients of the filters are normalized in order not to get an image with contrast or intensity discontinuities. For this purpose the coefficients are divided by the sum of the filter coefficients. In this way a filter which does not have any effect on a uniform region is obtained.

3 Implementation

For experimental studies, a system whose block diagram is shown in Figure 5 is built. The system is also a prototype to be used in the finger print identification system given in [5], [13], [14], [15].

Input to the system is 256 gray-level fingerprint images of size 512×512 that are scanned at 500-dpi resolution. Processing starts with extraction of a map of some features from the input image. The map is a two dimensional grid whose cells correspond to uniformly spaced square blocks tiling the fingerprint image. For each block, statistical or structural features are extracted. Then the blocks corresponding to background regions in the image are detected and a segmentation map is built. Filtering operation described in Section 2 follows background segmentation. The resulting image is then binarized. Binarization corresponds to ridge extraction.

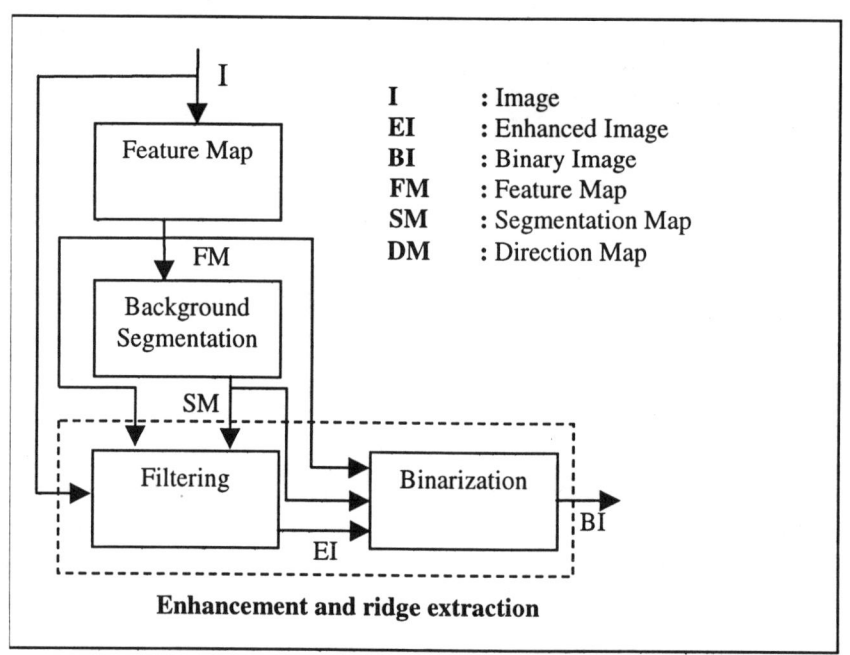

I	: Image
EI	: Enhanced Image
BI	: Binary Image
FM	: Feature Map
SM	: Segmentation Map
DM	: Direction Map

Figure 5. Block diagram of the ridge extraction system.

Each cell in the feature map corresponds to 8×8 blocks tiling the fingerprint image. The features for one cell are calculated using averages over the points in a neighborhood of size 16×16. The features can be listed as follows:

1. **Mean**: This is the mean value of the gray values of the pixels in the averaging box. Mean is used in binarization step.
2. **Variance**: This is the variance of the distribution of gray values in the region where the averages are calculated and is used in background segmentation.
3. **Orientation**: Determining the orientation with high certainty at a point needs examination of a large neighborhood around that point, and needs high computational load. Orientation is calculated as an average of uncertain orientations for each pixel as in [8], [16], [19]. For a single pixel an orientation vector is defined to be the vector with magnitude equal to the magnitude of the gradient vector and direction orthogonal to the gradient. The orientation vectors for the points in the 16×16 neighborhood are averaged. In averaging, the angles of the orientation vectors are doubled then a vector summation is performed and divided by the number of vectors. Half of the angle of the resulting vector gives the orientation angle.
4. **Orientation certainty**: For a single pixel, orientation certainty vector is defined to be the vector with unit magnitude and the same direction as the orientation vector. These vectors are also averaged like the orientation vectors. The magnitude of the resulting vector gives the orientation certainty of the calculated orientation angle. In high-curvature or low-quality regions, these orientation-certainty vectors point to different directions and cancel each other; therefore a vector with small magnitude is obtained. One problem is that it is possible to get high-certainty values in uniform regions, but these regions are supposed to be segmented in the background segmentation process.
5. **Ridge Spacing**: Ridge spacing is estimated using the projection of the image along a line segment perpendicular to the orientation vector and passing through the center point of the block. The projection is smoothed and the distances between consecutive peaks are calculated. The maximum distance gives the ridge spacing. Maximum possible ridge spacing is observed to be 16 pixels so for the projecting line to include at least two consecutive peaks in the interior section, its length is set to 40 tracing steps. Ridge spacing is assumed to be constant in most of the studies in literature [12], [18] but using constant ridge spacing forces one to choose a large bandwidth for the filters, which means a decrease in selectivity.

6. **Curvature**: The ridge curves has varying curvature at different regions of the fingerprint image. The curvature of the curves increases when they are close to core and delta points of the fingerprint. For each block, an approximate measure which gives an average curvature magnitude for the ridges nearby is calculated. The unit vector with an angle equal to the block orientation represents an average tangent vector for the ridges close to the block. For each block in the map the magnitude of the difference between its tangent vector and the tangent vectors of the blocks in the 3×3 neighborhood of the block is calculated. The maximum magnitude is assumed to be the curvature for the block. This feature is used for evaluation purposes only.

For a sample orientation map see Chapter 1. The gray-level pictures for the other calculated features are given in Figure 6.

After the calculation of features for all the blocks, the resulting measures are smoothed by a 3×3 averaging box filter. First, the orientation angle is smoothed. In smoothing of the orientation angle unit vectors in the direction of the orientation angle are averaged and each vector is weighted by the product of orientation certainty and the magnitude of the average orientation vector. The angle of the resulting vector is taken to be the orientation of the corresponding block. Then the other features, which are scalar numbers, are averaged with equal weights. The smoothing operation reduces estimation errors and guarantees that neighboring blocks are processed by similar filters.

The purpose of background segmentation is detecting the background regions in the image. These regions are discarded in the rest of the processing. The blocks with variance less than a threshold are segmented as background. Background regions have a very small variance and are detected with minor errors.

In the last processing stage of the system the image is binarized. Because of the feature selective nature of the filters, a very high output is expected for a ridge pixels and a very low output is expected for a valley pixel. Guided by this fact, the filtered images are thresholded in two levels. When the value of a pixel in the image at the filter output is higher than 255 it is set to 255 and when it is less than zero it is set to 0. The intermediate values are left intact. In this way a semi-binary image

(a)　　　　　　　　　　　　(b)

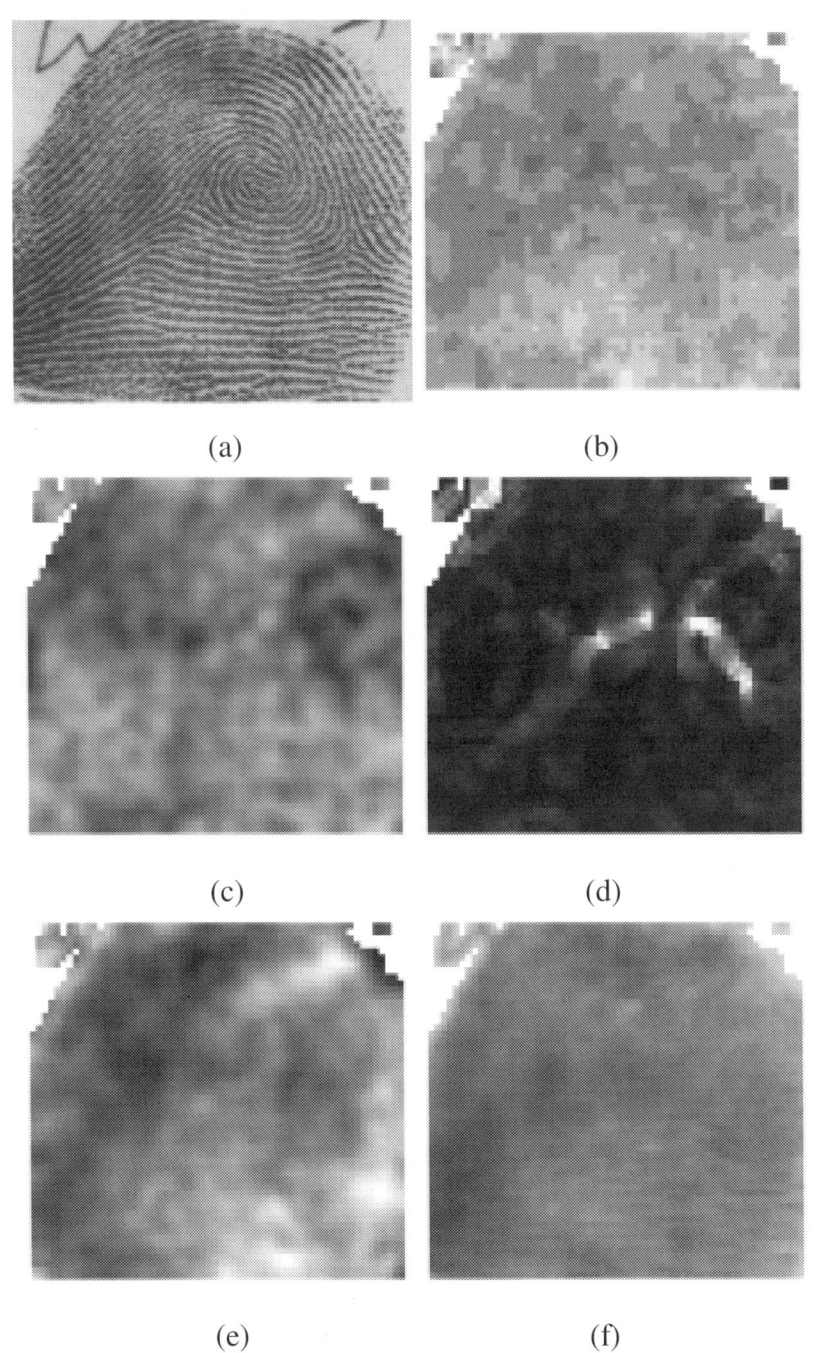

(c)　　　　　　　　　　　　(d)

(e)　　　　　　　　　　　　(f)

Figure 6. The feature map features except the orientation. (a) Original image, (b) ridge width, (c) orientation certainty, (d) curvature, (e) variance, (f) mean.

is obtained (See Figures 8–12). The remaining gray scale regions are thresholded using the local mean calculated in the feature map (See Figure 13). Pixels with values larger than the mean correspond to ridges and the lower valued pixels correspond to valleys.

4 Experiments

The images used in our experiments are taken from the National Institute of Standards Special Database 4 (NIST-4) which includes 4000 paired images from one finger of 2000 persons. The images are of size 512×512, have 256 gray scales, were scanned at approximately 500 dpi resolution and are all scanner images of rolled fingerprints.

In source fingerprint images high-intensity values actually correspond to the valleys on the fingerprint. Since these two features are duals of each other and high intensity surfaces representing the valleys are smoother than the ones in the negative image, the valleys are treated as ridges in all experiments given here. One can get the physically correct ridge pattern by negating the output image.

It is observed that the semi-binary image obtained as an intermediate step gives good clues about what has happened in the filtering stage; therefore the semi-binary images are used in evaluation of the algorithm. In Figure 7, a high-quality fingerprint image is shown. The obtained ridge structure is very clear almost everywhere. Notice that even the black line on the top disappears in some regions.

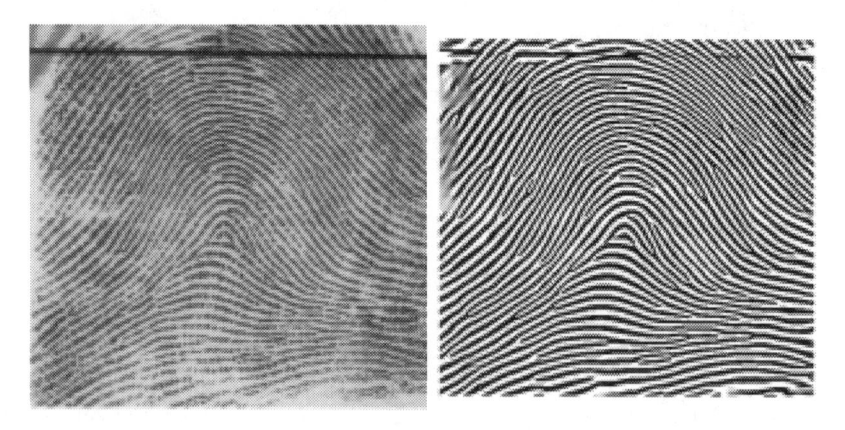

Figure 7. A high-quality finger print image and its enhanced form.

In Figure 8, the processing result of a fingerprint consisting mostly of high-quality regions is shown. It is observed that low-quality regions are not binarized. The orientation certainty map indicates that such regions have small certainty.

In Figure 9, a medium-quality image is processed. The source fingerprint has a finer ridge structure than the others. The same arguments in Figure 8 also apply here, but a positive observation is that although there is a finer and noisy ridge structure and visual minutiae detection is not easy in most of the fingerprint area, the algorithm adapts to the fine ridge structure and gives a clear output of ridges and valleys.

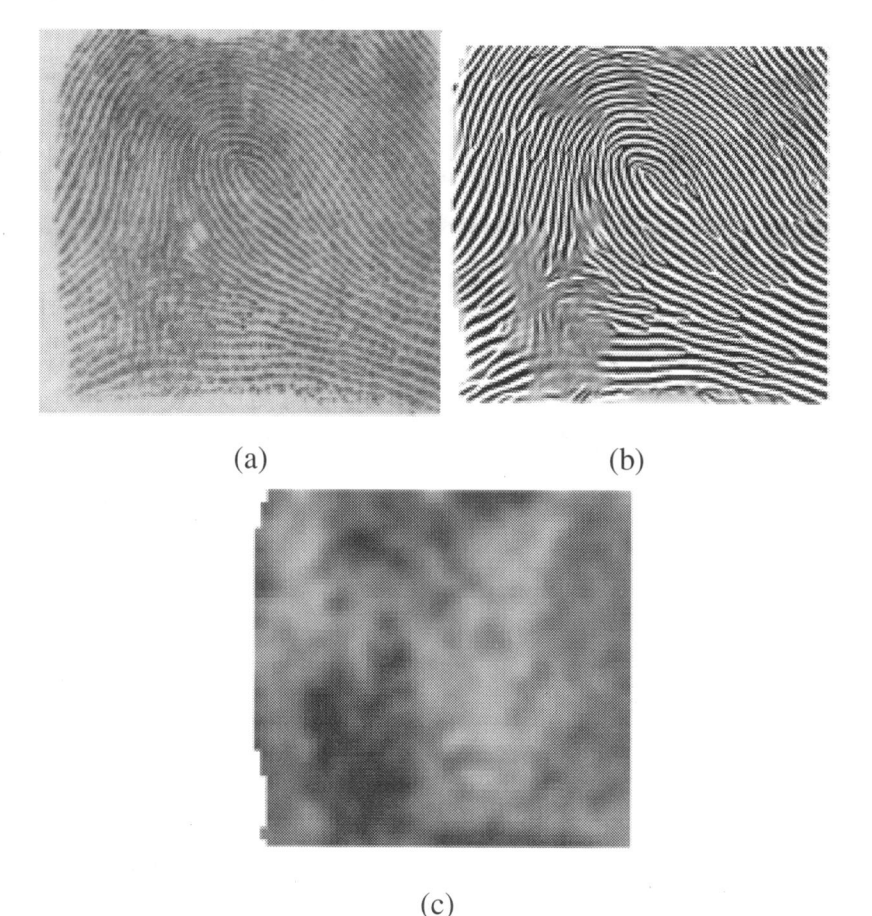

(a) (b)

(c)

Figure 8. The processing of a reasonable quality fingerprint image with defects, (a) the source image, (b) enhanced form which is the semi-binary image, (c) the orientation certainty map.

(a)

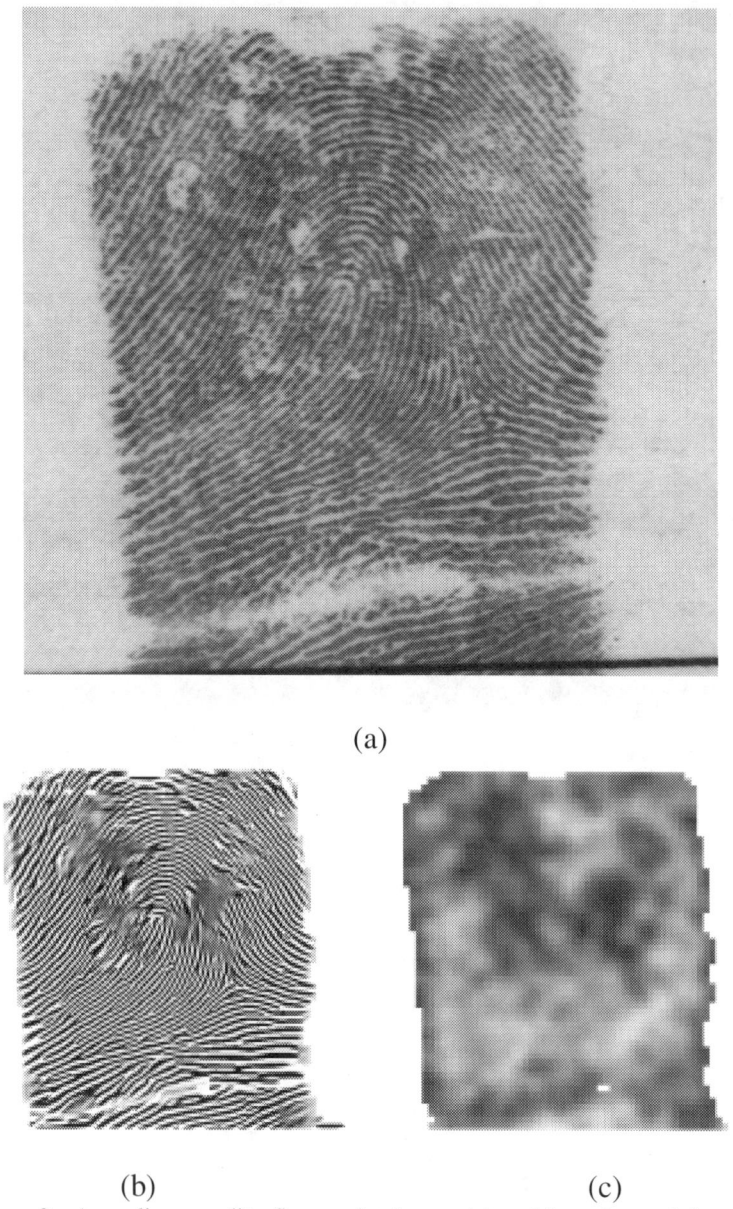

(b) (c)

Figure 9. A medium-quality fingerprint image (a) and its enhanced form (b); (c) is the certainty map. The source image is printed in larger size as it has a finer ridge structure than the others.

Figure 10 gives an example of an extremely low-quality fingerprint image. Actually such an image should be discarded in a minutiae-based fingerprint recognition system. The most important problem is that it will not be possible to estimate the filter adjustment parameters with high accuracy. The filters do the best they can and can still find ridges on the image but high certainties calculated in nearly uniform regions causes false ridges to appear.

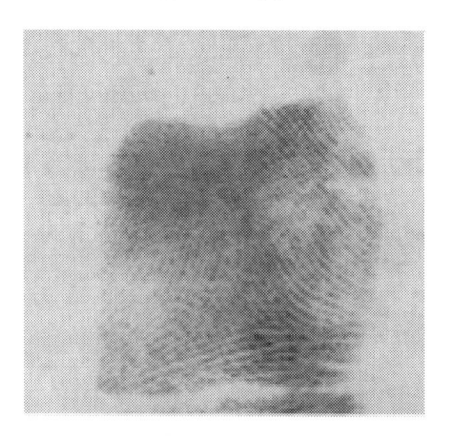

Figure 10. An extremely low-quality fingerprint image and its processed form.

In Figure 11 and Figure 12, two important problems with the algorithm are shown. In Figure 11 there are thin scars cutting the ridges on the right side of the fingerprint area. As they have high intensity, the orientation angles calculated in the corresponding blocks are completely wrong but a human can easily detect ridges in these blocks. In Figure 12, failure to detect ridges in high-curvature regions is observed. The curvature and certainty map indicates that low certainty values are calculated for the high curvature regions of the image. The resulting image still has gray level but clear signs of the ridges in these regions.

In Figure 13-a the semi-binary image obtained from the source image shown in Figure 1-a is given. Figure 13-b is the output of the binarization stage operating in its fully functional mode. The binary image is processed further and the thinned image shown in Figure 13-c is obtained. The processing results shown in Figure 1 are obtained by using the simplest possible binarization algorithm and it has been used in [14] and [15]. After elimination of uniform regions in the source image, it is smoothed by a smoothing mask and then ridge-valley binarization given in [20] is applied. If we compare the binary and thinned images in Figure 1and Figure 13, we see that the outputs of the system presented in this chapter are much better. The binary image in

Figure 13-b has clear and smooth ridge patterns while the one in Figure 1-b has a lot of broken and touching ridges. The thinned image in Figure 13-c is free from almost all of the bridges and breaks observed in Figure 1-c. The remaining ones are on the distorted or low quality regions of the fingerprint image.

The overall response on the images given in Section 0 shows that the filters are doing fine as long as they are supported by accurate measurement of block features and do not give false responses as long as the orientation certainty is calculated correctly and it still has enhancement capability in medium orientation certainty levels.

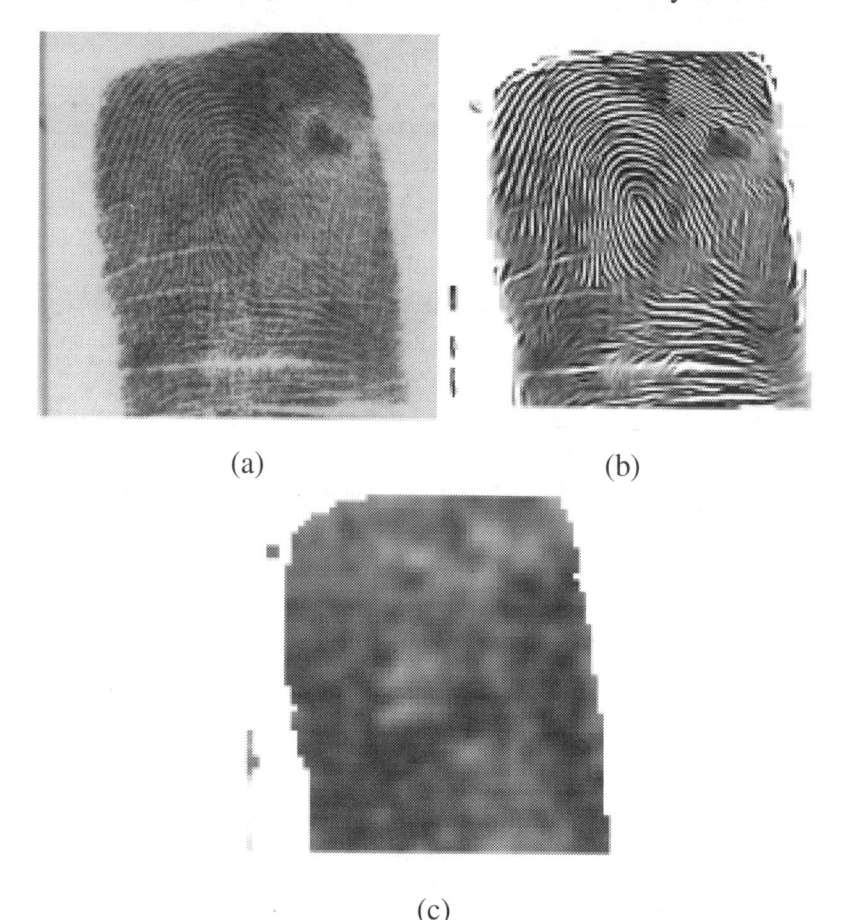

(a)   (b)

(c)

Figure 11. A fingerprint image and its processed form. (a) The source image, (b) enhanced form which is the semi binary-image, (c) the orientation certainty map.

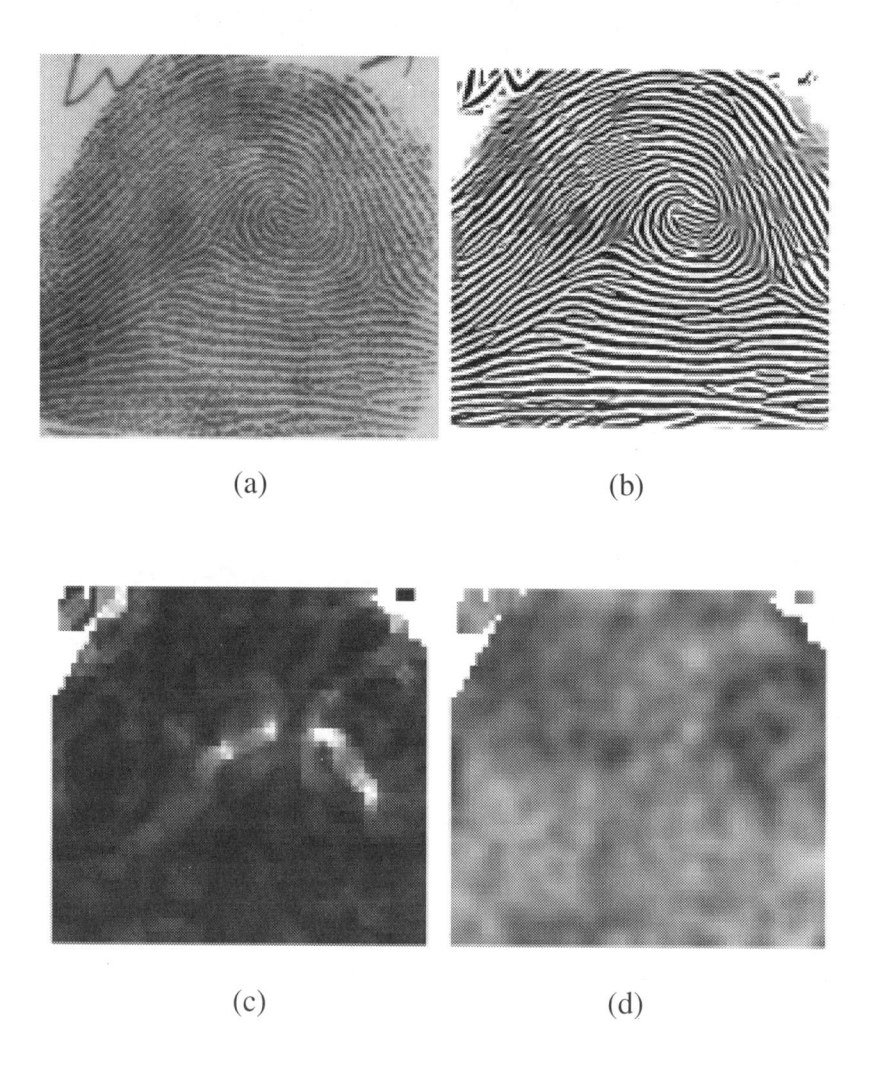

Feature Selective Filtering for Ridge Extraction

Figure 12. A fingerprint image and its processed form. (a) The source image, (b) enhanced form, which is the semi-binary image, (c) the curvature map, (d) the orientation certainty map

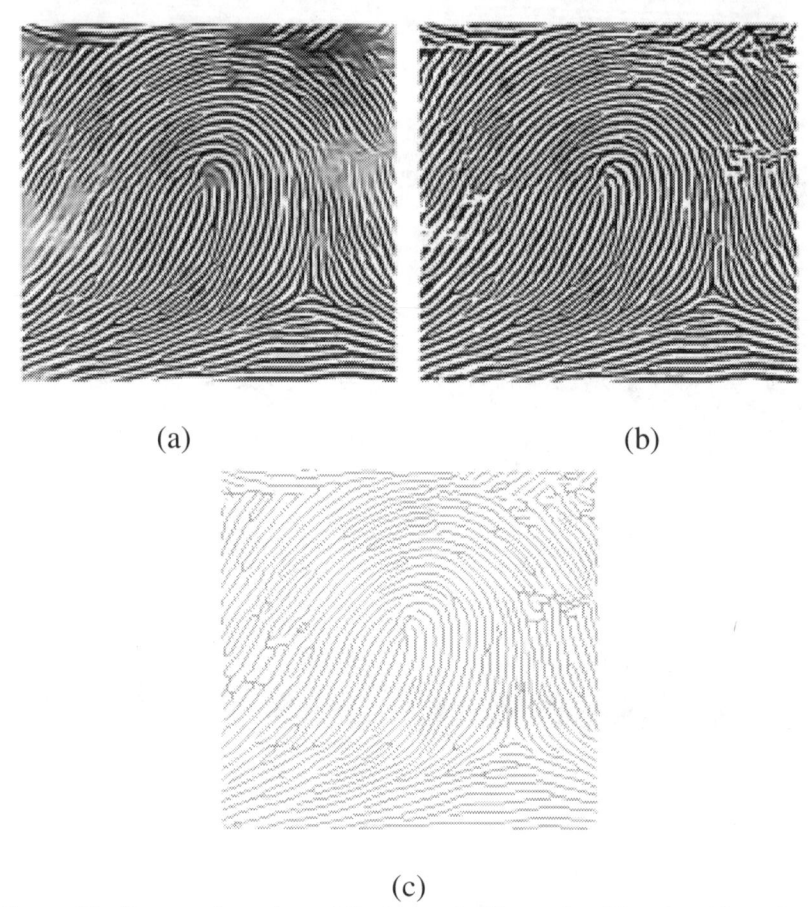

Figure 13. Processed version of the image in Figure 1-a. The algorithm given in this chapter is used. (a) Semi-binary image, (b) binary image, (c) thinned image.

5 Conclusion

The most important property of the enhancement system given in this chapter is its ridge detection capability. The usual binarization process in fingerprint recognition systems corresponds to detection of ridges. The overall process is like an edge detection process where the edges are modeled by regularly spaced, oriented tiny line segments.

As stated in Section 1, the usual approach to fingerprint enhancement is to filter an image to get an enhanced image where ridges are more clear and free from breaks [12], [18], [17]. Then simple methods like local thresholding or ridge valley binarization [20] is applied to get a binary image. A problem with linear filters is that they can give high responses although the condition they are supposed to give high output is not present. One level thresholding of linear filter outputs results in an image containing false ridges and valleys. This effect is also observed in our system for the gray-level regions in the semi-binary image and is demonstrated in Figure 13. However, extremely low and high values of a feature selective filter can be trusted and this fact is utilized by the two level thresholding mechanism in the binarization stage and it exactly corresponds to ridge or valley detection.

On the other hand, to be able to get filters with good selectivity, the filter parameters should be tuned with high precision and bandwidth should be minimized. In most of the studies on enhancement the ridge spacing parameter is assumed to be constant and this assumption forces one to use less selective high-bandwidth filters. However calculation of ridge spacing allows one to choose the center frequency and bandwidth of the filter in the method presented here.

A common approach in fingerprint processing is segmenting the low-quality regions of the fingerprint image using local features like variance, orientation certainty, or orientation histogram. The calculated features are simply thresholded for a decision of such regions. In our experiments it was not possible to find good threshold values that can give accurate results for a large population of fingerprint images. However one may get better results by an analysis of the semi-binary images obtained. The gray-level regions mostly correspond to distorted regions of the fingerprint and if the exceptional regions like high curvature regions or the ones shown in Figure 11 are detected or processed correctly a better thinned image with less false minutiae can be obtained.

As stated in Section 0, high-curvature regions where the orientation can not be estimated correctly form an important problem. The reason for this incorrect estimate is the use of a large neighborhood around the block; on the other hand, decreasing the window size produces noisy orientation data. Using a finer tiling of the image in these problematic regions may help to solve the problem.

References

[1] Moayer, B. and Fu, K.S. (1986), "A Tree System Approach for Fingerprint Recognition," *IEEE Transactions on Pattern Analysis and Machine Intelligence*, Vol. 9(3), pp. 376-387.

[2] Erol, A. (1998), *Automated Fingerprint Recognition*, Ph.D. Thesis Proposal Report, Dept. of Electrical and Electronics Eng., Middle East Technical University, Turkey.

[3] Federal Bureau of Investigation (1984), *The Science of Fingerprints: Classification and Uses*, Washington DC.

[4] Galton, F. (1892), *Finger Prints*, Macmillan, London.

[5] Halici, U. and Ongun, G. (1996), "Fingerprint Classification Through Self-Organizing Feature Maps Modified to Treat Uncertainties," *Proc. of the IEEE*, Vol. 84(10), pp. 1497-1512.

[6] Hrechak, A.K. and McHugh, J.A. (1990), "Automated Fingerprint Recognition Using Structural Matching," *Pattern Recognition*, Vol. 23, pp. 893-904.

[7] Hung, D.C.D. (1993), "Enhancement and Feature Purification of Fingerprint Images," *Pattern Recognition*, Vol. 26(11), pp. 1661-1671.

[8] Jain, A.K., Hong, L., and Bolle, R. (1997), "On-Line Fingerprint Verification," *IEEE Trans. on Pattern Analysis and Machine Intelligence*, Vol. 19(4), 302-314.

[9] Jain, A.K., Hong, L., Pankanti, S., and Bolle, R. (1997), "An Identity-Authentication System Using Fingerprints," *Proc. of the IEEE*, Vol. 85(9), pp. 1365-1388.

[10] Lee, H.C. and Gaenssley, R.E., (1991), *Advances in Fingerprint Technology*, Elsevier, New York.

[11] Maio, D. and Hanson, A.R. (1997), "Direct Gray-Scale Minutiae Detection in Fingerprints," *IEEE Trans. on Pattern Analysis and Machine Intelligence*, Vol. 19(1), pp. 27-40.

[12] O'Gorman, L. and Nickeron, J.V. (1989), "An Approach to Fingerprint Filter Design," *Pattern Recognition*, Vol. 22, pp. 29-38.

[13] Ongun, G. (1995), "An Automated Fingerprint Identification System Based on Self-Organizing Feature Maps Classifier," M.Sc. Thesis, Middle East Technical University, Turkey.

[14] Ongun, G., Halıcı, U., and Özkalaycı, E. (1995), "HALafis: An automated fingerprint identification system," *Proc. Bilisim 95*, Istanbul, pp. 95-100.

[15] Ongun, G., Halıcı, U., and Özkalaycı, E., "A Neural Network Based Fingerprint Identification System," *Proc. The Tenth International Symposium on Computer and Information Sciences*, Izmir, Turkey, pp. 715-722.

[16] Ratha, N.K., Karu, K., Shaoyun, C., and Jain, A.K. (1996), "A Real-Time Matching System for Large Fingerprint Databases," *IEEE Trans. Pattern. Analysis and Machine Intelligence*, Vol. 18(8), pp. 799-813.

[17] Sherlock, B.G., Monro, P.M., and Millard, K. (1992), "Algorithm for Enhancing Fingerprint Images," *Electronic Letters*, Vol. 8(18), pp. 1720-1721.

[18] Sherlock, B.G., Monro, P.M., and Millard, K. (1994), "Fingerprint Enhancement by Directional Fourier Filtering," *IEE Proc.-Vis. Image Signal Process.*, Vol. 141(2), pp. 87-94.

[19] Vizcaya, P.R. and Gerhardt, L.A. (1996), "A Non-Linear Orientation Model for Global Description of Fingerprints," *Pattern Recognition*, Vol. 29(7), pp. 1221-1231.

[20] Wilson, G.L., Candela, G., Grother, P.J., Watson, C.I., and Wilkinson, R.A. (1993), "Neural-Network Fingerprint Classification," *J. Artificial Neural Networks*, Vol. 1(2), pp. 1-25.

[21] Xiao, Q. and Raafat, H. (1991), "Fingerprint Image Post-processing: A Combined Statistical and Structural Approach," *Pattern Recognition*, Vol. 24, pp. 985-992.

Chapter 7:

Introduction to Face Recognition

INTRODUCTION TO FACE RECOGNITION

A. Jonathan Howell
School of Cognitive and Computing Sciences,
University of Sussex, Falmer,
Brighton BN1 9QH,
United Kingdom

This chapter introduces the general concepts concerning our face recognition task of automatic recognition of human faces in dynamic environments. By concentrating on face recognition, our work will cover only one of a larger set of techniques connected with the identification of people. The term 'biometrics' has come to be used for the study of automated methods for the identification or authorization of individuals using physiological or behavioral characteristics. Techniques such as speech recognition, iris scanning, hand geometry, fingerprint scanning and signature verification, as well as face recognition, can be combined to produce useful applications. In comparison to the other techniques, however, face recognition has the major advantage of being non-intrusive and requiring very little cooperation or modification of normal behavior on the part of the subjects in order to collect useful data.

The real-life problems to be tackled here concern identifying individuals and their intentions in everyday settings, such as offices or living-rooms. The dynamic, noisy data involved in this type of task is very different to that used in typical computer vision research, where specific constraints are used to limit variation. Historically, such limitations have been essential in order to limit the computational resources required to process, store and reason about the visual data. However, enormous improvements in computers in terms of speed of processing and size of storage media, accompanied by progress in statistical techniques and neurobiology, now allow more efficient handling of such data.

The development of intelligent environments has been highlighted recently by the 'Smart Rooms' projects [1] at the MIT Media Lab, which enable novel forms of interactive control for computer systems. Our particular focus within such an area is the role of adaptive learning techniques in recognizing the individuals and simple movement-based gestures like head rotation. Unfortunately, the relatively unconstrained appearance of faces of individuals in video scenes makes this a particularly difficult problem.

There is great commercial interest in logging and interpretation of activity within domestic or commercial environments. Applications include access control and personalization of domestic appliances such as computers, telephones and televisions. Burglar alarms could be improved so that they not only identify when unidentified people are in the house, but also record their activities for evidence. In addition, the logging of shoppers' interest and behavior patterns in shops would be of interest to marketing and consumer research groups. Although this latter task does not require explicit identification of individuals, short-term memory of what individuals in a room look like could be used to connect what a particular person does for the time they remain there.

We will first outline our task requirements. We then go on to survey general theories of object recognition, including a review of psychological evidence, and computational research within face recognition from the perspective of acquisition, representation and reasoning. The final section will apply our proposed face recognition scheme to a standard database, giving comparisons with published results for other approaches.

The particular face recognition task considered here concerns a known group of people in an indoor environment such as a domestic livingroom. Within such a task, it cannot be assumed that there will be clear frontal views of faces at all times. Therefore, it is important not to lose such vital information, which may only be present for a split-second if the subject is moving fast. To effectively tackle such a task requires the combination of three real-time processes: tracking of individuals as they move around the room, detection and localization of their faces, and recognition of the final, segmented face information. Each of these three processes currently occupies a large area of research within computer vision. It is very difficult to consider an overall solution to face recognition, therefore this introduction is mainly confined to the process of face recognition from video images, with the assumption that other processes, e.g., [2], [3], will provide suitable segmented face images and image sequences from our target environment.

In order for the system to be suitable for domestic environments, it needs to be as automatic and robust as possible. It cannot rely on monitoring or tuning from technical staff, nor should it constrain or require any particular actions from people in that environment. Any information collected should be from normal everyday behavior. This is in contrast to security access control systems using face recognition, where the users are often required to stand in front of the system for several seconds in controlled lighting and pose after giving an ID card for verification. The uncontrollable nature of our potential subjects means that the system needs to be able to collect and process data as quickly as possible (close to frame rate), and make reasonable 'guesses' where information is confusing or missing.

Applications suitable for mass-market domestic use require economical solutions. Computational techniques have to be simple enough to be accomplished on standard serial processors (such as used in PCs). Data collected with a simple, fixed camera system with low-cost frame grabber must be sufficient for the recognition process. This data may need to be monochrome at present, although color could soon be cheap enough to be used instead.

In terms of how many people the system should cope with, the maximum number distinguished does not have to be particularly high, as most family groups contain no more than 15 individuals, even including relatives. A face recognition system that could effectively discriminate a moderate number of individuals, for example around 40–50, could also be useful for monitoring other small groups, such as offices or small factories. If people are expected to stay in a room for at least a minute, for instance, a frame capture rate of 25 frames per second will provide 1500 test images for identification purposes during that minute. It is not required that all of these are correctly classified: if we can accurately discard 1499 of them which the system finds ambiguous and correctly identify the one remaining frame, that will be sufficient data for logging the activity of that person (assuming that occlusion or multiple people do not undermine identity constancy). If the person stays longer than a minute, then hypotheses about identity can be confirmed and confidence increased.

Limiting expectations as to how many people can be expected to be recognized makes explicit how different this *low precision, high discard* task is to *high precision, low discard* tasks, such as police records analysis. The former type of task is expected to distinguish a small number of people from a large amount of potentially ambiguous data, most of which it is allowed to throw away (if a clear classification cannot be made), whereas the latter require hundreds of thousands or even millions of people to be distinguished unequivocally, using very small amounts of image data (usually single frontal and/or profile views). However, such an application would be able to take advantage of the highly constrained nature of the face images (each having fixed pose and lighting) and almost unlimited computing time (in comparison to the inter-frame period required here).

A major difficulty in tackling identification of individuals in dynamic, real-life environments is that it is not known exactly how many people there will be in the picture or whether they will be standing, sitting, etc. Additionally, even if a person can be tracked and their head localized, their face could be facing in any direction. If a simple, single fixed camera system is being used, a high-resolution wide-angle view of an average domestic room, when heads have been localized, will provide fairly low-resolution data (certainly under 100×100 pixels, and generally around 50×50), but, as we will show, this is sufficient for identification of familiar individuals.

The lighting in normal, everyday environments is obviously less controlled than that encountered in laboratory conditions (the standard for most current research face recognition systems). Not only will the number of light sources be variable (often varying from one moment to next), but they will be of different types, such as natural light from a window (varying from direct sunlight to overcast diffuse light) and spotlights, and shadows and illumination will change due to reflections and people moving around the room.

There are many types of variations in facial appearance that can occur: some arise out of the location itself, such as variable direction and contrast in lighting, and occlusion from objects and other people. Variation due to pose will occur as the subjects are free to stand in whatever position relative to the camera that they like. Day-to-day changes will be encountered after the system has learnt a person's appearance, as details such as hair styles, beards or stubble, makeup and jewelery will all vary for each individual. In addition, if daylight is present, the lighting due to this will vary according to weather and time of day. Longer-term changes will also have to be tolerated by the system (or at least the system would have to 'relearn' people periodically). These include aging, weight change and facial changes, such as scars.

An important characteristic of the output is that it should be accompanied by a level of 'confidence' in that output, as it is essential to be aware of possible confusion in the classification process. 'Forced' classification, where a decision is given regardless of confidence, would not be appropriate or useful for our task, and the statistical validity of the approach is important so that classifications can be analysed effectively. This means that 'black box' solutions do not fare well as engineering solutions, since performance parameters will not be available, and it will not be known under what circumstances the system will be able to work.

In addition, the system ought to be capable of detecting if a viewed person is not from the group that it has learnt to identify. Such 'strangers' could then be monitored with temporary identities (to allow more than one at a time to be distinguished) which could then be assigned permanent titles if they needed to be 'remembered' for more than a short time (as determined by the user's requirements). Of course, deciding that someone is 'unknown' is a very much more difficult task than identification within a known group, where all classes will have explicit examples, as the system is trying to identify (as a member of a general group of classes, not as an individual) an almost infinite number of face classes that it has not previously seen. In a full system, a higher-level process would be required to monitor day-to-day events and allow some behavioral reasoning to help with ambiguous data. This could allow expectations of who is likely to be present at a particular time of day, and to assess the likelihood of encountering unknown people and conduct relearning of the database of distinct individuals (known and unknown) as required.

1 Task Requirements

The requirements for a useful, commercial face recognition and identity logging system for small groups of known individuals in busy, unconstrained environments, such as domestic living-rooms or offices, can be split into groups: there are *general requirements* that need to be satisfied by all parts of the system, *acquisition requirements* concerned with monitoring and extraction of useful information, *face recognition requirements* for the recognition stage and *identity requirements* which are concerned with how the recognition information is used.

1. General Requirements:

    (a) Computation involved possible on low-cost standard serial processor.

    (b) Robust performance with noisy, real-life data.

2. Acquisition Requirements:

    (a) Real-time tracking of individuals, with the ability to deal with multiple identities and occlusion.

    (b) Real-time detection and localization of faces.

3. Face Recognition Requirements:

    (a) Fast learning and real-time recognition of faces, with a minimum of tunable parameters, of a moderate number of individuals (under 50).

    (b) Ability to work with low-resolution (under 50×50 pixels) face images, segmented from a single, wide-angle view.

    (c) Invariance to typical variations in images in such an environment, including:

        i. Minor variation in shifted position and scale of class information (in this case, faces) in the segmented image (dependant on accuracy of Requirement 2b).

Introduction to Face Recognition 225

   ii. Moderate variation in lighting direction, contrast, brightness and spectral composition.
   iii. Minor occlusion by another object (self-occlusion is addressed in Requirement 3(d)ii).
   iv. Any variation in background areas of image.
  (d) Invariance to typical facial variations in such an environment, including:
   i. Moderate expression variation. This would include changes due to talking, eating or chewing, etc. but not extreme facial contortions.
   ii. Head pose orientation, within a range of angles that allow some facial area to still be seen in image (for example, not the back of the head). Note that this will need to accommodate self-occlusion.
   iii. Day-to-day facial differences due to glasses, makeup, skin tones, facial hair and head hair style. Note that this too may create some self-occlusion.
   iv. Long-term, permanent facial changes due to ageing, weight change, scars, etc.
  (e) Level of confidence in output available to allow discard of erratic or ambiguous data. Note this should be able to reduce 'false positive' results without creating a large proportion of 'false negatives'.
  (f) Ability to detect, but not recognize, unknown individuals (that is, people from outside the learned group).

4. Identity Requirements:

   Ability to adapt the known group of individuals (the new information coming from a mechanism to handling 'strangers'), including:

   (a) Learning a new individual.
   (b) Forgetting a currently known individual.
   (c) Learning the new appearance of a currently known individual.
   (d) Identifying different types of 'strangers' (people not previously encountered):

i. Authorized strangers, who are subsequently added to the known group and require an ID label from the user.
   ii. Authorized temporary strangers, who are, for instance, recognized for set period of time, such as the rest of the day, and then forgotten.
   iii. Unauthorized strangers, who have not been given permission to be in area.

2 Object Recognition

There are a number of introductions and surveys to the theoretical background of object recognition, such as [4] and [5], so it is not necessary, nor within the scope of the chapter, to reproduce such information here. We will summarize the basic common categories that have developed to describe different approaches.

First, it has to be noted that there are several levels at which an object can be 'recognized'. The most specific would be identifying a unique instance of an object, such as 'my diary', whereas the same object could also be categorized simultaneously as a type, 'a diary', and a more general category, 'a book'. Animals are often grouped into general categories based on their subparts, for instance, their number of legs, but can also be seen as mammals, reptiles, etc., and this process can be carried out even if we do not know the exact species the animal belongs to. For face recognition, specifically, identification of the face object class is generally termed 'face detection' (see Section 3.4), whereas what is termed 'face recognition' is the discrimination of subordinate identity classes within the general face class.

Recognition will always be dependant on context and expectations, and our division of the scene into individual parts can be subjective. For instance, whether we choose to see the apple tree as a tree or a source of apples depends on whether we wish simply to navigate around it or if we are hungry.

2.1 Approaches

This section will examine how computational object recognition can be tackled, and the contribution different approaches can make to a successful system. It should be appreciated that faces are a fairly specific subset of all possible objects, and that the task we are tackling is not of category classification (is it a chair or a table?), but of distinguishing very similar non-rigid, self-occluding objects (albeit from a smaller range of angles than is possible for most 3-D objects, as the face cannot be seen from the back of the head). It is assumed that all recognition requires a model which is matched to some form of representation, although this representation can be highly distributed or implicit.

Approaches to object recognition are divided into three broad areas [5], [6]: invariant properties, use of parts and structural descriptions, and alignment-based methods. The first two we will only be mentioning in passing. The use of *invariant properties* is based on the assumption that objects will have invariant properties that are common to all views of them, and was found to be useful for constrained recognition of flat, unoccluded objects. It is difficult to extend such an approach to more general applications, as it is unclear how invariant properties would be extracted from complex 3-D objects. [7] argued that even approximate metric invariants do not exist in the general case for 3-D recognition.

The use of *parts and structural descriptions* allows a pose invariance through the use of 3-D structural graphs based on generic parts and relationships, for instance, in hierarchical arrangements of generalized cylinders [8] or geometric 'geon' primitives ('recognition by components') [9]. This approach is discussed in depth in [10]. The use of parts is more useful for distinguishing general object classes than faces, for instance, as different facial identities will be constructed from same basic parts. A disadvantage of this approach is that it is difficult to construct 3-D models from the information present in 2-D images, due to the ambiguity of occluded surfaces. Visual systems incorporating 3-D models should exhibit complete pose invariance as any view is equally possible to compute, but this is not supported by psychophysical studies [11], [12].

2.1.1 Alignment-Based Methods

Alignment can be approached in two distinct ways: first, there is an explicit 3-D *alignment of pictorial feature descriptions*, where potential 3-D object models are transformed to maximize the degree of match between the image generated by the transformed model and the image of the input object. The calculation of appropriate transformations is dependant on establishing correspondence between the model and the image, often using edge-based features [6]. The disadvantage of this approach is the difficulty of selecting features common for all views of an object (which remain unique to that object): its reliability depends on extracting sufficient features for matching.

The second alignment approach, *alignment through the combination of views*, uses a combination of 2-D views of the object as the model. [13] were able to show that the linear combination of a small number of views was sufficient to express a wide range of views of object transformations, such as 3-D rotation, translation and scaling. The modes of variation have to be expressed though variations shown in example views, as the approach has little tolerance to orthogonal variation (see Figure 1). This is a view-based approach which relies on geometric relationships between the object, the image and the transformations between the two, using 2-D structural information, such as x, y coordinates and segment lengths and angles.

The general group of view-based methods encompasses other approaches that deal with photometric representations which are computed directly from image intensity values rather than trying to extract explicit structural information, although some approximate correspondences will be often established for normalization of the image. [14] gives an overview of the advantages of view-based recognition of 3-D objects from 2-D images over 3-D model-based approaches, citing robustness and efficiency as major issues.

Two major concepts in view-based approaches are: (1) the *canonical view*, [15] found that certain characteristic 'canonical views' allowed viewers to name an object much faster than other 'non-canonical views' and view-based methods can take advantage of such views to improve

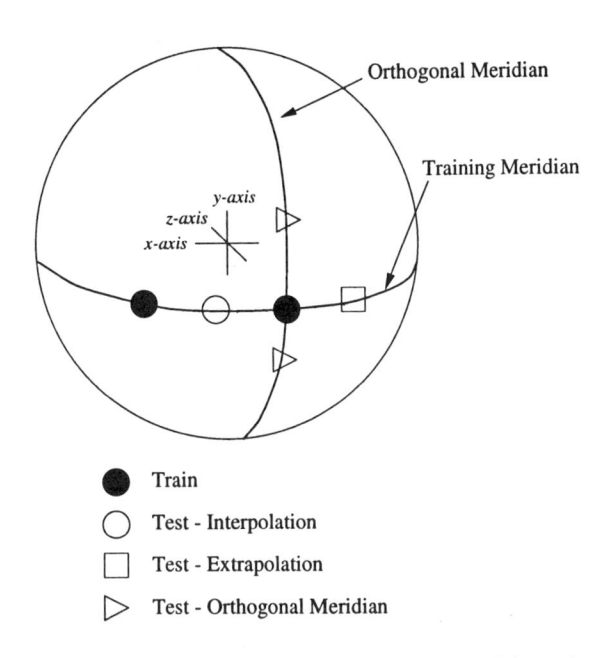

Figure 1. Illustration of view-sphere, demonstrating training view ranges and interpolating, extrapolation and orthogonal regions for test images in view-based recognition methods. Adapted from [11] with permission.

the efficiency of recognition, and (2) the *view-sphere*, which is the range of views from which the object can be seen (see Figure 1). A general 3-D object will have a view-sphere covering 360° movement in x-, y- and z-axes, although prior knowledge can allow this to be reduced. This is especially clear in the case of faces, as facial information is only visible on a human head from (roughly) the front ±120° of x- and y-axis movement, and z-axis movement is physiologically constrained to around ±20° (when standing or sitting). For example, [16] covered a facial view sphere of ±20° x-axis and ±30° y-axis with 15 example views.

2.1.2 Appearance-Based Methods

Another major approach that can be distinguished is what we have called *appearance-based methods*. This seeks to capture separately the essential visual characteristics of each object class to be recognized, and aims to create a viewpoint-independent representation. The appearance-based approach treats generalization across views as a function approximation problem [17], [18]. If an assumption of smoothness is made on the function, non-linear interpolation within example views [19], [20] can be used. Partially viewpoint-invariant representations of object classes can be formed by training an RBF network to interpolate a function in the space of all possible views of the objects [19]. Such representations can be seen as 'grandmother cells' (dismissed by [10] and reinstated by [21]!).

2.2 Psychological Evidence

We review psychological and psychophysical results here not to construct a 'biologically plausible' model of face recognition, but to look at issues central for 'engineering' a solution for our particular task.

Cognitive studies of the way human faces are perceived [22], [23], [24], [25] have contributed to our understanding of the problems for automating face processing. Psychological theories on the processing and recognition of objects and faces seem to point to different strategies for each; for a review, see [4], [26]. They suggest that the general level of object recognition may use edge-based (intensity discontinuity) information, whereas face recognition may use surface-based (texture, shading) 'holistic' information. However, if all visual recognition is seen as a spectrum from the most general to the most specific, this division can be seen as a consequence of comparing tasks from opposite ends of such a spectrum rather than some inherent difference in the nature of the recognition process. It is clear that the task of recognizing radically different general categories of object, such as trees and houses, requires different techniques than distinguishing those categories that are structurally very similar, such as faces from members of a family. Within familiar face recognition, there is evidence that we may use a kind of 'face-recognition unit' mechanism, where each is tuned to recognizing a known individual [22], [26].

The disproportionate effect on face recognition of *inversion* (making the photographic image negative) has been taken as support for special mechanisms in face processing [22], [24] separate from other other types of object. The inversion effect may be due to the type of object processing required for the task. The conclusion is that it is failure of the first-order feature information to distinguish class members that leads to use of second-order configural data, and it is the latter that is susceptible to inversion interference. However, [27] showed that this may be the effect of 'expertise', rather than something unique to human face processing, in that dog breeders who are good at identifying particular breeds of dog are more adversely affected by inversion than dog 'novices'.

The *'caricature advantage'*, where a face is recognized more efficiently from a caricature than from a veridical (undistorted) representation (noted in computer-generated line images [28] and computer-generated photographic-quality images [29]) also points to configural processing. In addition, [30] point out that elements of composite faces are recognized more easily in tests if misaligned than if they are correctly aligned, which suggests that when two different halves are exactly lined up, a new facial 'configuration' is created which is difficult to decompose into elements.

Within face processing research, there is also support for treating face classification as a task separate from others using facial information, such as expression interpretation. [23] describe evidence for separate mechanisms being present in human vision for facial recognition and facial expression recognition. This is shown most clearly in people with cognitive disorders such as *prosopagnosia*, where they cannot distinguish individual faces, but can usually still 'read' emotional states from expressions. [22] surveys neuropsychological disorders that indicate that familiar and unfamiliar face recognition also proceeds separately. In addition, PET brain scans [31] have provided further evidence for independent processing.

Psychophysical studies with monkeys [32] suggest that they use interpolation between 2-D, viewer-centred representations for object recognition. There is some psychophysical support for the appearance-based approach [11], [19], [33], since the results suggest that only partial view-invariance over the restricted view-sphere is found for pose in human vision. In contrast, 3-D models and linear combination of views predict full view invariance over either a full or restricted view-sphere, respectively.

[34] has been able to establish the minimum resolution required for humans to recognize facial identity in a image at around 18 pixels horizontally and 24 pixels vertically. This extremely low level of spatial information indicates that local feature processing is not being used. This confirms that there is sufficient information in low-resolution face images for effective recognition. Several researchers have found that different frequencies of facial information can be informative for different tasks. [35] found that finer scales of facial information, were related to familiarity, whilst coarser ones were related to distinctiveness. [36] used high-frequency receptive fields for detecting viewing conditions, such as illumination and pose, and low-frequency receptive fields for face identification.

[37] has compared eigenface and graph-based systems against human responses. Whilst finding that both approaches match the human results very closely, they concluded that graph matching was closer to face processing, as PCA was affected more by image variations, such as lighting.

In summary, three main points stand out from the psychological evidence: (1) a division of strategies for entry- (general object classification) and subordinate-level (face discrimination) object recognition, (2) that appearance-based methods are likely for human vision in at least some vision tasks, and (3) that a quite coarse image resolution is adequate for familiar face recognition.

2.3 Discussion

[5, pp. 34-35] stresses two important points about object recognition. First, practical research will not in general be heavily committed to specific abstract philosophies, as hybrid schemes can frequently be more effective. Secondly, the relative merits of techniques will be different depending on the target application, in other words, no single approach can be a general solution to object recognition.

We will now go on, in Sections 3–5, to isolate the important issues in face recognition and explore how they have been dealt with in previous research projects. *Acquisition* – what are the important factors in the way face information is collected? *Representation* – what is it in a face that can allow a face recognition system to remember it when it sees it again and distinguish from others? *Reasoning* – how can a face recognition system compare faces most effectively? This is followed, in Section 6, by a practical evaluation of some of these techniques to determine their suitability for our task requirements, as described in Section 1.

We will not be presenting an exhaustive discussion of all current face recognition research, as that is outside the scope of this chapter. The field of face recognition has developed quickly and even recent surveys [38], [39], [40] are perhaps already out of date. A more representative view of current research can be found in [41], [42] and the Face Recognition Home Page [43].

3 Face Acquisition

This section looks at how the original data is acquired before the issue of representation is raised. How many and what type of face images are needed? How much and what kind of variation should be present in the images? As we are assuming our face images are pre-segmented (Requirement 2b has been fulfilled), we are treating face detection as an acquisition issue.

3.1 Design of Face Databases

The environment and manner in which a database of face images is collected is vital to the success of any face recognition system in which it is used. Almost without exception, face recognition research has been carried out with highly constrained data [44], with variations due to lighting, expression and pose either fixed or within unrealistic limits (if compared to variations encountered in real-life data). Many variations important to long-term applications, such as how aging will affect recognition over months and years, are not allowed for in the system design. There are some exceptions, for instance, [45] make a conscious effort to make their data more varied, with large rotations and occlusions due to hands in front of faces. Despite these limitations, we need to use standard databases to obtain comparative results.

Two U.K. databases are of note. First, the Olivetti Research Laboratory (ORL) face database [46] is a small database of 40 people (400 images) showing some pose, lighting and expression variation. The usefulness of the ORL database lies in having a large number of comparative results from different groups, discussed in Section 6. Second, the Manchester face database [47] is larger (30 people, with 690 images). The training and test data have been deliberately kept separate to prevent systems using spurious environmental details, such as lighting or background features, to classify individuals, and have at least 3 weeks between their collection for each person to introduce more realistic variability into the data. The training images have fairly constrained lighting, whereas the test images are more variable. Two levels of difficulty of test data are present, with different levels of variability. The first is fairly easy as the images are quite similar to the training images, only allowing changes to hairstyle, background and the wearing of glasses. The second is much harder, featuring occlusion with hands, dark glasses and covered hair. Although the Manchester database tests face recognition methods more thoroughly than the ORL database, it does not have so many published comparative results.

The largest collection of face images to date is the FERET database [48], [49], currently still under development by the US Army Research Laboratory. This has made great advances in constructing a standard by which competing face recognition systems can be compared, and conducting independent testing of leading algorithms. They have allowed pose movements from frontal to profile and limited lighting variations within the group of images for each person, and have required that data is collected over a period of time to allow changes in the person's appearance, clothing and lighting. The database evolves from year to year, but at present, no attempt has been made to collect data as image sequences or with sound/vocal information. Its major disadvantage is its unavailability to non-U.S. institutions.

In summary, the ORL is the most useful standard database for our purposes, due to large amounts of comparative results. However, aspects of task requirements will remain that are not covered by such tests and where this is the case, it will be necessary to design and construct our own specific databases.

3.2 Multi-Modal Facial Information

Cues other than facial appearance play a part in human face recognition and could prove useful for automatic techniques. Speech patterns, body shape/height and posture/gait are all characteristic and easily collected alongside facial information [50].

The ATR Human Information Processing Research Laboratories in Japan are constructing a joint face and speech database of 60 people [51], however, the main face views are taken as still images rather than as continuous sequences. The European M2VTS multimodal face database [52] has been set up to test multimodal person verification strategies [53], [54]. At present, the database contains 35 people with several image sequences of each. Synchronized speech is provided with at least one sequence for each person.

3.3 Space-Variant Sampling

Most view-based approaches have used the rectangular aspect provided by standard video cameras, although this does not necessarily provide the most useful representation. There has been interest in space-variant pixel arrangements, such as radial, logarithmically sampled [55] or 'foveal' representations [56], where pixels are concentrated at the centre, and cover a progressively larger area as they spread out, as this gives a natural rotation invariance. [57] used a foveal grid arrangement for their iconic representation scheme, centred within a rough face boundary. It is not clear how useful such an approach is, as no comparisons were made with other sampling arrangements.

The disadvantages of spatially variant representations are: (1) the extra computation required to remap pixels (unless dedicated hardware is used), and (2) that peripheral detail is sacrificed (although angular extent of the field of view can be increased). This can be useful for areas of vision such as autonomous robots, for example, [58], where a constant level of detail is not required over the whole visual field. In the context of our particular face recognition task, however, this loss of detail is more of a problem, as the face is already localized and uniform sampling within this region seems to have very little disadvantage.

3.4 Face Detection and Segmentation

Detection of faces using specific facial features will not be possible for our task, due to the low resolution of the data (see Section 4.1). An alternative is to use the whole face pattern as a holistic representation, such as with eigenface information [59], [60]. A successful neural network face detector has been developed by [61], which also used receptive fields to give some translation and scale invariance. A bootstrapping algorithm is used to get around the problem of finding suitable 'non-faces' (negative examples) to train with by incorporating initial false-positives as subsequent training data. This use of only the most confusable near-face examples, rather than a potentially huge range from the whole spectrum of 'non-faces', can substantially reduce the size of training set required for good performance compared to earlier approaches.

Face detection in image sequences is very much easier, due to motion cues, than for single images and can be integrated into tracking techniques. Once a face been found in a frame, temporal correlations greatly reduce the search space in subsequent frames, for instance, [62] were able to combine motion detection by spatio-temporal filtering with face detection with a neural network based on [61]. More recently, they have been able to use color to further reduce computation and give greater invariance to rotations in depth and partial occlusions [3].

Face detection not only includes finding a face in an image, but also determines how much of the face and background is actually segmented for further testing. The approach taken to face segmentation is important when assessing performance, as transitory details, such as hair style and background details, if included in training data, may be used as the most effective distinguishing detail. For instance, if one person stands next to a plant for a picture, whilst another does not, it is very much easier to check for the presence of the plant rather than to compare subtle facial details. Some groups, such as [63], ignore higher performance of experiments conducted with face images with hair included, as this face representation is not seen as being sufficiently general for images taken over time, and prefer to cite poorer results for hair-free data. There is some psychological evidence that person-specific details such as hair may be used by humans for unfamiliar face recognition [37], however, so

Introduction to Face Recognition 237

the visual features that are used for recognition may well be dependent on the task.

In contrast, non-person-specific details such as background are more obviously spurious for recognition. [59] acknowledged that the background surrounding the faces in their database was a significant part of the image data used to classify the faces. Of course, this must severely limit generalization of such an approach when it is trained with data against one background and tested with data containing a different background.

3.5 Normalization and Vectorization of Images

Once a face has been localized and segmented within an image, the image itself must be standardized or normalized prior to further processing to improve the efficiency of matching. Sometimes such normalization is just an adjustment of gray-level intensity values, but here we are considering adjustments to the image shape. This could be as simple as a rescaling to some standard size, or as complex as remapping each pixel.

The normalization and vectorization of an image are approximately similar processes. Image normalization is generally taken to be a process of adjusting to allow particular areas in different images to line up when any two images are matched together. For example, face images are very commonly normalized via affine transform on the basis of the positioning of both eyes (and sometimes mouth or nose position). This can be taken further via the 'morphing' the face texture on the basis of a larger number of standard facial landmark positions. 'Dense correspondence' is the ultimate correspondence, where all elements of the image vector correspond to pixel information from the same object feature in scene, in other words, the process creates a feature-based representation from the pixel information (in the most abstract meaning of 'feature').

Two approaches have been taken to establish the required correspondence to a reference image for image vectorization: (1) *approximate correspondence*, either using a low number, often two or three, anchor points as features [63] or an intermediate number contained in an active shape model [47], [64], [65] and (2) *dense correspondence*, where each pixel is a feature point, which can be solved through optical flow algorithms [66].

The vectorized representation contains two vectors: shape and texture. The *shape vector* contains the feature coordinates, either in absolute terms (widely used, for instance [13], [19], [64], [67]) or relative to a standard reference shape [63], [68]. The *texture vector* can be the original image, geometrically normalized or warped to the standard reference shape [63], [67], [68], [69] or local texture areas [47].

The disadvantage of these approaches to the normalization of images is that they cannot be applied under wide variations in pose, and can be computationally expensive. Task Requirement 3(d)ii will not be satisfied through the use of simple 2-D affine transforms, as they treat as 2-D approximately, and anchor points will not be available at all views over large pose ranges. In addition, it is not clear how one would go about normalizing a profile view to match a frontal view. Our approach to tackling such problems is to use a 'nose-centering' technique, where the face images are centred on the tip of the nose, so that visible features on profiles, for instance, should be in roughly similar locations to those in frontal views of the same person.

4 Face Representation

For a face-recognition system to perform effectively, it is important to isolate and extract the salient features in the input data to represent the face in the most efficient way. The abstract elements of such a representation can be made up in a variety of ways, and it depends on the task which approach will be appropriate.

One of the main problems in computer vision, especially in face recognition, is dimensionality reduction to remove much of the redundant information in the original images. Simple mechanisms, such as subsampling, may give a rough reduction, but use of more specific prior knowledge to apply more sophisticated preprocessing techniques to an image is still required for the best results.

4.1 Simple Feature- and Template-Based Approaches

The feature-based approach requires the detection and measurement of salient facial points (see [38] for a survey). [70] used geometrical distances and angles between primary facial features such as eyes, nose and mouth to classify faces using an economic representation of the face where the elements are based on their relative positions and sizes. A disadvantage of such picture-plane measurements is that it is not obvious which features and configural information will categorize a face efficiently and accurately (especially if shape and texture variations are considered as part of a useful facial feature set), and so important data may be lost.

Automatic feature finding algorithms [71], [72], [73], [74] have been developed to locate facial 'key points'. However, this information has been more usefully used for normalization through transformations prior to recognition [67], [75] than for identification itself. A problem in using such techniques for our task is that the low resolution data which will be available would make the accurate identification and positioning of small facial areas very difficult, if not impossible.

Template matching, involves the use of pixel intensity information, either as original gray-level or processed to highlight specific aspects of the data. The templates can either be the entire face or regions corresponding to general feature locations, such as eyes or mouth. Cross-correlation of test images with all training images is used to identify the best match. [74] compared feature and template-based methods directly with the same database of frontal face views. Their template-matching strategy was based on the earlier work of [76], except that they automatically detected and used feature-based templates of the mouth, eyes and nose, in addition to whole face templates. These additional feature templates as well as the whole face image were used to give better performance. Geometrical alignment of the eyes to match test images with model views allows shift, scale and rotation normalization prior to the recognition process itself.

The use of raw pixel intensity values will make the representation very intolerant to lighting conditions and variability, so [74] compared several preprocessing techniques: none (plain gray-level values), intensity normalization (using a neighborhood average value) and the use of a gradient norm operator, which he found gave the highest recognition performance.

In summary, the simple use of templates or features will not be enough for real-life applications, but additional processes such as alignment [16], [36] or filtering may be able to improve on this (see Section 5.1).

4.2 Deformable Templates and Active Shape Models

A priori knowledge of face variations and the expected shape of geometric features can be used to construct deformable (flexible) templates [77], [78] to guide feature detection process. Size and shape parameters in such templates can be translated, rotated and deformed to fit the best representation of their shape present in the image and these variations give a feature description, allowing both detection and representation. Unfortunately, such approaches are critically dependant on appropriate starting positions for the template, and computationally expensive (5–10 minutes of Sun 4 CPU time to match one image was quoted). This use of hand-crafted templates, individually tailored for specific tasks, has been replaced by flexible shape models using a Point Distribution Model (PDM) [65], [79] which are learnt from examples. This, in turn, can be used together with a shape-free gray-level model, obtained by deforming and aligning each training face to the mean face, to give a combined face encoding scheme [80].

Statistical methods are useful for modeling shape and gray-level appearance of images, as they can give a compact encoding of permitted variability [47], [81]. Models containing prior structural knowledge of faces are learnt from a database of prototypical images. Such models can build flexible object representations (active shape models) [80], [82] through a linear combination of labeled examples, which can then be iteratively deformed to match image data. This requires under 100 parameters to describe each image with expression, lighting and limited pose ($\pm 15°$)

Introduction to Face Recognition 241

invariance to produce the PDM representation, and has been used for tracking in image sequences [83], [84].

In summary, the deformable templates are computationally expensive and not robust to everyday variation. Both they and the simpler active shape models will have problems establishing matches for model points on low-resolution images, such as provided in our task requirements.

4.3 Principal Components Analysis and 'Eigenfaces'

Principal components analysis (PCA), is a simple statistical dimensionality reducing technique that has perhaps become the most popular for face recognition. PCA, via the Kahunen-Loève transform, can extract the most statistically significant information for a set of images as a set of eigenvectors [85] (usually called 'eigenfaces' [59] when applied to faces), which can be used both to recognize and reconstruct face images. Eigenvectors can be regarded as a set of generalized features which characterize the image variations in the database. Once the face images are normalized for eye position, they can be treated as 1-D arrays of pixel values. Each image has an exact representation via a linear combination of these eigenvectors and an arbitrarily close approximation using the most significant eigenvectors (that is, those with the highest eigenvalues). The number of eigenvectors chosen determines the dimensionality of 'face space', and new images can be classified by a projection onto that face space. For example, [85] chose the 50 most significant eigenvectors. [86] and [87] compressed face images using a simple neural network, the weights and hidden unit activations representing eigenvectors and eigenvalues, and moderate success was made in recognizing novel images.

Comparisons can be made between pure image-based coding, which is effectively template matching with position and scale differences eliminated, and more extensive normalizations in which more shape variability was removed. [67] morphed faces to an average shape before applying PCA, as the 'shape-free' images give a more linear space for analysis. Such normalization of faces before extraction of eigenfaces is based on an assumption that faces lie within a low-dimensional manifold, linearly approximated by independent shape and shape-free texture. An eigenface coding of shape-free texture with manually coded landmarks has been found to be more effective for automatic recognition than correctly shaped faces, giving a higher-quality representation of the images in terms of facial variation [63]. Although earlier work concentrated only on frontal views, [88] extended this to encode wide pose ranges, both parametrically [89] (PCA calculated for all views together) and as modular view-spaces (PCA calculated separately for each view). [88] found a slight advantage for the latter approach. Assumptions have to be made about the suitability of the data before PCA is applied, hence the emphasis on normalization. [90] used data in the Fourier domain to gain shift invariance in subsequent PCA. Oriented Difference of Gaussians convolution [35] and Gabor wavelet transform [2] have also been performed before PCA to provide a greater level of invariance than found using gray-level pixel information.

In summary, PCA is a very efficient signal encoder, and designed specifically to characterize and encode variations rather than ignore them. Thus, it may find the optimal low-dimensional representation, but this may be more useful for reconstruction rather than recognition [91]. In addition, the eigenface method is not invariant to image transformations such as scaling, shift or rotation in its original form and requires complete relearning of the training data to add new individuals to the database. Instead, we prefer to overcome both image variation and the problem of picking out important information using receptive field functions and adaptive learning.

4.4 Receptive Field-Based Approaches

The receptive field (RF) of a visual neuron is the area of the visual field (image) where the stimulus can influence its response. For the different classes of these neurons, a receptive field function $f(x, y)$ can be defined. Precomputed filters can simulate such fields when applied to locations across the image. This type of preprocessing is more biologically motivated than simple edge detectors or intensity normalization, as there is psychophysical and physiological evidence for orientation and spatial frequency specific channels in biological visual systems [92].

4.4.1 Gaussian Receptive Fields

A simple dimensionality reduction strategy is to use receptive field (RF) responses. [93] used the responses of 75 asymmetrically positioned-oriented Gaussian RFs arranged around the image as input for the RBF classifier system which learnt from examples. [36] were able to use a similar arrangement to generalize from a single view of a face, using high-frequency RFs for detecting viewing conditions, such as illumination and pose, and low-frequency RFs for face identification. However, such simplistic filtering may not be making the input representation explicit enough. A Gaussian function will smooth the image at a given frequency, so it is good for removing noise. However, the disadvantages inherent to using raw pixel values will apply here, such as low tolerance to lighting conditions.

4.4.2 Natural Basis Functions

[57] used the dimensionality reducing properties of PCA via a fixed set of learned basis functions extracted from natural scenes, which appear to match V1 simple cell responses quite closely. Earlier, [94] had found that the eigenvectors of patches of real-world images were close approximations of derivative of Gaussian filters. These filters can be applied at different orientations and scales to provide feature jets, similar to Gabor jets (see below). The advantage of this approach over simple PCA on the dataset itself is that the basis functions do not need to be recalculated to accommodate new faces.

4.4.3 Laplacian/Difference of Gaussians

Retinal ganglion cells and lateral geniculate cells, early in biological visual processing, have receptive fields very similar to the Laplacian operator. This can be implemented as Difference of Gaussian (DoG) filters [95] which combine edge boundary detection with Gaussian smoothing. The output of this process is then in a suitable form for detecting *zero-crossings* – locations where the second derivative of the intensity values in the image undergo a sign change, such as used in the *primal sketch* representation [10], which can be useful for object segmentation. The idea of DoG-style *valley-detecting* convolution, where the 'width' scale is adjusted to be sensitive to face-sized features, has been proposed by [26] as being particularly useful for face processing. Scales and orientations can be introduced into the filtering process; for example, the Cresceptron network [96] used 8 directions of oriented zero-crossings at 2 scales for input representation.

The idea of edge information as a basic object representation is common, either unoriented as DoG filters, or oriented [97] (steerable filters) to give more specific information. [70] applied a Laplacian to binarize the grayscale values, but then used projection analysis to extract feature information. However, specific positions of 'edges' may be too precise for generalization, as matching will be 'brittle'. [93] found edge magnitude values from standard edge detection algorithms, such as the Canny operator, actually reduced performance when distinguishing faces. They thought such precise operations made generalization under pose and lighting variation difficult, and found a directional derivative more useful than either raw intensity values or intensity gradient magnitudes. For this reason, it may be more useful to use binarized gradient information rather than zero-crossings contours.

4.4.4 Gabor Wavelets

The receptive fields of the simple cells in the primary visual cortex (V1) of mammals are oriented and have characteristic spatial frequencies. [92] proposed that these could be modeled as complex 2-D Gabor filters, which have been found to be efficient in reducing image redundancy and robust to noise [98]. Such filters can be either convolved [99] or applied to a limited range of positions, such as for 'jets' [92], [100], [101], [102], where a region around a pixel is described by the responses of a set of Gabor filters of different frequencies and orientations, all centred on that pixel position.

[99] implemented a face recognition scheme based on Gabor wavelet input representations to imitate the human vision system. Unlike most preprocessing techniques which try to reduce the amount of input data,

Introduction to Face Recognition

the full convolution of the face image with a set of 64 Gabor functions (8 orientations and 8 scales) gives a very much larger representation than the original gray-level image. No learning algorithm was used to train the system: the test for recognition was based on simple comparisons of each image with all of the other images in the database. A successful match was where the closest (in terms of Gabor coefficients) to the test image was of the same person as the test image. The calculation of the Gabor coefficients as a complete convolution (rather than sparse sampling) of the image was reported as extremely computationally expensive, processing of the 64 Gabor functions for a single image taking about half an hour on a fast workstation. The approach was extended to use a Kohonen-style self-organizing network used as classifier [103]. Gabor coefficients can, in addition, be used as data for PCA to provide a greater level of illumination invariance than found using gray-level pixel information [2].

4.4.5 Summary

Filter-based preprocessing of the images is an important intermediate step in image-based techniques, as the input representation contributes a great deal to the learnability of the task. It is important to highlight relevant parts of the information (leading to reduction in the dimensionality of input) and provide moderate invariance to normal environmental illumination [95]. This is in contrast to tackling strong, incidental lighting, which is very much more difficult [104], but luckily not expected in domestic environments. The approach can both suppress variation that is not important for the task, such as illumination variability, and highlight those variations that are useful, via, for example, orientations and scales used for Gabor filters.

4.5 Dynamic Link Graphs

The dynamic link approach to object recognition can be seen in two lights. First, as a theoretical model of biological vision [105], [106], and second, as an algorithmic form which has been shown to perform extremely well on the standard databases [107]. The important features of the approach are labeled graphs containing layers of Gabor feature jets and the dynamic links within the graphs that establish the image/model correspondence match. The process can be extremely computationally expensive, taking 10-15 minutes of SPARC 10 CPU time to recognize one face from a gallery of 111 models [106].

Objects can be described by both shape and texture information using elastic graphs [100], [101], [102] of local features. This process uses a rectangular graph laid over the training images, the graph edges represent the distances between features (the geometric data), and the graph vertices hold coefficients from Gabor 'jets' (see Section 4.4) applied to the image at the feature locations. An alternative to the rectangular grid is to use manually constructed 'face bunch graphs' that are specific to faces, using fiducial landmarks, such as eyes, mouth, etc. [107]. A coarse match of the graph onto the test image is made first with fixed parameters, followed by finer matching using a cost function to offset graph distortion against object distortion. This approach has some similarities to flexible templates (Section 4.2), as the matching algorithm is in terms of geometrical deformation and similarity of Gabor coefficients.

Some pose invariance for the elastic graph models can be gained by global transformations to the feature jets to account for changes in view [108], [109]. This accounted for rotation up to half-profile (45°), but separate, manually-designed face grids and graphs have to be used to cope with self-occlusion at greater pose ranges [107], [110]. Matching times for a single image for this type of approach was reported as 15–20 seconds [108].

However, these highly specialized representations clearly illustrate that the boundary between representational issues and reasoning issues is hard to define. They seem too committed and computationally expensive for our purposes.

4.6 Discussion

In general, approaches relying on simple templates or features alone will not be sufficiently robust under pose and lighting variations for our requirements, especially as the extraction of common features under all poses will be hindered by self-occlusion. This means that a photometric view-based or appearance-based representation is likely to be the most useful for our task.

Methods that rely on locating specific facial features, either for classification or normalization, may turn out not to be efficient when applied to low resolution images. As mentioned earlier, although standardizing face images (especially to an average shape) can be an extremely efficient representation for frontal views, it is not clear how such a process could be carried out for large pose variations, as there are no common features for all facial views.

PCA is an efficient way of reducing dimensionality, but has the drawback of being more sensitive to image variations than to facial characteristics. Its performance is dependent on the accuracy of normalization, and the process has no inherent invariance to translation, scale or rotation. Lighting can severely disrupt matching [37], and although pose can be dealt with, it cannot be accommodated easily.

Other representations approaches to improving generalization through learning other aspects of the task are possible, such as low-dimensional object representations from examples [111], class-based image transformations [36], or specific invariances [112], but their computational expense made them unattractive for our specific task.

The way representations are devised is primarily led by the need to reduce dimensionality to reduce complexity and computation. There is an implicit assumption that much of image data is redundant or irrelevant. Obviously, the dimensions discarded should be from this category in order to emphasize the useful data left over. This is the major reason for using filter-based representations, as one can specify the nature of feature that should be extracted. This has been observed in biological systems, where parallel processes can deal with different aspects of the images which were specifically extracted at at an early stage.

We will take a filter-based approach, which is fast and yet fairly general. We regard this filtering as an early stage of representation for identity, which we will develop further using adaptive learning in the next section about reasoning. Finally, in Section 6 we provide a comparison of other techniques to our proposed approach with a standard database.

5 Face Reasoning

Once a database has been collected and a representation decided upon for the images, the method of comparison between exemplar and test faces has to be determined. This reasoning can be simple matching if the representation extracted is extremely face specific or can be very adaptive if a more generalized representation (not very discriminable) is chosen. It can be seen that the type of representation has determined a 'face-space' in which distance comparisons can be made. Standard distance metrics, such as Euclidean or Mahalanobis (for eigenvector spaces) [63], can be used for matching, whereas simple weighted sums may be more suitable for internal 'hidden' representations.

Learning is an important factor in any useful application, to avoid the 'brittleness' commonly found in manually extracted rule systems. Even simple vision tasks are of such complexity that original assumptions in manual systems turn out not to be valid or only partially valid in certain circumstances. In addition, such an approach is neither scalable nor modifiable in day-to-day operation. For example, if the task changes from the original specification due to different people or rooms being involved, the system should be able to automatically relearn the task, rather than require an operator to reprogram new rules to cover the changed circumstances.

5.1 Matching Techniques

Traditionally, matching has been found very useful in low-level vision for localizing and identifying patterns, based on simple correlation of the image vectors. Such simple approaches were discussed in Section 4.1.

A different approach to matching is taken by [16] who uses examples of faces in varying pose to learn a pose-invariant face description in terms of shape and texture vectors. This is essentially an alignment-based ex-

Introduction to Face Recognition

tension to traditional template-matching approaches [74], [76], but the model can solve the correspondence problem between face images in different poses, which can be used both for face image analysis and synthesis. Two methods were developed, an interpolating multiple view recognizer, and a virtual views approach. For the latter, [66] described how such synthetic views could complete a view-sphere for example-based learning where insufficient real views were available. Affine transform and optical flow were used to bring image templates into registration (they termed this process 'vectorization'), and normalized correlation determined the best match. Despite good results, this approach is fairly slow and can take several seconds of processing time per test image.

[16] used simple normalized correlation with example templates of eyes, nose and mouth, following a two-stage geometrical registration step. This was originally done with 15 example views of each person to be recognized, but this was adapted to work with virtual views. This 'analysis through synthesis' approach, recognizing faces from one original and several, synthetic views [66], [113] is extremely useful where data is very sparse. This is, however, extremely intensive computationally, taking up to half an hour to analyse one image, and therefore not applicable to our task. This low data, high computation is quite the opposite to our high-data, low-computation task requirements.

5.2 Early Connectionist Approaches

Neural networks have a long history of being used for face recognition, though computational limitations of the time seem to have restricted the amount of testing that was possible.

The Kohonen associative networks [114], [115] were able to demonstrate quite early on one of the main advantages of the distributed processing in neural networks, which is a tolerance to noisy or incomplete test data. They could classify gray-level images of faces when a forcing stimulus (the desired output activity) was provided along with the stimulus pattern (the input data). These values were clamped until a steady state of activations was reached. The idea was that, when unclamped, the network would converge when given the original input to give the desired output values. It could also generalize in classifying new views of learnt faces by interpolating within the range of angles already seen, but could not extrapolate to images outside this area. [116] used a Kohonen memory model to encode zero-crossing edge segments rather than gray-level values. The results, though better, are difficult to assess, as a greater pixel area was used to extract edge segments than was used for the pixel intensity values [26, p. 107].

WISARD is a pattern classifier system that uses a neural network-like approach. It has a single layer local adaptive network with an n-tuple selection mechanism which is used to recognize human faces and expressions, and is able to distinguish between smiling and frowning faces [117]. WISARD was trained with many binary exemplar images of each face, input to the system from real-time video until a sufficiently high recall was achieved. This gave reasonable results, although it was intolerant to scale, 3-D rotation, and lighting or background variation. WISARD was not a distributed model, as trained concepts (individual faces) were held locally. The output for a particular image was a numerical representation of the detector responses, rather than a classification against trained input. This could form a personal identification code, either for confirmation of known faces or for matching instances of unknown faces. This approach has been used again for face recognition recently [118], see Section 6.

5.2.1 Multi-Layer Perceptrons and Associative Networks

The Multi-Layer Perceptron (MLP), commonly trained using gradient descent with error back-propagation, is capable of good generalization for difficult problems, but is notoriously difficult to ensure global convergence under all training runs, as the non-linearity of the hidden units and the nature of the input-output mapping lead to a large number of local minima, and training times can typically be long. [86], [87] used multi-layer networks with target output equal to input (auto-association) in order to compress photographic images. The network was trained on random patches of image. The compressed signal could be taken from the hidden layer of units (these values were effectively eigenvalues, the eigenvectors, called 'holons' here, being contained in the weight values between the unit layers), and these values could, in turn, be put back in to decode or uncompress the original image as output values.

[86] found that the non-linear arrangement of their multi-layer network did not actually improve the compression of images when compared to networks using linear units. For this reason, all following networks used for PCA, such as [59] for instance, have used simpler linear associative networks. However, [39] suggests that while linear associative networks and MLPs using back-propagation which calculate PCA can be effective for single-viewed classification tasks, they may not be as effective as HyperBF networks [19], [20] in a nonlinear mapping task, for example the classification of people with varying head pose (see Section 5.4).

5.3 Hierarchical Neural Networks

The Cognitron [119] and Neocognitron [120] were biologically-inspired self-organizing hierarchical approach to object recognition. The neural network structure had successive layers of cells of increasingly large receptive fields with a cascaded grouping of features, which allowed it to become invariant to scale, rotation, and translation. The recognition of analogue input has been developed by [121], but training still requires binary patterns, and the approach has only been used on 2-D objects, such as numerals and digits, so it is not known how such an approach would behave with 3-D variations.

The Cresceptron [96] had a similar retinotopic structure to the Neocognitron, but differed in that its configuration could be automatically determined during learning. The higher-level layers of units can be regarded as increasingly complex receptive fields, in that they become more and more specific to the training objects. Information can be 'grown' incrementally, with new network units being added as new concepts are detected. It was trained on complex images containing faces from TV news programs, and appeared quite robust to expression and minor pose variation (greater pose ranges could be explicitly learnt as different instances of the same object). However, the approach is computationally expensive and has not been tested with large numbers of objects or under large image variations, such as illumination.

Neurophysiological evidence has come from [122] for image-based coding in face-sensitive neurons in the macaque STS area of the brain, showing a viewer-based, rather than object-based, representation for faces. The view-invariance seen in some of the face cells has been supported by work on high order cortical sensory areas by [123]. [124] have also created a neural network simulation, 'VisNet', for learning spatio-temporally invariant object representations based on observed responses of temporal cortical visual neurons. Hierarchical layers of competitive networks are used, with short range mutual inhibition within each layer. This multi-stage feed-forward architecture was able to learn invariant representations of objects, including faces. A wide range of invariances have been observed, including spatial-temporal, translation and view, using a modified Hebb-style training rule incorporating a temporal 'trace' of each cell's previous activity [125]. This approach is useful for simulation purposes, but its complexity would not make it suitable currently for real-time applications.

The hierarchical style of network structure has had considerable success in overcoming rotation and scale differences, but this type of processing requires considerable computational effort even to train with small amounts of data, due to the large numbers of layers the information has to passed through. It is clear that 3-D objects can be invariantly represented in such structures [124], [126], but at present the computational load precludes them from real-time applications. They would be suitable for a parallel process, but the specialized hardware required would exclude them from task suitability this time through cost (Task Requirement 1a).

5.4 Radial Basis Function Networks

One can implicitly model a view-based recognition task using linear combinations of 2-D views [13] to represent any 2-D view of an object. A simpler approach is for the system to use view interpolation techniques [19], [20] to learn the task explicitly. Radial basis function (RBF) neural networks have been identified as valuable adaptive learning model by a wide range of researchers [18], [127], [128], [129], [130], [131] for such tasks. Their main advantages are computational simplicity, supported by

Introduction to Face Recognition

well-developed mathematical theory, and robust generalization, powerful enough for real-time real-life tasks [132], [133]. They are seen as ideal for practical vision applications by [134] as they are good at handling sparse, high-dimensional data and because they use approximation to handle noisy, real-life data. The nonlinear decision boundaries of the RBF network make it better in general for function approximation than the hyperplanes created by the multi-layer perceptron (MLP) with sigmoid units [18], and they provide a guaranteed, globally optimal solution via simple, linear optimization. The RBF network is a poor extrapolator (compared to the MLP) and this behavior can give it useful low false-positive rates in classification problems. This is because its basis functions cover only small localized regions, unlike sigmoidal basis functions which are nonzero over an infinitely large region of the input space.

Regularization Networks are based on mathematical regularization theory and include RBF and HyperBF (HBF) networks in configurations where the networks have an equal number of hidden units and training examples [135]. They can be seen as performing generalization through non-linear view approximation [11], which has the advantage over linear interpolation (linear combination of views) [13] in that it is less affected by variation orthogonal to learnt variation, see Figure 1. The RBF network can be considered as a special case of the more general HBF network [18].

Once training examples have been collected as input-output pairs, that is, with the target class attached to each image, tasks can be simply learnt directly by the system. This type of supervised learning can be seen in mathematical terms as approximating a multivariate function, so that estimations of function values can be made for previously unseen test data where actual values are not known. This process can be undertaken by the RBF network using a linear combination of basis functions, one for every training example, because of the smoothness of the manifold formed by the example views of objects in a space of all possible views of that object [19].

Although [73] used a simple nearest neighbor classifier to discriminate feature vectors, their success with their HBF networks for object recognition [20] led them to conclude that an HBF network would be a more effective solution to their template matching scheme [74] for face recognition. Template matching is related to RBF and HBF Network schemes, with the difference that Gaussian, non-linear functions are applied to the correlation coefficients. The HBF network allows the use of non-radial basis functions and may find a more optimal solution than the RBF [20], as more precision is available in the choice of basis function. They are less attractive for real-time applications, however, as the calculations for the higher-order centre functions are computationally more intensive than the simple Gaussian function used by the RBF network. The ability of such networks to train according to very specific tasks is shown by [136], where the HyperBF architecture was used to identity gender information from geometrical descriptions very similar to those used in [20].

[130] trained a variety of nets to recognize stationary hand gestures from computer-generated 2-D polar coordinates of fingertips (not actual images). They achieved good generalization in 3-D orientation and their system was able to cope well even when much of the data was missing. Their standard test data was best handled by a back-propagation net, but this performed badly with missing or uncertain (noisy) features, suffering a serious fall-off in performance as more elements were lost. They show, however, that a Gaussian RBF net can cope well with this type of data, with a success rate over 90% even with 50% of the features missing. This indicates that the RBF network would be suitable for learning the 3-D transformations and occlusion found in faces under large variations in head pose.

[36] used RBF networks in two separate stages, to capture pose and lighting parameters (using high-frequency filters) and to classify individuals (using low-frequency filters). In between the two networks, a face class specific transformation was applied to the original image (using parameters from the first network) to align the test image with a single standard view for all trained identities (discriminated by the second network). This is related to the analysis by synthesis approach [66], in that there is only one training prototype of each class. The difference is that the method attempts to transform the test image to the single canonical view, rather than relying on interpolation between several views. Of course, this is not using the valuable view interpolation ability of the RBF net-

work, and it is not clear how intensive such transformation are computationally.

A major advantage of the RBF over other network models, such as the MLP, is that a direct level of confidence is reflected in the level of each output unit. This is because regions in input space that are far from training vectors are always mapped to low values due to the local nature of the hidden units receptive fields, so that 'novel' input will give a low activation. This is in contrast to the global function approximation of the sigmoidal hidden units in the MLP, which can have spurious high output in similar regions of input space, allowing high confidence output. In addition, the normalization of RBF hidden unit activities allow their output to represent probability values for the presence of their class [137]. In light of the probabilistic nature of the RBF network's output, we will be using a discard measure in our work to exclude low-confidence output and reduce false positives.

5.5 Committees and Ensemble-Based Networks

Committees of networks can be used to give a consensus opinion where each network is trained with different parameters or data examples. The combination of results from all the networks may be better than the use of the one that works best on test data, which may not generalize most efficiently [131].

An ensemble network scheme was used by [93], who had a series of RBF networks, one for each person each trained on several images, with single output units. Output from each network was combined and used as input for a second stage ensemble RBF network which coordinated a final 'winner take all' classification. Each network only had one output, which signified the strength of classification for a particular individual.

An ensemble of RBF networks was used by [138] to identify faces, each network in the ensemble using different numbers of clusters and amounts of overlap. The number of hidden units was not related to training examples, so training is more computationally intensive than approaches using one unit per example, as extra effort is required to cluster the centre vectors. However, subsequent classification may be less intensive, as the system will have fewer hidden units. Although the FERET database used for testing includes a wide range of different pose views, only results for the frontal views were presented. The system has been updated [139] to use a decision tree to coordinate the output from the ensemble for a contents-based image retrieval task. The decision tree component improved performance, but it is possible that a coordinating RBF network (such as used by [93]) could achieve similar results more efficiently.

The interactive activation and competition (IAC) network model [140], [141] has an ensemble style of organization which can be used to account for psychological phenomena in face recognition [142], based on the [22] face recognition perceptual framework model. 'Pools' of units, representing face recognition units (FRUs), person identity nodes (PINs) and semantic information units (SIUs), have inhibitory connections between their constituent units. These units then have excitatory links with specific units in other pools to allow the activation in a FRU, for instance, to activate that individuals PIN, to signal familiarity, and SIU, to allow more specific information about that person. The FRUs are view-independent, so that the unit can become active from any view of a particular face. This has been developed into interactive activation and competition with learning (IACL) [143] to allow unknown faces to learn their own FRUs. The model was not intended to be used as a functional system, but as a tool to confirm theoretical expectations.

5.5.1 Use of Negative Examples

Ensembles of networks, such as used by [93], rely on a second stage network to utilize implicit negative knowledge, where if one input has a large value it will act as a negative influence on all the other input units, because only one class can be present at any one time. This approach may be made more accurate if explicit negative (non-class) examples are learnt alongside the positive class examples. This technique has been shown to be of critical importance in building robust face detection systems [60], [61], [144]. The selection of prototypical non-face training examples can be very difficult, as they have to represent the entire class space of non-face images, which is considerably larger than the class space of face images. Most successfully so far, [61] has trained networks with a 'bootstrapping' algorithm which adds previous false detections to the training set as the training progresses, which reduces the number of negative examples required. This shows that prior knowledge of confusion in the distinction of classes can be used to guide the choice of appropriate training examples. This issue is developed in Chapter 5, where we introduce the 'Face Unit' RBF network model, which uses this type of positive and negative evidence to signal the presence of one particular class.

5.6 Temporal Networks

Representations and reasoning only concerned with data from single points in time are ignoring potentially useful information occurring through time, including significant temporal correlations in image sequences. Study of the statistical properties of static images [94], [145] has shown some regularities, see Section 4.4. This has been extended to image sequences [146] to show a high level of spatio-temporal correlation, showing that natural time-varying images do not change randomly over space or time. If our data source provides information over time, we would do well to take advantage of this.

Recognizing simple temporal behaviors is an important capability in computer vision applications such as visual surveillance [147] and biomedical sequence understanding [148]. Dynamic neural networks for such tasks can be constructed by adding recurrent connections to form a contextual memory for prediction in time [149], [150], [151]. These partially recurrent neural networks (RNNs) can be trained using back-propagation but there may be problems with stability and very long training times when using dynamic representations.

A limited alternative is to use a simple Time Delay structure which can provide fast, robust solutions. The Time-Delay Neural Network (TDNN) model (for an introduction, see [152]), incorporates the concept of time-delays in order to process temporal context, and has been successfully applied to speech and handwriting recognition tasks [153]. Its structured design allows it to specialize on spatio-temporal tasks, but, as in weight-sharing networks, the reduction of trainable parameters can increase generalization [154] and give some shift invariance when used as convolutional networks (CN) [155]. A time delay variant of the RBF network, the TDRBF network, has more recently been developed for speech recognition [156]. This has the benefits of ordinary RBF networks over other models, such as low numbers of tunable parameters and fast training times.

An even simpler approach than Time-Delays for processing image sequences was used by [157] to create an ensemble of single emotion RBF networks. A decay constant was applied to encode temporal facial optical flow information from several frames into a single frame, so that earlier frames had less value than later ones when summed into the composite input frame. Although this reduces the representation size, it creates ambiguities between spatial and temporal changes as both have to be shown in the same space.

The capacity of temporal cortex for making associations [158] has led research into using temporal relationships in patterns for learning, for instance, different face views [159], [160], using competitive Hebbian learning with a temporal 'trace rule' originally proposed by [161]. In contrast to previous models, these temporal learning rules use differences over time, rather than simple time windows, to directly learn those temporal relationships required for specific tasks.

5.7 Discussion

The limitations imposed by the requirements for the face recognition task prevents most of the computationally-intensive techniques from being used here. In particular, the most 'biologically-plausible' approaches, such as the VisNet network model [124], IAC model [143] and the full Gabor processing of [99], are the slowest in operation.

An example-based view interpolation learning approach using Regularization Networks, especially RBF networks, is very attractive as a face recognition technique, due to its simplicity and ease of training. In ad-

Introduction to Face Recognition

dition, they provide fast and robust operation. We noted that there is evidence that we use some kind of 'face recognition unit' to recognize familiar faces [22], [26]. In addition, primate vision systems seem to use some kind of view-based representations for recognition [32], [122], [162]. These ideas are partially captured by the RBF network where the first layer of the network maps the inputs with a hidden unit devoted to each view of the face to be classified. The second layer is then trained, to combine the views so that a single output unit corresponds to the individual person. If we regarded filter-based preprocessing as an early stage of representation of identity, we can now regard hidden unit output from an RBF network as a later stage of that representation, which has been transformed into a space of considerably fewer dimensions.

Up until this point, we have deliberately left out specific performance figures from our discussion of face recognition research, as it is extremely difficult to compare work from different groups when each chooses their own recognition task. Even current 'state of the art' tests, such as the FERET database, can be seen as too easy if their test images are compared to those encountered in realistic situations. Without dealing with expression, pose and lighting as confounding variables, a system can appear to work well, but turn out not to be useful in a practical application. As discussed in Section 1, truly robust systems would need to account for many other non-trivial aspects, such as temporal behaviors, occlusion and speech-related facial changes.

We believe that the best approach to satisfying our specific task requirements is to learn face classes over a wide range of poses with an RBF network. A preliminary test of the suitability of our approach is given in the following section, where we compare our results directly with published experimental results for other approaches using a common database.

6 Comparing Face Recognition Techniques

In the earlier sections of this chapter, we suggested that RBF network view interpolation with filtering preprocessing was a good way forward to meet our task requirements. The RBF network has been shown to provide robust classification even where data is noisy or partially missing [130]. Our original question [163] was whether this ability can be used with complex 3-D objects such as faces, where the data varies in lighting, expression and pose. Here, we compare the RBF techniques with other methods using a standard database.

It is particularly important to establish that the RBF network is able to distinguish a useful number of face classes, as this will indicate its potential as a practical technique for future applications. A suitable source of data to test this is the Olivetti Research Laboratory (ORL) database of faces. This contains 400 images of 40 people, which is sufficient to satisfy Task Requirement 3a (see Section 1). It should be noted that comparisons with these separately published results is a quick and simple compromise, as the test results presented here were collected on different systems and under different testing regimes. However, they give a rough indication of comparative performance and suitability for our task requirements.

6.1 Results

Table 1 summarizes the results from several published papers, plus our own tests. Test generalization performance for systems with differing numbers of training images are given, together with times for the train and test (classification) stages (where available).

6.1.1 Hidden Markov Models

[46] initially developed and experimented with the ORL database, using conventional Hidden Markov Models (HMMs) as a graphical probabilistic approach to encoding feature information. This approach used several subjective parameter selections, and gave a top performance around 87% for a system trained with 200 images. Further work using pseudo 2-D HMMs [164] was able to improve this to 95%, but the computational complexity of this approach seems to count this out as a useful real-time technique, as 4 minutes is a long time to wait for a classification.

Introduction to Face Recognition

Table 1. Test generalization (% correct) and processing times for various face recognition techniques used by various researchers using ORL Face Database of 40 people, averaged over several selections.

Group	Technique	\multicolumn{5}{c}{Images per Person}	\multicolumn{2}{c}{Processing Time}
		1	2	3	4	5	Train	Classify
Samaria	HMM	?	?	?	?	87	?	?
& Harter	pseudo 2-D HMM	?	?	?	?	95	?	4min
Lawrence	Eigenfaces	61	79	82	85	89	?	?
et al.	PCA + MLP	?	?	?	?	59	?	?
	SOM + MLP	?	?	?	?	60	?	?
	PCA + CN	66	83	87	88	92	?	?
	SOM + CN	70	83	88	93	96	4hr	<0.5sec
Lin et al.	PDBNN	?	?	?	?	96	20min	<0.1sec
Lucas	n-tuple	54	68	75	78	81	0.9sec	0.025sec
	cont n-tuple	73	84	90	93	95	0.9sec	0.33sec
	1-NN	?	?	?	?	97	0sec	1sec
Howell	RBF before discard	49	65	72	80	86	8sec	0.01sec
& Buxton	after discard	84	90	91	95	95	8sec	0.01sec

6.1.2 Eigenfaces

Both [46] and [155] tested the ORL database with the 'eigenface' [59], [85] approach. Both report performance of around 90%, though the latter found that they could only get this by using separate training vectors for each image. This is in contrast to [59], who averaged the eigenfaces for all images of each person in their tests. When tested with ORL data, this latter approach gave 74% for 5 training images per person [155]. That this is much lower than the MIT results (where results over 90% are common) would seem to indicate that the ORL database represents a much harder task than the MIT face database (assuming the implementations were equivalent).

6.1.3 Convolutional Networks

[155] used a self-organizing map (SOM) to reduce the dimensions of the input representation, and a five-layer convolutional network (CN) to give translation and deformation invariance. This was faster than the previous HMM approach and performed equally well, but still required several hours training time.

They compared the dimensionality-reducing abilities of the SOM with principal components analysis (PCA), and the CN with a multi-layer perceptron (MLP). This latter approach gave very poor results, especially when several hidden layers were used. It should be noted that the figures for these approaches in Table 1 show the single best results from all combinations (which came from a MLP with one hidden layer) rather than average results (which are given for the other approaches from other groups).

6.1.4 Probabilistic Decision-Based Neural Networks

[165] used a probabilistic, decision-based neural network (PDBNN) a modular network structure with non-linear basis functions (each sub-network similar to a HyperBF (HBF) network [18]). This was able to train and classify much faster than the CN approach of [155], while reaching a similar level of performance.

6.1.5 Continuous n-tuple Classifiers

The continuous n-tuple classifier [118] is an updated version of earlier n-tuple classifiers, such as WISARD (see Section 5.2). The updating refers to speed and storage efficiency, so it is likely that this technique would suffer the same problems with image variations such as pose or lighting in real-world tests. However, the approach does train and classify quickly and provide a high level of performance.

The figures shown are for tests with 200 3-tuples (600 values) per image. Using 500 4-tuples (2000 values) per image improved recognition to 86% and 97% for the n-tuple and continuous n-tuple classifier, respectively.

6.1.6 Nearest Neighbor Classifiers

[118] was also able to achieve very high performance using a simple 1-nearest-neighbor (1-NN) classifier using a City-Block distance measure. The success of simple matching indicates how constrained the database is in terms of lighting and pose, as such techniques will not be invariant to such factors.

Introduction to Face Recognition 263

6.1.7 RBF Networks

To use the ORL data with the RBF network, we subsampled each image to 25×25 and applied a four scales, three orientation sparse-sampling Gabor filter preprocessing process (termed 'A3' in [166]). A simple discard measure, based on the relative magnitudes of the output units, was used to remove low confidence classifications (these being those where the highest output value was less than a certain ratio below the next highest). Each training example was used as a centre vector for a hidden unit.

The RBF network approach was fast in training and the fastest in classification of all the published techniques. Our experiments were conducted on a moderately fast Sun SPARC 20 workstation. Test generalization before discard was fairly poor in comparison to the other approaches, though the results were well above random (2.5%). For 5 training examples per person, discarding 39% of results allowed performance to be improved from 84% to 95%, which was comparable with the best of the other techniques. The results after discard for the RBF network were especially good where lower numbers of training examples per person were provided.

6.2 Discussion

Table 1 shows that although, in pure generalization terms, our RBF network approach is not the overall top performer, it does have a sufficient level of performance (95% after discard) for our target application where it will have to deal with image sequences. In this type of application, training data is relatively sparse (compared to the large range of variation in real-life images) and test data is abundant. The success of the RBF discard measure, which makes it the top performer where low numbers of training examples are available, highlights its efficiency in interpolating between even small number of views for reliable classification. Although discarding does reduce the number of useful classifications, a significant amount of data will remain when such techniques are used with image sequences.

The ORL database is a highly constrained database and not designed to meet our task requirements. Thus, success in recognizing the ORL faces does not necessarily indicate a suitability for our less constrained face recognition task. For example, the constant lighting conditions do not require an invariance to illumination to be developed, and thus no consideration has been made by the other approaches to the issue of preprocessing. We are not able to know how the other techniques would perform in the presence of variable lighting, but we believe that real-life applications would require some type of preprocessing to overcome this kind of variability.

A particularly important point is that all the other face recognition techniques gave processing times which are very much slower than the RBF network in classification of the ORL data. It is apparent that the RBF network provides a solution which can process test images in inter-frame periods on low-cost processors (Task Requirement 1a).

7 Summary

This chapter started by introducing our face recognition task and establishing its requirements. This was followed by a discussion of previous face recognition approaches, in terms of acquisition, representation and reasoning. We have evaluated a wide variety of general approaches with respect to our specific task requirements. It is perhaps not surprising that many of the approaches to face recognition discussed in this chapter do not fulfill these task requirements for the simple reason that they were not designed with such an application in mind. We stress that success of an application will be determined by relevance of the approach to the task.

The combination of Task Requirements 1a and 1b to be robust in the face of noisy and variable data, and yet fast enough to give results in inter-frame periods with standard processors is demanding, but not impossible. Simple filter-based preprocessing can give some invariance for the input representation, and RBF networks will give speed and robust performance for the recognition itself. The suitability of the RBF network approach for handling occlusion, covered by Task Requirements 3(c)iii and 3(d)ii, has been shown by results in [21] and [130].

In summary, we can see that our proposed filter-based RBF view interpolation scheme appears to be very suitable for our target task: it combines fast training and testing times with the ability to cope with complex 3-D transformations. Results using the ORL standard database indicate that the network can discriminate useful numbers of face classes.

References

[1] Pentland, A. (1996), "Smart rooms," *Scientific American*, 274(4):68–76.

[2] McKenna, S.J., Gong, S., and Collins, J.J. (1996), "Face tracking and pose representation," in R.B. Fisher and E.Trucco (Eds.), *Proceedings of British Machine Vision Conference*, pp. 755–764, Edinburgh. BMVA Press.

[3] McKenna, S.J., Gong, S., and Raja, Y. (1997), "Face recognition in dynamic scenes," in A.F. Clark (Ed.), *Proceedings of British Machine Vision Conference*, pp. 140–151, Colchester, U.K. BMVA Press.

[4] Bruce, V. and Humphreys, G.W. (1994), "Recognising objects and faces," *Visual Cognition*, 1:141–180.

[5] Ullman, S. (1996), *High-level Vision*, MIT Press, Cambridge, MA.

[6] Ullman, S. (1989), "Aligning pictorial descriptions: an approach to object recognition," *Cognition*, 32:193–254.

[7] Moses, Y. and Ullman, S. (1992), "Limitations of non model-based recognition schemes," in G. Sandini (Ed.), *Proceedings of European Conference on Computer Vision, Lecture Notes in Computer Science*, Vol. 588, pp. 820–828, Santa Margherita Ligure, Italy. Springer-Verlag.

[8] Marr, D. and Nishihara, H.K. (1978), "Representation and recognition of the spatial organization of three dimensional structure," *Proceedings of Royal Society London, Series B*, 200:269–294.

[9] Biederman, I. (1987), "Recognition by components: a theory of human image understanding," *Psychological Review*, 94:115–147.

[10] Marr, D. (1982), *Vision*, Freeman, San Francisco, CA.

[11] Bülthoff, H.H. and Edelman, S. (1992), "Psychophysical support for a 2-D view interpolation theory of object recognition," *Proceedings of the National Academy of Sciences, USA*, 89:60–64.

[12] Bülthoff, H.H., Edelman, S., and Tarr, M. (1995), "How are three-dimensional objects represented in the brain?" *Cerebral Cortex*, 5:247–260.

[13] Ullman, S. and Basri, R. (1991), "Recognition by linear combinations of models," *IEEE Transactions on Pattern Analysis & Machine Intelligence*, 13:992–1006.

[14] Breuel, T.M. (1992), "Geometric aspects of visual object recognition," Technical Report 1374, AI Lab, MIT, Cambridge, MA.

[15] Palmer, S.E., Rosch, E., and Chase, P. (1981), "Canonical perspective and the perception of objects," in J. Long and A. Baddeley (Eds.), *Attention and Performance IX*, pp. 135–151. Erlbaum, Hillsdale, NJ.

[16] Beymer, D.J. (1994), "Face recognition under varying pose," *Proceedings of IEEE Conference on Computer Vision & Pattern Recognition*, pp. 756–761, Seattle, WA. IEEE Computer Society Press.

[17] Poggio, T. and Girosi, F. (1990), "Networks for approximation and learning," *Proceedings of IEEE*, Vol. 78, pp. 1481–1497.

[18] Poggio, T. and Girosi, F. (1990), "Regularization algorithms for learning that are equivalent to multilayer networks," *Science*, 247:978–982.

[19] Poggio, T. and Edelman, S. (1990), "A network that learns to recognize three-dimensional objects," *Nature*, 343:263–266.

[20] Brunelli, R. and Poggio, T. (1991), "HyperBF networks for real object recognition," in J. Myopoulos and R. Reiter (Eds.), *Proceedings of International Joint Conference on Artificial Intelligence*, pp. 1278–1284, Sydney, Australia. Morgan Kaufmann.

[21] Edelman, S. and Poggio, T. (1992), "Bringing the grandmother back into the picture: a memory-based view of object recognition," *International Journal of Pattern Recognition & Artificial Intelligence*, 6:37–62.

[22] Bruce, V. and Young, A. (1986), "Understanding face recognition," *British Journal of Psychology*, 77:305–327.

[23] Ellis, H.D. and Young, A.W. (1989), "Are faces special?" in A.W. Young and H.D. Ellis (Eds.), *Handbook of Research on Face Processing*, North-Holland, Amsterdam, The Netherlands.

[24] Hay, D.C. and Young, A. (1982), "The human face," in H.D. Ellis (Ed.), *Normality and Pathology in Cognitive Functions*, Academic Press, San Diego, CA.

[25] Hay, D.C., Young, A., and Ellis, A.W. (1991), "Routes through the face recognition system," *Quarterly Journal of Experimental Psychology: Human Experimental Psychology*, 43:761–791.

[26] Bruce, V. (1988), *Recognising Faces*, Lawrence Erlbaum Associates, London.

[27] Diamond, R. and Carey, S. (1986), "Why faces are, and are not special: An effect of expertise," *Journal of Experimental Psychology: General*, 115:107–117.

[28] Rhodes, G. and McLean, I.G. (1990), "Distinctiveness and expertise effects with homogeneous stimuli: Towards a model of configural coding," *Perception*, 19:773–794.

[29] Benson, P.J. and Perrett, D.I. (1994), "Visual processing of facial distinctiveness," *Perception*, 23:75–93.

[30] Bruce, V. and Green, P. (1990), *Visual Perception*, Lawrence Erlbaum Associates, London.

[31] Sergent, J., Ohta, S., MacDonald, B., and Zuck, E. (1994), "Segregated processing of facial identity and emotion in the human brain: a PET study," *Visual Cognition*, 1:349–370.

[32] Logothetis, N.K., Pauls, J., Bülthoff, H.H., and Poggio, T. (1994), "View-dependent object recognition by monkeys," *Current Biology*, 4:401–414.

[33] Sinha, P. and Poggio, T. (1996), "Role of learning in three-dimensional form perception," *Nature*, 384:460–463.

[34] Bachmann, T. (1991), "Identification of spatially quantised tachistoscopic images of faces: how many pixels does it take to carry identity?" *European Journal of Cognitive Psychology*, 3:87–103.

[35] Hancock, P.J.B., Burton, A.M., and Bruce, V. (1995), "Preprocessing images of faces: correlations with human perceptions of distinctiveness and familiarity," *Proceedings of IEE Fifth International Conference on Image Processing and its Applications*, Edinburgh, Scotland.

[36] Lando, M. and Edelman, S. (1995), "Receptive field spaces and class-based generalization from a single view in face recognition," *Network: Computation in Neural Systems*, 6:551–576.

[37] Hancock, P.J.B., Bruce, V., and Burton, A.M. (1997), "A comparison of two computer-based face recognition systems with human perceptions of faces," *Vision Research*, (Submitted).

[38] Samal, A. and Iyengar, P.A. (1992), "Automatic recognition and analysis of human faces and facial expressions: a survey," *Pattern Recognition*, 15:65–77.

[39] Valentin, D., Abdi, H., O'Toole, A.J., and Cottrell, G.W. (1994), "Connectionist models of face processing: A survey," *Pattern Recognition*, 27:1208–1230.

[40] Chellappa, R., Wilson, C.L., and Sirohey, S. (1995), "Human and machine recognition of faces: A survey," *Proceedings of IEEE*, Vol. 83, pp. 705–740.

[41] Bichsel, M. (Ed.) (1995), *Proceedings of International Workshop on Automatic Face & Gesture Recognition*, Zurich, Switzerland. University of Zurich.

[42] Essa, I. (Ed.) (1996), *Proceedings of International Conference on Automatic Face & Gesture Recognition*, Killington, VT. IEEE Computer Society Press.

[43] Kruizinga, P. (1995), The Face Recognition Home Page, web page: http://www.cs.rug.nl/~peterkr/FACE/face.html.

[44] Robertson, G. and Craw, I. (1994), "Testing face recognition systems," *Image & Vision Computing*, 12:609–614.

[45] Bouattour, H., Fogelman-Soulié, F., and Viennet, E. (1992), "Solving the human face recognition task using neural nets," in I. Aleksander and J. Taylor (Eds.), *Proceedings of International Conference on Artificial Neural Networks*, pp. 1595–1598, Brighton, U.K. Elsevier Science Publishers.

[46] Samaria, F.S. and Harter, A.C. (1994), "Parameterisation of a stochastic model for human face identification," *Proceedings of 2nd IEEE Workshop on Applications of Computer Vision*, Sarasota, FL.

[47] Lanitis, A., Taylor, C.J., and Cootes, T.F. (1995), "An automatic face identification system using flexible appearance models," *Image & Vision Computing*, 13:393–401.

[48] Phillips, P.J., Rauss, P.J., and Der, S.Z. (1996), *FERET (face recognition technology) recognition algorithm development and test results*, Technical Report ARL-TR-995, US Army Research Laboratory, Adelphi, MD.

[49] Phillips, P.J., Moon, H., Rauss, P.J., and Rizvi, S.A. (1997), "The FERET September 1996 database and evaluation procedure," *Proceedings of 1st International Conference on Audio & Video-based Biometric Person Authentication, Lecture Notes in Computer Science*, Crans-Montana, Switzerland, Vol. 1206, pp. 395–402, Springer Verlag.

[50] Brunelli, R. and Falavigna, D. (1995), Person recognition using multiple cues. *IEEE Transactions on Pattern Analysis & Machine Intelligence*, 17:955–966, 1995.

[51] ATR (1996), "Cognitive and representative models of visual image information: Interaction of multi-modal information relayed by face," web page: `http://www.hip.atr.co.jp/departments/Dept2/Reports_FP.html`.

[52] Pigeon, S. and Vandendorpe, L. (1997), "The M2VTS multi-modal face database (release 1.00)," *Proceedings of 1st International Conference on Audio & Video-based Biometric Person Authentication, Lecture Notes in Computer Science*, Crans-Montana, Switzerland, Vol. 1206, pp. 403–410, Springer Verlag.

[53] Duc, B., Maître, G., Fischer, S., and Bigün, J. (1997), "Person authentication by fusing face and speech information," *Proceedings of 1st International Conference on Audio & Video-based Biometric Person Authentication, Lecture Notes in Computer Science*, Crans-Montana, Switzerland, Vol. 1206, pp. 311–318, Springer Verlag.

[54] Kittler, J., Li, Y.P., Matas, J., and Ramos Sánchez, M.U. (1997), "Combining evidence in multimodal personal identity recognition systems," *Proceedings of 1st International Conference on Audio & Video-based Biometric Person Authentication, Lecture Notes in Computer Science*, Crans-Montana, Switzerland, Vol. 1206, pp. 327–334, Springer Verlag.

[55] Young, D.S. (1987), *Representing images for computer vision*, Technical Report CSRP 96, School of Cognitive and Computing Sciences, University of Sussex, U.K.

[56] Tistarelli, M. (1994), "Recognition by using an active/space-variant sensor," *Proceedings of IEEE Conference on Computer Vision & Pattern Recognition*, pp. 833–837, Seattle, WA. IEEE Computer Society Press.

[57] Rao, R.P.N. and Ballard, D.H. (1995), "Natural basis functions and topographic memory for face recognition," in C.S. Mellish

(Ed.), *Proceedings of International Joint Conference on Artificial Intelligence*, pp. 10–17, Montréal, Canada. Morgan Kaufmann.

[58] Cliff, D.T. and Bullock, S.G. (1993), "Adding 'foveal vision' to Wilson's Animat," *Adaptive Behaviour*, 2:47–70.

[59] Turk, M. and Pentland, A. (1991), "Eigenfaces for recognition," *Journal of Cognitive Neuroscience*, 3:71–86.

[60] Moghaddam, B. and Pentland, A. (1995), "Probabilistic visual learning for object detection," *Proceedings of International Conference on Computer Vision*, pp. 786–793, Cambridge, MA. IEEE Computer Society Press.

[61] Rowley, H.A., Baluja, S., and Kanade, T. (1996), "Human face detection in visual scenes," in D.S. Touretzky, M.C. Mozer, and M.E. Hasselmo (Eds.), *Advances in Neural Information Processing Systems*, Vol. 8, pp. 875–881, Cambridge, MA. MIT Press.

[62] McKenna, S.J. and Gong, S. (1997), "Non-intrusive person authentication for access control by visual tracking and face recognition," *Proceedings of 1st International Conference on Audio & Video-based Biometric Person Authentication, Lecture Notes in Computer Science*, Crans-Montana, Switzerland, Vol. 1206, pp. 177–184, Springer Verlag.

[63] Craw, I., Costen, N., Kato, T., Robertson, G., and Akamatsu, S. (1995), "Automatic face recognition: combining configuration and texture," in M. Bichsel (Ed.), *Proceedings of International Workshop on Automatic Face & Gesture Recognition*, Zurich, Switzerland, pp. 53–58. University of Zurich.

[64] Cootes, T.F. and Taylor, C.J. (1992), "Active shape models - 'smart snakes'," in D. Hogg and R. Boyle (Eds.), *Proceedings of British Machine Vision Conference*, pp. 266–275, Leeds, U.K. Springer-Verlag.

[65] Cootes, T.F., Taylor, C.J., Lanitis, A., Cooper, D.H., and Graham, J. (1993), "Building and using flexible models incorporating grey-level information," *Proceedings of International Conference on*

Computer Vision, pp. 242–246, Berlin, Germany. IEEE Computer Society Press.

[66] Beymer, D.J. and Poggio, T. (1995), "Face recognition from one example view," *Proceedings of International Conference on Computer Vision*, pp. 500–507, Cambridge, MA. IEEE Computer Society Press.

[67] Craw, I. and Cameron, P. (1991), "Parameterising images for recognition and reconstruction," in P. Mowforth (Ed.), *Proceedings of British Machine Vision Conference*, pp. 367–370. Springer-Verlag.

[68] Beymer, D.J. (1995), *Vectorizing face images by interleaving shape and texture computations*, Technical Report 1537, AI Lab, MIT, Cambridge, MA.

[69] Bichsel, M. and Pentland, A. (1994), "Human face recognition and the face image set's topology," *Computer Vision, Graphics, and Image Processing: Image Understanding*, 59:254–261.

[70] Kanade, T. (1973), *Picture processing by computer complex and recognition of human faces*, Technical report, Department of Information Science, Kyoto University, Japan.

[71] Bennett, A. and Craw, I. (1991), "Finding image features using deformable templates and detailed prior statistical knowledge," in P. Mowforth (Ed.), *Proceedings of British Machine Vision Conference*, pp. 233–239. Springer-Verlag.

[72] Craw, I., Tock, D., and Bennett, A. (1992), "Finding face features," in G. Sandini (Ed.), *Proceedings of European Conference on Computer Vision, Lecture Notes in Computer Science*, Vol. 588, pp. 92–96, Santa Margherita Ligure, Italy. Springer-Verlag.

[73] Brunelli, R. and Poggio, T. (1992), "Face recognition through geometrical features," in G. Sandini (Ed.), *Proceedings of European Conference on Computer Vision, Lecture Notes in Computer Science*, Vol. 588, pp. 792–800, Santa Margherita Ligure, Italy. Springer-Verlag.

[74] Brunelli, R. and Poggio, T. (1993), "Face recognition: Features versus templates," *IEEE Transactions on Pattern Analysis & Machine Intelligence*, 15:1042–1052.

[75] Craw, I. and Cameron, P. (1992), "Face recognition by computer," in D. Hogg and R. Boyle (Eds.), *Proceedings of British Machine Vision Conference*, pp. 498–507, Leeds, U.K. Springer-Verlag.

[76] Baron, R.J. (1981), "Mechanisms of human facial recognition," *International Journal of Man-Machine Studies*, 15:137–178.

[77] Yuille, A.L. (1991), "Deformable templates for face recognition," *Journal of Cognitive Neuroscience*, 3:59–70.

[78] Yuille, A.L., Hallinan, P.W., and Cohen, D.S. (1992), "Feature extraction from faces using deformable templates," *International Journal of Computer Vision*, 8:99–111.

[79] Cootes, T.F., Taylor, C.J., Cooper, D.H., and Graham, J. (1992), "Training models of shape from sets of examples," in D. Hogg and R. Boyle (Eds.), *Proceedings of British Machine Vision Conference*, pp. 9–18, Leeds, U.K. Springer-Verlag.

[80] Lanitis, A., Taylor, C.J., and Cootes, T.F. (1995), "A unified approach to coding and interpreting face images," *Proceedings of International Conference on Computer Vision*, pp. 368–373, Cambridge, MA. IEEE Computer Society Press.

[81] Vetter, T. and Poggio, T. (1996), "Image synthesis from a single example image," in B. Buxton and R. Cipolla (Eds.), *Proceedings of European Conference on Computer Vision, Lecture Notes in Computer Science*, Vol. 1065, pp. 652–659, Cambridge, U.K. Springer-Verlag.

[82] Lanitis, A., Taylor, C.J., and Cootes, T.F. (1997), "Automatic interpretation and coding of face images using flexible models," *IEEE Transactions on Pattern Analysis & Machine Intelligence*, 19:743–756.

[83] Edwards, G.J., Lanitis, A., Taylor, C.J., and Cootes, T.F. (1996), "Modelling the variability in face images," *Proceedings of International Conference on Automatic Face & Gesture Recognition*, pp. 328–333, Killington, VT. IEEE Computer Society Press.

[84] Edwards, G.J., Taylor, C.J., and Cootes, T.F. (1997), "Learning to identify and track faces in image sequences," in A.F. Clark (Ed.), *Proceedings of British Machine Vision Conference*, pp. 130–139, Colchester, U.K. BMVA Press.

[85] Kirby, M. and Sirovich, L. (1990), "Application of the Karhunen-Loève procedure for the characterization of human faces," *IEEE Transactions on Pattern Analysis & Machine Intelligence*, 12:103–108.

[86] Cottrell, G.W., Munro, P., and Zipser, D. (1987), "Learning internal representations from gray-scale images: An example of extensional programming," *Proceedings of Annual Conference of the Cognitive Science Society*, pp. 461–473, Seattle, WA. Lawrence Erlbaum Associates.

[87] Fleming, M.K. and Cottrell, G.W. (1990), "Categorization of faces using unsupervised feature extraction," *Proceedings of International Joint Conference on Neural Networks*, pp. 65–70, San Diego, CA.

[88] Pentland, A., Moghaddam, B., and Starner, T. (1994), "View-based and modular eigenspaces for face recognition," *Proceedings of IEEE Conference on Computer Vision & Pattern Recognition*, pp. 84–91, Seattle, WA. IEEE Computer Society Press.

[89] Murase, H. and Nayar, S. (1995), "Visual learning and recognition of 3-D objects from appearance," *International Journal of Computer Vision*, 14:5–24.

[90] Akamatsu, S., Sasaki, T., Fukamachi, H., Masui, N., and Suenaga, Y. (1992), "An accurate and robust face identification scheme," *Proceedings of International Conference on Pattern Recognition*, pp. 217–220, The Hague, Netherlands.

[91] O'Toole, A.J., Abdi, H., Deffenbacher, K.A., and Valentin, D. (1993), "A low-dimensional representation of faces in high dimensions of the space," *Journal of the Optical Society of America A*, 10:405–410.

[92] Daugman, J.G. (1988), "Complete discrete 2-D Gabor transforms by neural networks for image analysis and compression," *IEEE Transactions on Acoustics, Speech, & Signal Processing*, 36:1169–1179.

[93] Edelman, S., Reisfeld, D., and Yeshurun, Y. (1992), "Learning to recognize faces from examples," in G. Sandini (Ed.), *Proceedings of European Conference on Computer Vision, Lecture Notes in Computer Science*, Vol. 588, pp. 787–791, Santa Margherita Ligure, Italy. Springer-Verlag.

[94] Hancock, P.J.B., Baddeley, R.J., and Smith, L.S. (1992), "The principal components of natural images," *Network: Computation in Neural Systems*, 3:61–70.

[95] Marr, D. and Hildreth, E. (1980), "Theory of edge detection," *Proceedings of Royal Society London, Series B*, 207:187–217.

[96] Weng, J.J., Ahuja, N., and Huang, T.S. (1993), "Learning recognition and segmentation of 3-D objects from 2-D images," *Proceedings of International Conference on Computer Vision*, pp. 121–128, Berlin, Germany. IEEE Computer Society Press.

[97] Ballard, D.H. and Rao, R.P.N. (1994), "Seeing behind occlusions," in J.O. Eklundh (Ed.), *Proceedings of European Conference on Computer Vision, Lecture Notes in Computer Science*, Vol. 800, pp. 274–285, Stockholm, Sweden. Springer-Verlag.

[98] Bossomaier, T.R.J. (1989), "Efficient image representation by Gabor functions – an information theory approach," in J.J. Kulikowski, C.M. Dickinson, and I.J. Murray (Eds.), *Seeing Contour and Colour*, pp. 698–704. Pergamon Press, Oxford, U.K.

[99] Petkov, N., Kruizinga, P., and Lourens, T. (1993), "Biologically motivated approach to face recognition," in J. Mira, J. Cabestany,

and A. Prieto (Eds.), *New Trends in Neural Computation, Proceedings of International Workshop on Artificial Neural Networks, Lecture Notes in Computer Science*, Vol. 686, pp. 68–77, Sitges, Spain. Springer-Verlag.

[100] Manjunath, B.S., Chellappa, R., and Von der Malsburg, C. (1992), "A feature based approach to face recognition," *Proceedings of IEEE Conference on Computer Vision & Pattern Recognition*, pp. 373–378, Champaign, IL. IEEE Computer Society Press.

[101] Würtz, R.P. (1994), *Multilayer Dynamic Link Networks for Establishing Image Point Correspondences and Visual Object Recognition*, PhD thesis, Bochum University, Germany.

[102] Konen, W. and Schulze-Krüger, E. (1995), "ZN-Face: a system for access control using automatic face recognition," in M. Bichsel (Ed.), *Proceedings of International Workshop on Automatic Face & Gesture Recognition*, pp. 18–23, Zurich, Switzerland. University of Zurich.

[103] Petkov, N. (1995), "Biologically motivated computationally intensive approaches to image pattern recognition," *Future Generation Computer Systems*, 11:451–465.

[104] Moses, Y., Adini, Y., and Ullman, S. (1994), "Face recognition: the problem of compensating for illumination changes," in J.O. Eklundh (Ed.), *Proceedings of European Conference on Computer Vision, Lecture Notes in Computer Science*, Vol. 800, pp. 286–296, Stockholm, Sweden. Springer-Verlag.

[105] Lades, M., Vorbruggen, J.C., Buhmann, J., Lange, J., Von der Malsburg, C., Würtz, R.P., and Konen, W. (1993), "Distortion invariant object recognition in the dynamic link architecture," *IEEE Transactions on Computers*, 42:300–311.

[106] Wiskott, L. and Von der Malsburg, C. (1996), *Face recognition by dynamic link matching*, Technical Report IR-INI 96-05, Institut für Neuroinformatik, Ruhr-Universität Bochum, Bochum, Germany.

[107] Wiskott, L., Fellous, J.M., Krüger, N., and Von der Malsburg, C. (1997), "Face recognition by elastic bunch graph matching," *IEEE Transactions on Pattern Analysis & Machine Intelligence*, 19:775–779.

[108] Maurer, T. and Von der Malsburg, C. (1995), "Learning feature transformations to recognize faces rotated in depth," in F. Fogelman-Soulié and P. Gallinari (Eds.), *Proceedings of International Conference on Artificial Neural Networks*, Vol. 1, pp. 353–358, Paris, France. EC2 & Cie.

[109] Maurer, T. and Von der Malsburg, C. (1995), "Single-view based recognition of faces rotated in depth," in M. Bichsel (Ed.), *Proceedings of International Workshop on Automatic Face & Gesture Recognition*, pp. 248–253, Zurich, Switzerland. University of Zurich.

[110] Krüger, N., Pötzsch, M., Maurer, T., and Rinne, M. (1996), "Estimation of face position and pose with labeled graphs," in R.B. Fisher and E. Trucco (Eds.), *Proceedings of British Machine Vision Conference*, pp. 735–743, Edinburgh. BMVA Press.

[111] Edelman, S. and Intrator, N. (1997), "Learning as extraction of low-dimensional representations," in D. Medin, R. Goldstone, and P. Schyns (Eds.), *Mechanisms of Perceptual Learning*. Academic Press, San Diego, CA. In press.

[112] Simard, P., Victorri, B., Le Cun, Y., and Denker, J. (1992), "Tangent prop - a formalism for specifying selected invariances in an adaptive network," in J.E. Moody, R.P. Lippman, and S.J. Hanson (Eds.), *Advances in Neural Information Processing Systems*, Vol. 4, pp. 895–903, San Mateo, CA. Morgan Kaufmann.

[113] Ezzat, T. and Poggio, T. (1996), "Facial analysis and synthesis using image-based models," *Proceedings of International Conference on Automatic Face & Gesture Recognition*, pp. 116–121, Killington, VT. IEEE Computer Society Press.

[114] Kohonen, T., Oja, E., and Lehtiö, P. (1981), "Storage and processing of information in distributed associative memory systems," in

G.E. Hinton and J.A. Anderson (Eds.), *Parallel Models of Associative Memory*, pp. 105–143. Lawrence Erlbaum Associates, Hillsdale, NJ.

[115] Kohonen, T. (1989), *Self-Organization and Associative Memory*, Springer-Verlag, Berlin, Germany, 3rd edition.

[116] Millward, R. and O'Toole, A. (1986), "Recognition memory transfer between spatial-frequency analysed faces," in H.D. Ellis, M.A. Jeeves, F. Newcombe, and A.W. Young (Eds.), *Aspects of Face Processing*, pp. 34–44. Nijhoff, Dordrecht, The Netherlands.

[117] Stonham, T.J. (1986), "Practical face recognition and verification with WISARD. in H.D. Ellis, M.A. Jeeves, F. Newcombe, and A.W. Young (Eds.), *Aspects of Face Processing*, Martinus Nijhoff, Dordrecht, The Netherlands.

[118] Lucas, S.M. (1997), "Face recognition with the continuous n-tuple classifier," in A.F. Clark (Ed.), *Proceedings of British Machine Vision Conference*, pp. 222–231, Colchester, U.K. BMVA Press.

[119] Fukushima, K. (1975), "Cognitron: A self-organizing multilayered neural network," *Biological Cybernetics*, 20:121–136.

[120] Fukushima, K. (1988), "Neocognitron: A hierarchical neural network capable of visual pattern recognition," *Neural Networks*, 1:119–130.

[121] Ting, C. and Chuang, K.-C. (1993), "An adaptive algorithm for Neocognitron to recognize analog images," *Neural Networks*, 6:285–299.

[122] Perrett, D.I., Mistlin, A.J., and Chitty, A.J. (1989), "Visual neurons responsive to faces," *Trends In Neurosciences*, 10:358–364.

[123] Rolls, E.T. (1994), "Brain mechanisms for invariant visual recognition and learning," *Behavioural Processes*, 33:113–138.

[124] Wallis, G. and Rolls, E.T. (1997), "A model of invariant face and object recognition in the visual system," *Progress in Neurobiology*, 51:167–194.

[125] Wallis, G., Rolls, E.T., and Földiák, P. (1993), "Learning invariant responses to the natural transformations of objects," *Proceedings of International Joint Conference on Neural Networks*, pp. 1087–1090.

[126] Rolls, E.T. (1995), "Learning mechanisms in the temporal lobe visual cortex," *Behavioural Brain Research*, 66:177–185.

[127] Moody, J. and Darken, C. (1988), "Learning with localized receptive fields," in D. Touretzky, G. Hinton, and T. Sejnowski (Eds.), *Proceedings of 1988 Connectionist Models Summer School*, pp. 133–143, Pittsburgh, PA. Morgan Kaufmann.

[128] Broomhead, D.S. and Lowe, D. (1988), "Multivariable functional interpolation and adaptive networks," *Complex Systems*, 2:321–355.

[129] Musavi, M.T., Ahmad, W., Chan, K.H., Faris, K.B., and Hummels, D.M. (1992), "On the training of radial basis function classifiers," *Neural Networks*, 5:595–603.

[130] Ahmad, S. and Tresp, V. (1993), "Some solutions to the missing feature problem in vision," in S.J. Hanson, J.D. Cowan, and C.L. Giles (Eds.), *Advances in Neural Information Processing Systems*, Vol. 5, pp. 393–400, San Mateo, CA. Morgan Kaufmann.

[131] Bishop, C.M. (1995), *Neural Networks for Pattern Recognition*, Oxford University Press, Oxford, U.K.

[132] Pomerleau, D.A. (1989), "ALVINN: An autonomous land vehicle in a neural network," in D.S. Touretzky (Ed.), *Advances in Neural Information Processing Systems*, Vol. 1, pp. 305–313, San Mateo, CA. Morgan Kaufmann.

[133] Rosenblum, M. and Davis, L.S. (1996), "An improved radial basis function network for autonomous road-following," *IEEE Transactions on Neural Networks*, 7:1111–1120.

[134] Girosi, F. (1992), "Some extensions of radial basis functions and their applications in artificial intelligence," *Computers & Mathematics with Applications*, 24(12):61–80.

[135] Girosi, F., Jones, M., and Poggio, T., "Regularization theory and neural networks architectures," *Neural Computation*, 7:219–269.

[136] Brunelli, R. and Poggio, T. (1992), "HyperBF networks for gender classification," *Proceedings of DARPA Image Understanding Workshop*, pp. 311–314, San Diego, CA.

[137] Moody, J. and Darken, C. (1989), "Fast learning in networks of locally-tuned processing units," *Neural Computation*, 1:281–294.

[138] Gutta, S., Huang, J., Singh, D., Shah, I., Takacs, B., and Wechsler, H. (1995), "Benchmark studies on face recognition," in M. Bichsel (Ed.), *Proceedings of International Workshop on Automatic Face & Gesture Recognition*, pp. 227–231, Zurich, Switzerland. University of Zurich.

[139] Gutta, S. and Wechsler, H. (1997), "Face recognition using hybrid classifiers," *Pattern Recognition*, 30:539–553.

[140] McClelland, J.L. and Rumelhart, D.E. (1981), "An interactive activation model of context effects in letter perception: Part 1. an account of basic findings," *Psychological Review*, 88:375–407.

[141] Rumelhart, D.E. and McClelland, J.L. (1982), "An interactive activation model of context effects in letter perception: Part 2. the contextual enhancement effect and some tests and extensions of the model," *Psychological Review*, 89:60–94.

[142] Burton, A.M., Bruce, V., and Johnston, R.A. (1990), "Understanding face recognition with an interactive activation model," *British Journal of Psychology*, 81:361–380.

[143] Burton, A.M. (1994), "Learning new faces an interactive activation and competition model," *Visual Cognition*, 1:313–348.

[144] Sung, K.-K. and Poggio, T. (1994), *Example-based learning for view-based human face detection*, Technical Report 1521/CBCL paper 112, AI Lab, MIT, Cambridge, MA.

[145] Field, D.J. (1987), "Relations between the statistics of natural images and the response properties of cortical cells," *Journal of the Optical Society of America A*, 4:2379–2394.

[146] Dong, D.W. and Atick, J.J. (1995), "Statistics of natural time-varying images," *Network: Computation in Neural Systems,* 6:345–358.

[147] Buxton, H. and Gong, S. (1995), "Visual surveillance in a dynamic and uncertain world," *Artificial Intelligence,* 78:431–459.

[148] Psarrou, A. and Buxton, H. (1993), "Hybrid architecture for understanding motion sequences," *Neurocomputing,* 5:221–241.

[149] Jordan, M.I. (1989), "Serial order: A parallel, distributed processing approach," in J.L. Elman and D.E. Rumelhart (Eds.), *Advances in Connectionist Theory: Speech.* Lawrence Erlbaum, Hillsdale, NJ.

[150] Elman, J. (1990), "Finding structure in time," *Cognitive Science,* 14:179–211.

[151] Mozer, M.C. (1994), "Neural net architectures for temporal sequence processing," in A.S. Weigend and N.A. Gershenfeld (Eds.), *Time Series Prediction: Predicting the Future and Understanding the Past,* pp. 243–264. Addison-Wesley, Redwood City, CA.

[152] Hertz, J.A., Krogh, A., and Palmer, R.G. (1991), *Introduction to the Theory of Neural Computation,* Addison-Wesley, Redwood City CA.

[153] Waibel, A., Hanazawa, T., Hinton, G., Shikano, K., and Lang, K. (1989), "Phoneme recognition using time-delay neural networks," *IEEE Transactions on Acoustics, Speech, & Signal Processing,* 37:328–339.

[154] Le Cun, Y., Boser, B., Denker, J.S., Henderson, D., Howard, R.E., Hubbard, W., and Jackel, L.D. (1989), "Backpropagation applied to handwritten zip code recognition," *Neural Computation,* 1:541–551.

[155] Lawrence, S., Giles, C.L., Tsoi, A.C., and Back, A.D. (1997), "Face recognition: A convolutional neural network approach," *IEEE Transactions on Neural Networks,* 8:98–113.

[156] Berthold, M.R. (1994), "A Time Delay radial basis function network for phoneme recognition," *Proceedings of IEEE International Conference on Neural Networks*, Vol. 7, pp. 4470–4473, Orlando, FL. IEEE Computer Society Press.

[157] Rosenblum, M., Yacoob, Y., and Davis, L.S. (1996), "Human emotion recognition from motion using a radial basis function network architecture," *IEEE Transactions on Neural Networks*, 7:1121–1138.

[158] Stryker, M.P. (1991), "Temporal associations," *Nature*, 354:108–109.

[159] Bartlett, M.S. and Sejnowski, T.J. (1996), "Unsupervised learning of invariant representations of faces through temporal association," in J.M. Bower (Ed.), *Computational Neuroscience: Trends in Research 1995*, pp. 317–322. Academic Press, San Diego, CA.

[160] Bartlett, M.S. and Sejnowski, T.J. (1997), "Viewpoint invariant face recognition using independent component analysis and attractor networks," in M. Mozer, M. Jordan, and T. Petsche (Eds.), *Advances in Neural Information Processing Systems*, Vol. 9, Cambridge, MA. MIT Press.

[161] Földiák, P. (1991), "Learning invariance from transformation sequences," *Neural Computation*, 3:194–200.

[162] Perrett, D.I. and Oram, M.W. (1993), "Neurophysiology of shape processing," *Image & Vision Computing*, 11:317–333.

[163] Howell, A.J. and Buxton, H. (1995), "Invariance in radial basis function neural networks in human face classification," *Neural Processing Letters*, 2(3):26–30.

[164] Samaria, F.S. (1994), *Face Recognition using Hidden Markov Models*, PhD thesis, Cambridge University, U.K.

[165] Lin, S.-H., Kung, S.-Y., and Lin, L.-J. (1997), "Face recognition/detection by probabilistic decision-based neural network," *IEEE Transactions on Neural Networks*, 8:114–132.

[166] Howell, A.J. (1997), *Automatic face recognition using radial basis function networks*, PhD thesis, University of Sussex, U.K.

Chapter 8:

Neural Networks for Face Recognition

NEURAL NETWORKS FOR FACE RECOGNITION

A.S. Pandya
Computer Science and Computer Engineering Department
Florida Atlantic University
P.O. Box 3091, Boca Raton, FL 33431-0991
U.S.A.

R.R. Szabo
School of Computer and Information Sciences
Nova Southeastern University
3100 S. W. 9th Avenue, Fort Lauderdale, FL 33315
U.S.A.

This chapter focuses on a neural network approach for pattern recognition. Section 2 contains an overview of the different attempts to apply ANN techniques to a variety of face recognition tasks. It also describes critical elements of a typical face recognition system. Design of a neural network based transformation-invarient face recognition system is presented as an example in Section 3. This example describes the use of a higher-order neural network based pattern classifier for the classification of gradient (edge-extracted) facial images.

1 Introduction

Pattern recognition typically deals with a search for structure (regularity) in data. In 1973, Duda and Hart characterized it as "a field concerned with machine recognition of meaningful regularities in noisy or complex environment" [9]. The field of pattern recognition encompasses a wide range of information processing problems, from speech recognition and classification of handwritten characters, to fault detection in machinery and medical diagnosis. Humans solve these problems in a seemingly effortless fashion. However, the solution using computers has, in many cases, proven to be very difficult.

In general, a pattern can be referred to as a quantitative or structural measure of an object. A set of patterns that share some common properties can be regarded as a pattern class. A pattern-recognition system performs a classification function on its input. Input represents a set of measurements called the pattern vector. For each input vector the system must decide the class to which it belongs. Object classification is closely related to recognition. Patterson views it as, "the process of establishing a close match between some new stimulus and previously stored stimulus pattern" [42].

The ability to classify or group objects according to some commonly shared features is a form of object recognition. The subject matter of pattern recognition by machine deals with techniques for assigning patterns to their respective classes, automatically and with as little human interaction as possible. For example, the machine, automatically sorting mail based on 5-digit zip at the post office is required to recognize numerals. In this case there are ten pattern classes, one for each of the 10 digits. The function of the machine is to identify geometric patterns as being a member of one of the available classes.

Broadly speaking, pattern recognition involves the partitioning or assignments of measurements into meaningful categories by extracting significant attributes of the input data from the background of (seemingly unrelated) irrelevant details. For example, speech recognition system maps a waveform into words. When viewed as a decision-making systems, pattern recognition includes applications, such as medical imaging, risk assessment in financial services, signature verification, face and fingerprint recognition from pixel-maps, and detection of explosive and hostile threats (military applications).

1.1 Pattern Recognition in Practice

1.1.1 Approaches Used in Pattern Recognition

The major approaches for designing pattern recognition systems are (1) statistical, (2) syntactic or structural, and (3) artificial neural networks.

The most general framework in which to formulate solutions to pattern recognition problems is a statistical approach [5]. Statistical methods developed during the last 30 years offer many powerful techniques in handling complex pattern-recognition problems. They rely on the discrimination of an appropriate combination of feature values that provide measures for discriminating between classes. However, in some cases, the features are not important in themselves. Rather the critical information regarding pattern class, or pattern attributes, is contained in the structural relationship among the features. Applications involving recognition of pictorial patterns (which are recognizable by shapes) such as character recognition, chromosome identification, and elementary particle collision photographs fall into this category.

Classic books available on this subject are [9], [12], [58], [62], whereas [2], [3], [14], [33], [39], and [41] are an excellent source of the most relevant neural network techniques from the perspective of practical applications. In applications involving patterns that can be represented meaningfully by using vector notations the statistical pattern recognition approach is ideal. However, this approach lacks a suitable formalism for handling pattern structures and their relationships. In the syntactic approach [15] the patterns are represented in a hierarchical fashion, and are viewed as being composed of sub-patterns. A set of rules governing the syntax is called a grammar, and used for the generation of strings (sentences) from the given hierarchy. For multi-class pattern recognition at least one grammar per class needs to be identified. The subject of syntactic pattern recognition deals with this type of pattern recognition problems, since it possesses the structure-handling capability lacked by the statistical pattern recognition approach. Many of the techniques in this field draw from earlier work in mathematical linguistics and results of research in computer languages. A large body of literature exists in this field, such as [12], [16], [62].

Despite the existence of a number of good statistical, syntactic (grammar-based), and graphical approaches to pattern recognition, we limit the scope of this chapter to the applications based on neural networks. Although, it should not be overlooked that neural network recognizers may and have been used in combination with other types of recognition engines, or optimization techniques [4], and [25]. Those techniques may include fuzzy logic, genetic algorithm, and rough sets.

1.1.2 Neural Network Approach

Neural networks have been applied as pattern classifiers in a variety of fields including signal processing [17], speech recognition [21], image recognition [49], character recognition [39] and dynamic (time-varying) systems [40]. This chapter begins by examining the basic issues that arise in the applications of neural networks for pattern recognition.

Several characteristics of neural network technology set them apart from conventional computing and artificial intelligence approaches. Unlike a traditional computer, neural networks do not execute a series of instructions, but, rather, respond to the variety of inputs presented to them. A conventional artificial intelligence's techniques (predicate logic and symbol manipulation) used to model a human reasoning in some narrow areas of expertise if explicit knowledge is available. Therefore, they are not the right tools to deal with the real world problems: imprecise, uncertain, and incomplete information [25].

A neural network can be described as an information processing system composed of a large number of interconnected processing elements [10], [39], and [50]. Each processing element (also called a node, P.E., cell, neuron) calculates its activity locally on the basis of the activities of the cells to which it is connected. The strengths of its connections (also called weights) are changed according to some transfer function that explicitly determines the cell's output, given its input. The connectivity between the processing elements (network architecture), the transfer function, and the learning algorithm determine the functionality and the performance of the neural network system.

The ability of neural networks to learn from their experience is the key element in the problem solving strategy of a pattern-recognition task. Differences in neural network models are mostly due to the number of processing elements and their interconnections, and learning laws. In other words, artificial neural networks are cellular systems which can acquire, store, and utilize experiential knowledge. These structural differences identify learning paradigms by which neural networks are classified. Many networks are good at recognizing patterns; some perform better as classifier builders, and others are good for feature extraction.

The main advantages neural networks have over other technologies can be listed as follows: (1) no need for a prior knowledge on statistical distribution of data, (2) inherent parallelism, (3) fault tolerance, (4) adaptability, and (5) fast classification. Their disadvantages include: (1) a lack of explanation of a neural network learning process, (2) time-consuming training, and (3) lack of the uniformity in setting parameters of the chosen paradigm.

A comprehensive study on recent neural network methods and techniques used in pattern recognition field can be found in [7], [8], [39]. These books describe both classical and recent contributions towards understanding autonomous patterns, vision, speech and languages, cognitive information processing, recognition learning, reinforcement learning, associate mapping, adaptive timing, adaptive sensory-motor control, and the self-organization of temporally organized plans.

2 Face Recognition

2.1 Introduction

Face recognition has been studied for many years and has practical applications in areas such as security systems, identification of criminals and assistance with speech recognition systems. Face recognition is important to humans because the face plays a major role in social intercourse, conveying emotions and feelings. Humans are very adept at recognizing faces and can do so with ease even under a range of different physical conditions [20]. However, developing an artificial system to mimic the human ability has proven to be a very difficult and computationally complex task [11].

There have been numerous studies exploring various concepts and problems in the face recognition process and many efforts in designing human face recognition systems [3], [6], and [51]. Some of the systems have employed artificial neural networks [19], [34], [36], [44], [51],

[52], while the others, a variety of approaches [13], [27], [35], [63], such as:

(i) template matching of isodensity lines of subjects' faces;
(ii) comparison of sizes/relative distances of facial features (nose, eyes, mouth) of subjects' facial images.

However, the performance of a majority of proposed schemes has been (generally) sensitive to 2-D coordinate transformations of the image (e.g. scaling, translation) [22]. In this chapter we will focus on a system that has the ability to exploit the three dimensional nature of the human face.

2.2 A Typical Face Recognition System

2.2.1 Design Process

In general, the design process of a face recognition system is a multi-staged process. These stages are essentially the same steps as performed by humans in order to identify or classify objects. According to Patterson [42], and Fischler and Firschein [11], the major steps of face-recognition process are (refer to Figure 1):

Step 1. Sensory phase
Studies are conducted in order to identify stimuli produced by object attributes (size, shape, color, texture, etc.). The values of those attributes and their relations are used to characterize an object in the form of a pattern vector. The range of characteristic attribute values is known as the measurement space. For current technology a set of values for sampling and digitizing an image is typically a 256×256 grid with 256 possible intensity levels at each such a grid point. An important issue is the fidelity with which the actual scene appearance is captured by the pattern vector. While converting analog signal into digital, a sufficient number of levels of quantization needs to be determined, in order to insure that the introduced noise is below the desired level (threshold), and no fundamental loss of information has occurred.

Step 2. Preprocessing phase
A subset of attributes whose values provide cohesive object grouping or clustering, consistent with a given task are selected. Preprocessing

Neural Networks for Face Recognition

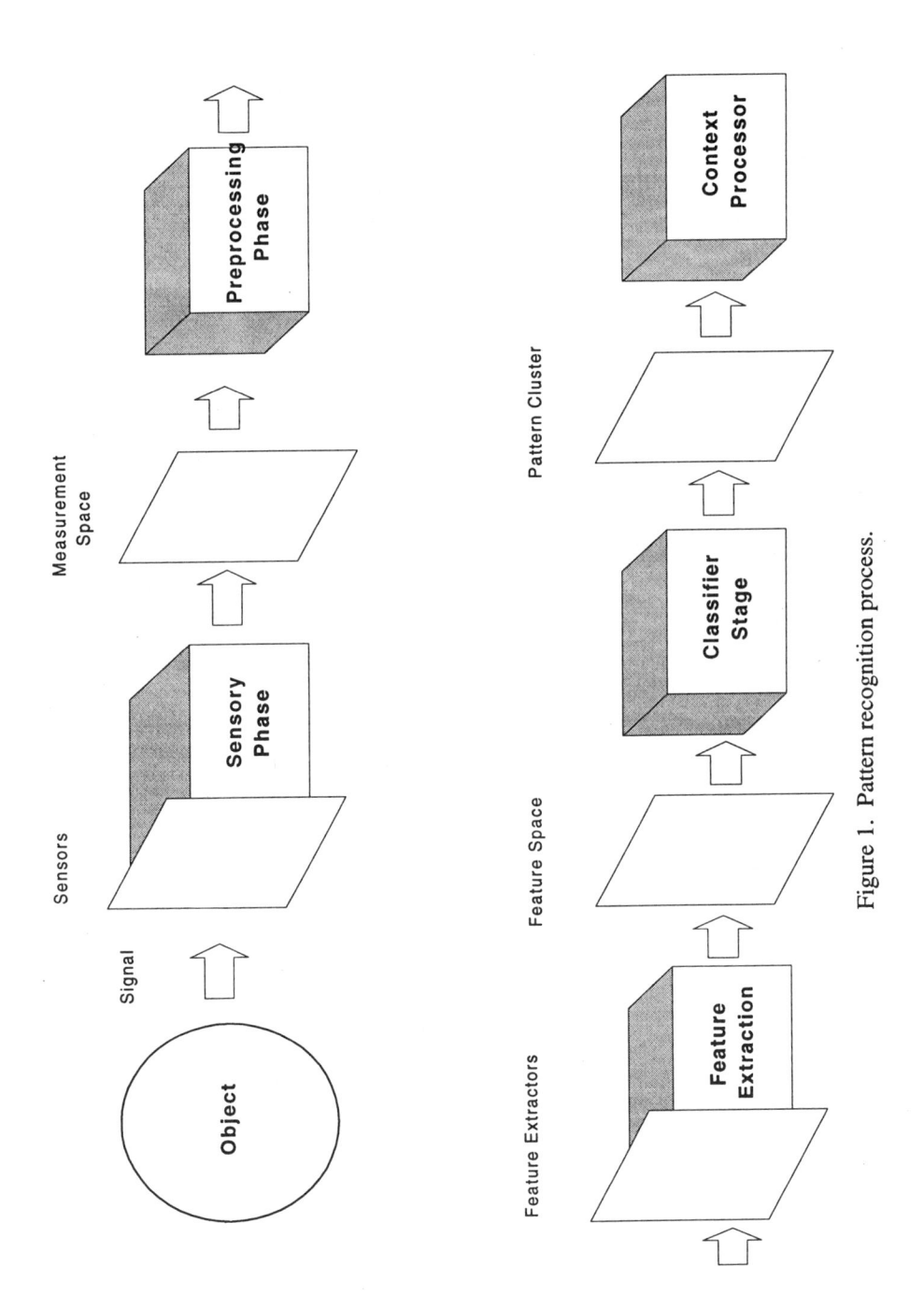

Figure 1. Pattern recognition process.

stage is indented to remove noise, enhance certain aspect of image, and induce other changes that will simplify the higher level processing steps. In order to avoid the elimination of existing edges or introduce false edges, typical operations are utilized: thresholding and smoothing. Thresholding transforms a gray-level image into a binary image in which each pixel is either black or white. Image smoothing is utilized to reduce noise, to enhance selected image features, and to degrade unwanted image details. Therefore, the image is partitioned into isolated objects. The range of the subset of attribute values is known as the feature space.

Step 3. Feature extraction

Features of attributes of individual patterns are refined further to facilitate recognition. This refinement process may take into account the following attributes: color, texture, intensity, distance, motion, etc.

Step 4. Classifier stage

Using the selected attribute values, object or class characterization models are learned by forming generalized prototype descriptions, classification rules, or decision functions. The model identifies the category to which the pattern belongs by identifying the attributes associated with the pattern.

Step 5. Context processor

The context processor increases the recognition accuracy by providing relevant information regarding the environment surrounding the object (by matching of object features with the stored models, or by further classification based on additional criteria).

2.2.2 Experimental Design Issues

Success of a face recognition application often depends on how information (signal) is presented to the neural network. Depending on the application, appropriate preprocessing should be done in order to produce a signal representation that accentuates information most relevant to the given application. In the context of neural network classifiers two constraints are noteworthy:

- Classifier dimension (number of degrees of freedom) which is associated with the number of connections in a neural network.
- Number of training set samples (size of data set).

In general, the data set size should exceed the number of weights by an order of magnitude. Increasing the number of input neurons and the number of training samples helps the generalization capability of the neural network. Practically, however, some compromises are required since increasing the network size (number of neurons, layers, and weights) beyond certain limits adversely affects the network's convergence behavior.

Thus, in applications involving image processing, some decision must be made to reduce the size of the pattern vector to be presented to the neural network. In applications involving time series data (i.e. speech recognition, sonar signal processing, forecasting), frequently, transformation of time series to a spectral domain is performed. The dimension of input pattern can be reduced using a relatively small number of spectral bins. Pandya and Macy [39] reported on an application where the neural network trained more quickly when it was presented with the power spectrum as the input. However, the neural network performed better as a classifier when it was trained with raw data. This is a classic example of the trade-off between neural network size and pattern vector size. The power spectrum may eliminate any signal phase information, which may carry important classification information.

Wavelet's transform [48] is another popular technique used in preprocessing, where a transient signal is modeled in terms of a finite number of scaled basis functions:

$$X(t) = \sum_j \sum_k C_{jk} h_{jk}(t) \tag{1}$$

where, $X(t)$ is the original time series, C_{jk} are the wavelet coefficients, $h_{jk}(t)$ are the wavelet kernels, index j is the temporal position of the kernel, and index k is the scale of the kernel. After transformation the wavelet coefficients C_{jk} are used as inputs to a neural network for training.

Fitting the signal properties to match the neural network's dynamic range is another important step in designing a face recognition classifier. The relationship between the maximum and minimum values of a paradigm's nonlinearity is identified by its threshold function. This range limitation coupled which the dynamics of the algorithm identifies the efficiency of the neural network classifier. As a result, input patterns must be normalized accordingly in order to fully utilize this dynamic range.

Generally, the neural networks provide an adaptive mechanism for combining a set of nonlinearities in order to approximate a given non-linear transformation [17]. The majority of applications in face recognition have been implemented using first-order neural networks (traditional multi-layer perceptron). While in training, the optimization algorithms of this category of neural networks attempt to reduce the training time and eliminate local minima.

The nonlinear kernel classifiers, which are interpolative versions of the nearest-neighbor classifiers, constitute the second most popular category of neural networks in face recognition. They require little training, but most contend with balancing the number of basic functions with memory and classification speed constraints.

2.3 Various Face Recognition Systems

Face recognition process can be decomposed into two major tasks:

1. finding a face (or faces) in an image; and
2. recognizing the identity of that face.

Task 1. Finding a face: Finding a face in an image is also known as face registration or face localization. Figure 2 shows that the degree of difficulty of this task depends on several factors, such as:

a. the control one has over lighting and background conditions,
b. whether the images are color or monochrome, and whether the images are still or video.

If the background lighting can be controlled, then it might be possible to extract the head very simply as one of the main sources of brightness in the image. Hence, depending on the control, one has over these factors, face localization can be a hard or an easy problem. In Section 3 of this chapter we discuss an approach which normalizes images to provide invariance against background illumination.

Task 2. Face recognition and verification: There are two versions of the face recognition problem: the recognition and the verification problem. In the case of verification, one only has to test the likelihood that the face is that of who it claims to be, hence, this involves testing the quality of match of an image against a single model. In the case of face recognition, the problem is to find the best match of an unknown image against a database of face models or to determine whether it does not match any of them well. The practical importance of this distinction is the speed required; generally, if there are N subjects (i.e., people) in the database, then the recognition process will be N times slower than the verification process. This may place practical limits on the algorithm used.

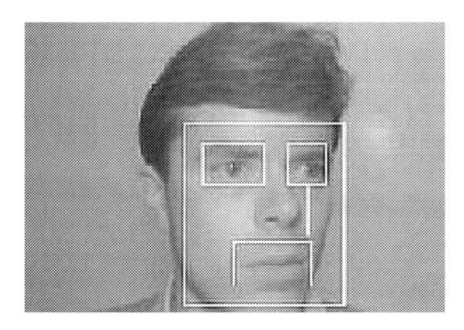

Figure 2. Face Registration and Feature Identification.

Lucas [32] describes a novel approach called the continuous n-tuple classifier: a type of n-tuple classifier that is ideally suited to problems where the input is continuous or multi-level rather than binary. Results on a widely used face database show the continuous n-tuple classifier to be as accurate as any method reported in the literature, while having the advantages of speed and simplicity over other methods.

The Olivetti Research Laboratory (ORL) in U.K. has developed a facial database which can serve as a good benchmark for testing a face recognition system for consistent performance. A large number of previous methods for face recognition reported in the literature have used the Olivetti Research Laboratories face-database. This facial database (refer to Figure 3) is available at: http://www.cam-orl.co.uk/facesataglance.html

298 Intelligent Biometric Techniques in Fingerprint and Face Recognition

The FERET program, is one of the major efforts on face recognition in the USA. FERET is a three- year, $900K program sponsored by ARPA. The program is now in its second year. ARPA is interested in application of face recognition technologies to surveillance databases for both immigration and drug interdiction programs. In September 1996, the final round of FERET tests was administered by the US Army Research Laboratory which consisted of a large gallery test containing nearly 3,800 frontal images. For more information on the FERET database and the results of the competition contact Jonathan Phillips at email:jphillip@nvl.army.mil.

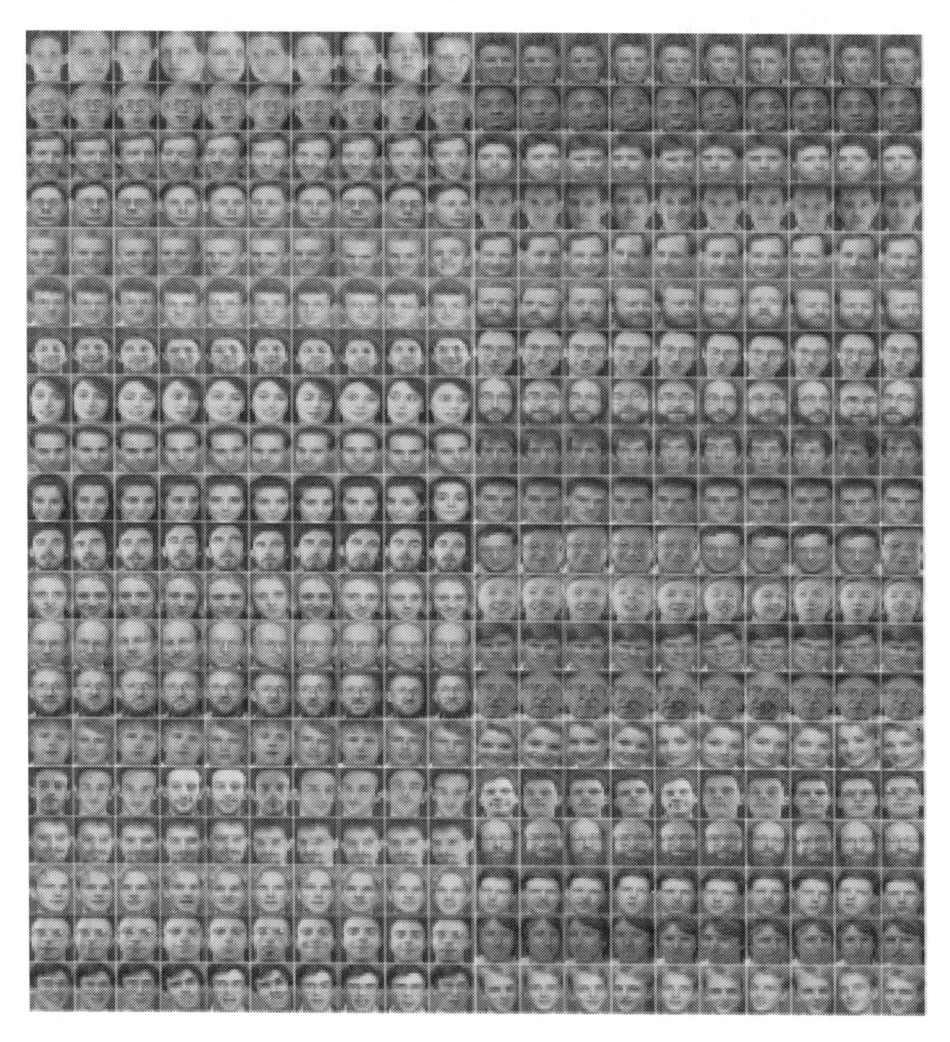

Figure 3. A database for testing face recognition systems.

Most of the work in the field of face recognition is focused on 2-D images, like the ones shown in Figure 3. However it is important to take into consideration the 3-D nature of the human face. Several researchers have developed paradigms that deal with 3-D images. Harashima and his colleagues [1] have created a system capable of animating a face using text or speech as input data. The system which uses a neural network for voice to image conversion has several application including:

1. an intelligent human-machine interface;
2. an intelligent communication system;
3. an automatic animation production system; and
4. a simple computer interface for the handicapped.

Their ultimate aim is to design a system based on visual input. The system could become an intelligent user interface that could read the user's face and interpret the speech, rather than using a keyboard or mouse. The screen displays a synthesized face image display and has synthesized speech to talk to the user. Thus the system provides a much friendlier and more natural interface. The system uses a 3-D wire frame model and maps a 2-D texture onto it. Points of importance on the 3-D face are matched up to corresponding points on the texture map using an affine-transformation. As shown in Figure 4, the authors set up 17 phoneme positions for the face. The model also includes teeth, and the movement of the teeth followed directly from the jaw movement.

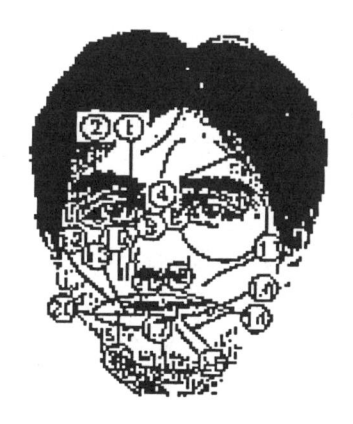

Figure 4. A face with various phonemes mapped on it, including the teeth and jaws.

Two methods of voice to image conversion are used. The first is vector quantization, and the other is synthesis using a neural network. The output of each of these converters becomes the input to the image synthesis system. Four sub-processes are involved in the image display function:

1. facial movement calculation using eight parameters;
2. transformation of the wire-frame model;
3. texture mapping for each polygon; and,
4. output to the screen.

Radiologists at Washington University [18] have developed techniques for 3-D surface imaging to obtain the image description of a human face. The digitized 3-D data can be used in a variety of medical applications. Various techniques have been developed using Image Processing, Computer Graphics, Computer Aided Design and Engineering and other Software Packages to refine and manipulate the facial data. Figure 5 illustrates several images that were created with this Surface Imaging System designed to digitize human head forms.

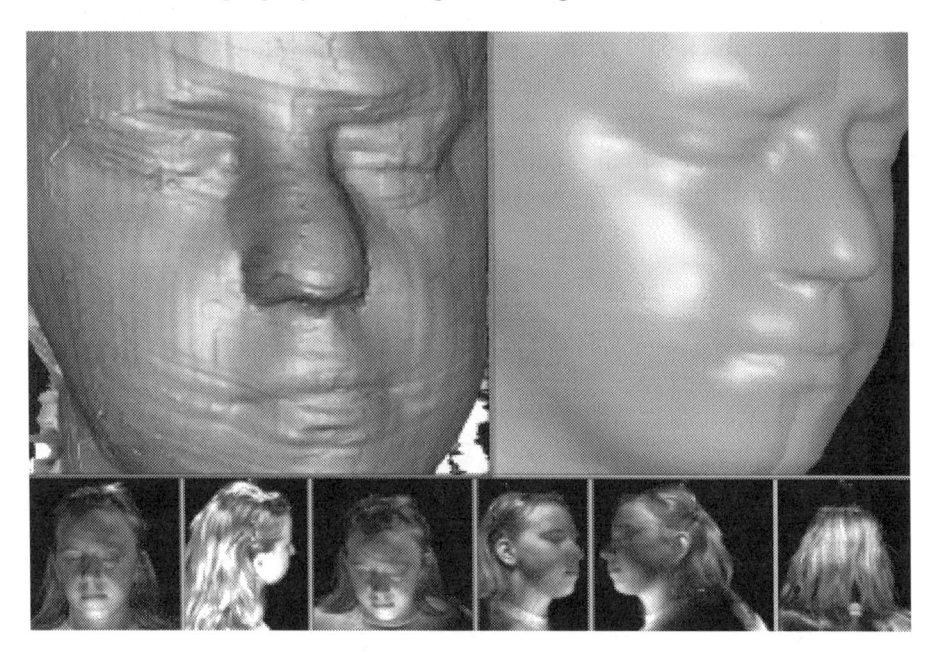

Figure 5. Images created by Bhatia and Vannier using the Surface Imaging System.

3 Example: Transformation-Invariant Face Recognition

3.1 Introduction

In face recognition an input image must be recognized regardless of its position, size, and angular orientation. Therefore, pattern recognition requires the nonlinear subdivision of the pattern space into subsets representing the objects to be identified. From the perceptron family, only multi-layer first-order networks and higher-order networks can perform such a discriminatory classification.

A key property of neural networks is their ability to recognize invariances and extract essential parameters from complex high-dimensional data. The most significant advantage of the higher-order neural networks (HONN) over first order networks is that such invariances to geometric transformations can be incorporated into the network and need not be learned through iterative weight updates. The weights of the higher-order net are configured in such a way as to provide position-, scale-, and rotation-invariant pattern recognition and then trained on binary gradient images. In many image processing applications, the use of gradient images yields superior neural network performance in certain cases over the use of ordinary (full-bodied) images.

In this section we discuss an application involving face recognition. This example describes the use of a higher-order neural net (HONN) based pattern classifier for the classification of gradient (edge-extracted) facial images. Concepts of coarse coding and isodensity regions are also discussed since such preprocessing often helps in reducing the size of the neural network.

The use of higher-order networks for invariant pattern recognition has been explored and applied to the classification of simple 2-D binary images by several researchers [53], and [61]. Higher-order nets can be designed to be invariant to a 2-D coordinate transformation of images by adjusting their weights accordingly. A second-order version of such a network can be made insensitive to translation and scale distortions. A third-order network can be used to perform translation, scale and rotation-invariant object recognition with a significant reduction in training time over other neural network paradigms, such as back propagation. Spirkovska and Reid [54] were able to achieve a 100% recognition rate for binary, edge-only aircraft images; their test suite comprised of objects translated, scaled down to 38% of the original size, and rotated through arbitrary angles.

In order to achieve desired invariance to these three transformations, appropriate equivalence classes need to be constructed using *relative* coordinates of the pixels involved. Thus, the architecture is constrained in such a way that all combinations of three pixels (triples) that define similar triangles are connected to the output with the same weights. The internal angles of these triangles remain unchanged under translation, rotation and scale transformations.

3.2 Edge Extraction

There are three basic different types of discontinuities in a digital image: points, lines, and edges. A common way to look for these discontinuities is to run a mask through the image. This involves computing the sum of products between the mask coefficients and the intensities of the pixels under the mask at a specific location of the image (refer to Figure 6). The mask is then moved to the next pixel location and the process is repeated. This continues until all pixel locations have been covered. An edge is the boundary between two regions with relatively distinct gray-level properties. Most edge-detection techniques are based on the computation of a local derivative operator: the first derivative at any point in an image is obtained by using the magnitude of the gradient at that point.

z_1	z_2	z_3
z_4	z_5	z_6
z_7	z_8	z_9

Figure 6. Sobel mask.

It is well established, that for an image $f(x, y)$, the gradient vector points in the direction of maximum rate of a change of f at (x, y) [16]. The gradient of the image at the location (x, y) is given by the vector:

$$\nabla f = \begin{bmatrix} Gx \\ Gy \end{bmatrix} = \begin{bmatrix} \partial f / \partial x \\ \partial f / \partial x \end{bmatrix} \quad (2)$$

The magnitude of this vector, ∇f, is given by:

$$\text{Magnitude } (\nabla f) = [G_x^2 + G_y^2]^{1/2} \quad (3)$$

The gradient can then be approximated by using absolute values:

$$\nabla f \approx |G_x| + |G_y| \quad (4)$$

Computation of the gradient of an image is based on obtaining the partial derivatives $\partial f/\partial x$ and $\partial f/\partial x$ at every pixel location (also called Sobel operators). Derivatives may be implemented in digital form in several ways using various derivatives – approximating operators. However, the Sobel operators have the advantage of providing both a differentiating and a smoothing. Because derivatives enhance noise, the smoothing effect is a particularly attractive feature of the Sobel operators. The Sobel mask is a 3×3-pixel structure (refer to Figure 6), and the operator derivatives are given by:

$G_x = (z_7 + 2z_8 + z_9) - (z_1 + 2z_2 + z_3),$
$G_y = (x_3 + 2z_6 + z_9) - (z_1 + 2z_4 + z_7),$

Where the z_s represent gray-level values on a 0-255 scale.

The corresponding masks are shown in Figures 7 and 8.

-1	-2	-1
0	0	0
1	2	1

-1	0	1
-2	0	2
-1	0	1

Figure 7. X-axis.   Figure 8. Y-axis.

3.3 Isodensity regions

Many schemes for face recognition have been proposed and implemented by researchers, however performance of most of the proposed schemes have generally been sensitive to 2-D coordinate transformations of the image (rotation, scaling, translation). Various approaches to invariant 2-D pattern recognition have been implemented, such as use of Fourier descriptors [30], [45], [47] auto-regressive models [26], circular harmonic expansion [55], and moment invariants [24], [28], [31], [43], [46], [56], [57].

The human face is a 3-D surface; therefore, it seems reasonable to base the face recognition algorithm on its three-dimensional characteristics. The perceived light intensity when viewing a small area of an illuminated 3-D object depends on various factors including shape and depth of the surface of the object [23], [29]. A 3-D structure of each face in the database was captured by identifying and isolating regions of constant intensity (also called isodensity regions) [35], and then by transforming them to binary images. The binary images of the isodensity regions were later processed and used as the training set for an artificial neural network recognition system. The use of binary-valued data reduces storage requirements. In the case of a human face, isodensity regions do not provide an exact 3-D reconstruction of the image because the reflection coefficient is not constant over the whole face. This phenomenon (fact) is due to the variations in color between different regions of the face (eyes, eyebrows, skin, lips, pupils, and stubble), and the presence of perspiration on the skin. However, a reasonable amount of 3-D information is still present in isodensity regions thereby justifying this approach.

An isodensity image is a binary (monochrome) image created from the original gray-scale image that displays the desired intensity levels in white (gray-scale level 255) and all other pixels in black (gray-scale level 0). A major drawback in using isodensity information is the fact that this method is very sensitive to lighting conditions for the facial image. A simplistic isodensity analysis is very susceptible to changes in lighting conditions because the decision boundaries (thresholds) are fixed at certain intervals. Therefore, for an isodensity method to be viable, it needs to be impervious to such disturbances. The 8-bit gray-scale range (000 to 255) was arbitrarily divided up into four regions by

3.4 Coarse Coding

A serious limitation of high-order nets is the rapid increase in the number of weights required with the input image size. The number of weights increases factorially with the image dimensions. For example, in an 8×8-pixel image, the number of possible triplets is 64 over 3 or 41,664. Quadrupling the image area to a 16×16-pixel image yields 2,763,520 triplets, an increase by a factor 66. In order to obtain images with a reasonable amount of detail 128×128 images have been used. In this case, the number of interconnections without the use of encoding, would have become prohibitively large (7.3×10^{11}). Coarse coding [53], [54] is a technique whereby such a high-resolution image is transformed into a series of superimposed lower resolution images. For example, a 128×128 image can be represented by a series of eight 16×16 coarse images overlaid on each other with an offset equivalent to the dimensions of a pixel in the original high resolution image (refer to Figure 9) without any loss of information. This transformation reduces the problem to a manageable size since we now only require a 64-input neural network to process the 16×16 images as opposed to the 16,384-input (128×128) net needed for an unencoded image. Of course, the number of training samples increases from 1 to 8, but such an increase in training time is tolerable in this application.

Assume a binary 2-D image $f(x,y)$, composed of binary-valued pixels is to be coarse-coded. The following condition describes the relationship between the image and its nth coarse-coded field $g_n(a,b)$:

$$\text{IF } f(x,y) = 1 \text{ THEN } g_n((x + \Delta_n)/S_x, (y + \Delta_n)/S_y) = 1, \quad (5)$$

where Δ_n = offset (in number of pixels) of nth course image origin from $f(x,y)$ origin reference, and S_x, S_y = 2-D scaling factor that relates coarse image dimensions to original image dimensions.

A coarse pixel is activated if a corresponding fine pixel lies within its boundaries, thus each original pixel is represented by a unique set of coarse pixels. In the case of Figure 9, since $f(3,6)$ is activated and belongs to both $g_1(3,2)$ and $g_2(1,3)$, they too are both activated.

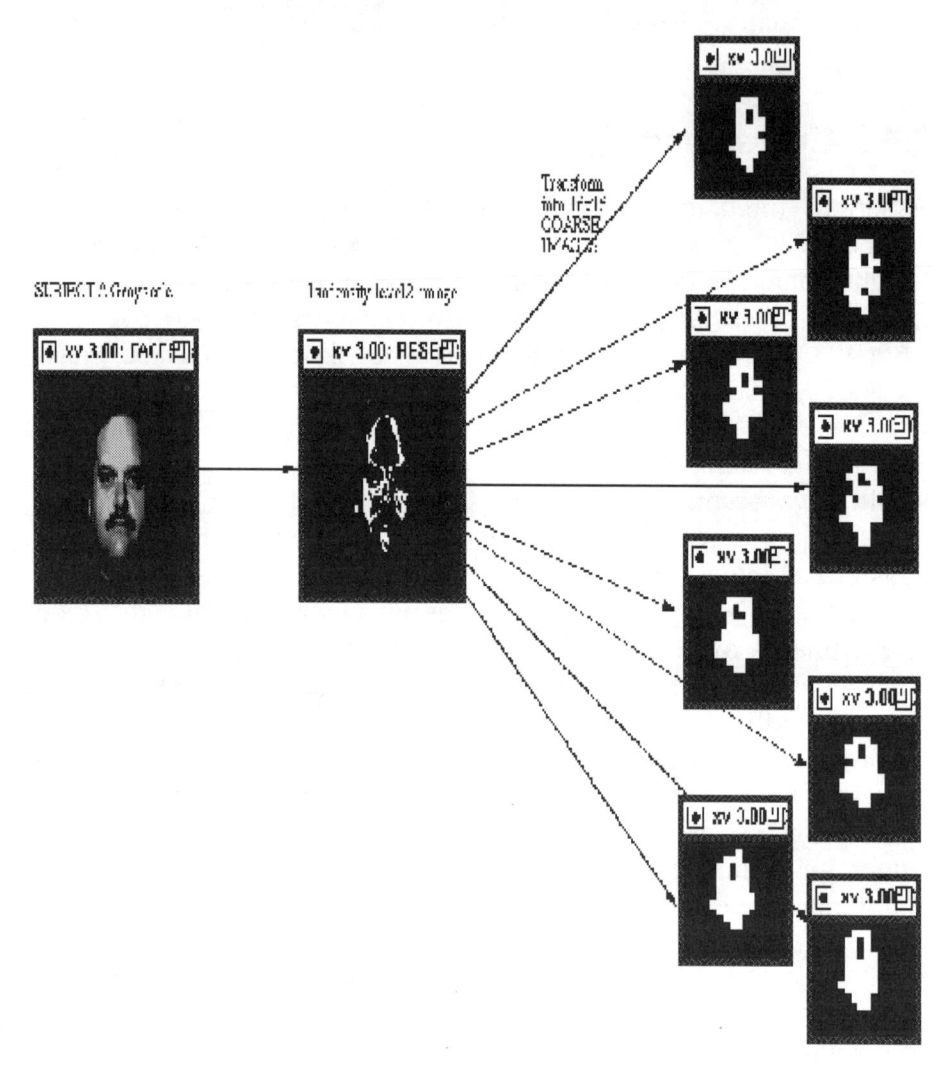

Figure 9. 128×128 Image represented by a series of coarse 16×16 images.

3.5 Results

Each image used was subjected to a 4-level isodensity decomposition whereby four images were generated per gray-scale image. Each of the images was coarse-coded then to produce a set of eight coarse images per isodensity image. The neural network was then trained on two sets of coarse images at a time: one set per isodensity image per subject. The training was performed using various pairs of facial images using both gradient coarse images and ordinary coarse images and the results compared. The test suite for each pair of training images comprised of the training subjects under different facial expressions subjected to a combination of geometric transformations (rotation, scale and translation).

Figure 9 shows the base set of the gray-scale facial images used (with various expressions and geometric transformations). Each gray-scale image was broken down into four isodensity images by applying four isodensity thresholds. A classification was regarded as "correct" when the corresponding images of at least 3 of these 4 levels classified correctly. Cases of network non-convergence were discarded and were not regarded as either correct or incorrect classification so as not to affect the overall network classification decision. The network parameters were set as follows: learning rate = 2.0; threshold = 0.01.

In-plane rotation-, scale- and translation-invariant pattern recognition has been achieved using human faces as image data on a third-order neural network. Invariance to facial expressions was also achieved as the test facial images had a range of different facial expressions. The system designed is to be trained on a pair of facial images at a time and tested with a single image at a time. It must then categorize the test face into one of the two available templates.

Out of 23 facial image pairs that have been trained and classified correctly under the gradient scheme, 8 have been classified correctly under the ordinary (non-gradient) scheme [60]. The success rate for classification of facial images of the system can be increased in several ways, one of which is by increasing the size of the coarse-coded images thereby allowing a more faithful reproduction of the original images. It has been reported [55] that a high-order net can achieve 100% correct classification for simple geometric patterns. The success was partially due to the obvious differences between the training patterns: the two shapes considered were very different geometrically.

4 Summary

In the case of human faces, it becomes more difficult to make a distinction between subjects' faces as all faces have many very similar features (characteristics). For example, the shape of the face is generally oval, there are two eyes, a nose, a mouth etc., and for each individual, these occupy the same relative positions on the oval face. Therefore, as the image resolution is reduced, such as is the case in coarse-coding, many of the salient features that could be used to distinguish the target patterns are lost as they coalesce into larger pixels. Increasing the coarse image dimensions will alleviate this problem but more memory will be required to store the new weight values. As discussed in [60], [61] the space required increases combinatorially with the number of image pixels and may pose a problem for the computer system.

The transformation of images to their gradient equivalents prior to network training results in superior performance. The superior response is expected in the case of scaled images due to the fact that when an image is scaled, the image surface area (i.e., the number of pixels) is altered. Use of a gradient image minimizes this change in the number of pixels since only the contour pixels are involved in the transformation.

References

[1] Aizawa, K., Harashima, H., and Saito, T. (1989), "Model-Based Analysis; Synthesis Coding System for a Person's Face," *Signal Processing: Image Communication*, Vol. 1, pp.139-152.

[2] Anderson, J.A. and Rosenfield, E. (1986), *Neurocomputing, Foundation of Research*, The MIT Press, Cambridge MA.

[3] Baron, R.J. (1981), "Mechanisms of Human Facial Recognition," *Intl. Journal of Man-Machine Studies*, Vol. 15, pp. 137-178.

[4] Bezdek, J.C. (1996), "A Review of Probabilistic, Fuzzy, and Neural Models for Pattern Recognition," in C.H. Chen (Ed.), *Fuzzy Logic and Neural Network Handbook*, McGraw-Hill Series on Computer Engineering.

[5] Bishop, C.M. (1996), *Neural Networks for Pattern Recognition*, Clarendon Press, Oxford.

[6] Bruce, V. and Young, A. (1986), "Understanding Face Recognition," *British Journal of Psychology*, Vol. 77, pp. 305-326.

[7] Carpenter, G. and Grossberg, S. (1991), *Pattern Recognition by Self-Organizing Neural Networks*, MIT Press.

[8] Chen, C.H. (Ed.) (1996), *Fuzzy Logic and Neural Network Handbook*, McGraw-Hill Series on Computer Engineering.

[9] Duda, R.O. and Hart, P.E. (1973), *Pattern Classification and Scene Analysis*, John Wiley & Sons, New York.

[10] Fausett, L. (1991), *Fundamentals of Neural Networks, Architecture, Algorithms, and Applications*, Prentice Hall.

[11] Fischler, M. and Firschein, O. (1987), *Intelligence. The Eye, the Brain, and the Computer*, Addison-Wesley.

[12] Fu, K.S. (1974), *Syntactic Methods in Pattern Recognition: Applications*, Springer-Verlag.

[13] Fuchs, A. and Haken, H. (1988), "Pattern Recognition as Dynamical Processes in a Synergetic System," *Biological Cybernetics*, Vol. 60, pp. 17-22, 1988.

[14] Gelsema, E.S. and Kanal, L.N. (1994), *Pattern Recognition in Practice IV: Multiple Paradigms, Comparative Studies and Hybrid Systems*, Elsevier Publisher.

[15] Gonzales, R.C. and Thomason, M.G. (1978), *Syntactic Methods in Pattern Recognition*, Addison-Wesley, Reading, MA.

[16] Gonzalez, R.C. and Woods, R.E. (1992), *Digital Image Processing*, Addison-Wesley, Reading, MA.

[17] Gorman, R.P. (1996), "Sonar Signal Processing and Classification using Neural Networks," in C.H. Chen (Ed.), *Fuzzy Logic and Neural Network Handbook*, McGraw-Hill Series on Computer Engineering.

[18] Gulab, B., Vannier, M.W., Pilgram, T., Brunsden, B., and Commean, P. (1991), "Facial Surface Scanner," *IEEE Computer Graphics and Applications*, Vol. 11, No. 6, pp. 72-80.

[19] Gupta, M. and Knopf, G.K. (Ed.) (1993), *Neuro-Vision Systems. Principles and Applications*, IEEE Press.

[20] Harinon, L.D. (1973), "The Recognition of Faces," *Scientific American*, October, pp. 70-82.

[21] Haton, J.P. (1996), "Neural Networks for Speech Recognition," in C.H. Chen (Ed.), *Fuzzy Logic and Neural Network Handbook*, McGraw-Hill Series on Computer Engineering.

[22] Hoe, Y. and Kashyap, R.L. (1991), "3-D Shape from a Shaded and Textural Surface Image," *IEEE Trans. on Pattern Analysis and Machine Intelligence*, Vol. 13, No. 9, pp. 907-919.

[23] Horn, B. (1975), *Obtaining Shape from Shading. The Psychology of Computer Vision*, McGraw-Hill.

[24] Hu, M. (1962), Visual Pattern Recognition by Moment Invariants, *IRE Trans. on Information Theory*, Vol. 4, pp. 179-187.

[25] Jang, J.-S.R., Sun, C.-T., and Mizutani, E. (1997), *Neuro-Fuzzy and Soft Computing. A Computational Approach to Learning and Machine Intelligence*, Prentice Hall.

[26] Kashyap, R.L. and Chellappa, R. (1981), "Stochastic Models for Closed Boundary Analysis: Representation and Reconstruction," *IEEE Transactions on Information Theory*, Vol., IT-27, pp. 627-637.

[27] Kaufman, G.J. and Breeding, K.J. (1976), "The Automatic Recognition of Human Faces from Profile Silhouettes," *IEEE Transactions: Systems, Man and Cybernetics*, Vol. 6, pp. 113 -121.

[28] Khotanzad, A. and Huong, Y.H. (1990), "Invariant Image Recognition by Zemike Moments," *IEEE Trans. on Pattern Analysis and Machine Intelligence*, Vol. 12, No. 5, pp. 489-497.

[29] Kim, B. and Burger, P. (1991), "Depth and Shape from Shading Using the Photometric Stereo Method," *CVGIP Image Understanding*, Vol. 54, No. 3, pp. 416-427

[30] Krzyzak, S., Leung, S.Y., and Suen, C.Y. (1988), "Reconstruction of Two-Dimensional Patterns by Fourier Descriptors, *Proceedings of the 9th ICPR*, Rome, pp. 55-58.

[31] Lo, C.H. and Don, H.S. (1989), "3-D Moment Forms: Their Construction and Application to Object Identification and Positioning," *IEEE Transactions: Pattern Analysis and Machine Intelligence*, Vol. I 1, No. 10.

[32] Lucas (1998), *Electronics Letters*, Vol. 33, pp. 1676-1678.

[33] Masters, T. (1995), *Advanced Algorithms for Neural Networks. A C++ Sourcebook*, John Wiley & Sins, Inc.

[34] Midorikawa, H. (1988), "The Face Pattern Identification by Backpropagation Learning Procedure," *Abstracts of the First Annual INNS Meeting*, Boston, p. 515.

[35] Nakamura, O., Mathur, S., and Minami, T. (1991), "Identification of Human Faces Based on Isodensity Maps," *Pattern Recognition*, Vol. 24, No. 3, pp. 263-272.

[36] O'Toole, A.J., Milward, R.B. and Anderson, J.A. (1988), "A Physical System Approach to Recognition Memory for Spatially Transformed Faces," *Neural Networks*, Vol. 1, pp. 179-199.

[37] Packard, N.H., Crutchfield, J.P., Farmer, J.D., and Shaw, R.S. (1980), "Geometry From Time Series," *Physics Review Letters*, Vol. 45, pp. 712-716.

[38] Pandya, A.S. and Szabo, R.R. (1991), "A Fast Algorithm for Neural Network Applications," *International IEEE/SMC Conference*, Charlottesville, pp. 1569-1573.

[39] Pandya, A.S. and Macy, R.B. (1996), *Pattern Recognition with Neural Networks in C++*, CRC Press & IEEE Press.

[40] Pandya, A.S., Kulkarni, D.R., and Parikh, J.C. (1997), "Study of Time Series Prediction under Noisy Environment," *Proceedings of SPIE Conference on Applications and Science of Artificial Neural Networks III*, Orlando, pp. 116-126.

[41] Pao, Y. (1989), *Adaptive Pattern Recognition and Neural Networks*, Addison-Wesley, Reading, MA.

[42] Patterson, D.W. (1990), *Introduction to Artificial Intelligence & Expert Systems*, Prentice Hall.

[43] Perantonis, S.J. and Lisboa, P.J. (1992), "Translation, Rotation, and Scale Invariant Pattern Recognition by High-Order Neural Networks and Moment Classifiers," *IEEE Trans. on Neural Networks*, Vol. 3, No. 2, pp. 241-251.

[44] Perry, J.L. and Carney, J.M. (1990), "Human Face Recognition Using a Multilayer Perceptron," *Proceedings: International Conference on Neural Networks*, Washington D.C., Vol. 2, pp. 4-13.

[45] Person, E. and Fu, K.S. (1979), "Shape Discrimination Using Fourier Descriptors," *IEEE Transactions on Systems, Man and Cybernetics*, Vol. SMC-7, pp. 170-179.

[46] Reddi, S.S. (1981), "Radial and Angular Moment Invariants for Image Identification," *IEEE Trans. on Pattern Analysis and Machine Intelligence*, Vol. PAM I-3, No. 2, pp. 240-242.

[47] Reeves, A.P., Prokop, R.J., Andrews, S.E., and Kuhl, F. (1988), "Three-Dimensional Shape Analysis Using Moments and Fourier Descriptors," *IEEE Transactions on Pattern Analysis and Machine Intelligence*, Vol. 10, No. 6, pp. 937-943.

[48] Rioul, O. and Vetterli, M. (1991), "Wavelets and Signal Processing," *IEEE Signal Processing Magazine*, Vol. 8, No. 4, pp. 14-38.

[49] Roli, F. and Serpico, B. (1996), "Neural Networks for Classification of Remotely Sensed Images," in C.H. Chen (Ed.), *Fuzzy Logic and Neural Network Handbook*, McGraw-Hill Series on Computer Engineering.

[50] Russel, S. and Norvig, P. (1995), *Artificial Intelligence. A Modern Approach*, Prentice Hall.

[51] Samal, A. and Lyengar, P. (1992), "Automatic Recognition and Analysis of Human Faces and Facial Expressions: A Survey," *Pattern Recognition*, Vol. 25, No. 1, pp. 65-77.

[52] Solheim, I., Paync, T., and Castain, R. (1992), "The Potential in Using Backpropagation Neural Networks for Facial Verification Systems," *Simulation*, Vol. 58, No. 5, pp. 306-310.

[53] Spirkovska, L. and Reid, M.B. (1992), "Higher Order Neural Networks in Position, Scale, and Rotation Invariant Object Recognition," in Soucek B. & the IRIS Group (Eds.), *Fast Learning and Invariant Object Recognition*, John Wiley & Sons Inc.

[54] Spirkovska, L. and Reid, M.B. (1993), "Coarse-Coded Higher-Order Neural Networks for PSRI Object Recognition," *IEEE Transactions on Neural Networks*, Vol. 4, No. 2, pp. 276-283.

[55] Su, Y.N., Arsenault, H.H., and April, G. (1982), "Rotational Invariant Digital Pattern Recognition Using Circular Harmonic Expansion," *Applied Optics*, Vol. 21, pp. 4012-4015.

[56] Teague, M.R. (1980), "Image Analysis via a General Theory of Moments," *Journal of the Optical Society of America*, Vol. 70, No 8., pp. 920-930.

[57] Teh, C. and Chin, R.T. (1988), "On Image Analysis by the Method of Moments," *IEEE Trans. on Pattern Analysis and Machine Intelligence*, Vol. 10, No. 4, pp. 496-512.

[58] Tou, J.T. and Gonzalez, R.C. (1974), *Pattern Recognition Principles*, Addison-Wesley, Reading, MA.

[59] Troxel, S.E., Rogers S.K., and Kabrisky, M. (1988), "The Use of Neural Networks in PSRI Recognition," *Proceedings of the Joint International Conference on Neural Networks*, San Diego, CA, pp. 593-600.

[60] Uwechue, O.A. and Pandya, A.S. (1997), *Human Face Recognition Using Third-Order Synthetic Neural Networks*, Kluwer Academic Publishers.

[61] Uwechue, O.A., Pandya, A.S., and Szabo, P. (1995), "High-Order Neural Networks for Image Recognition," *S.P.I.E. Proceedings*, San Diego, Vol. 2568, pp. 252-263.

[62] Watanabe, S. (1985), *Pattern Recognition: Human and Mechanical*, John Wiley & Sons, New York.

Chapter 9:

Face Unit Radial Basis Function Networks

FACE UNIT RADIAL BASIS FUNCTION NETWORKS

A. Jonathan Howell
School of Cognitive and Computing Sciences,
University of Sussex, Falmer,
Brighton BN1 9QH,
United Kingdom

This chapter introduces a different way of learning the face recognition task through the reorganization of the standard radial basis function (RBF) networks [1] into a group of smaller 'face recognition units', each trained to recognize a single person. This type of system organization allows flexible scaling up which could be used either by itself or in conjunction with a standard RBF network trained on all classes where the combined decisions might give greater reliability.

The concept of *face recognition units* was suggested in the perceptual frameworks for human face processing proposed by [2] and [3]. We are adopting this face unit concept as a useful way of developing a modular, scalable architecture, creating fast small RBF networks trained with examples of views of the person to be recognized. The face unit network uses these views of the person to be recognized as positive evidence together with selected confusable views of other people as the negative evidence, which are linked to just 2 outputs corresponding to 'yes' or 'no' decisions for the individual. This training using explicit negative examples is in contrast to the HyperBF network scheme used by [4], who preferred to use implicit negative evidence in their study (see Chapter 7, Section 5.5).

For each individual, an RBF network is trained to discriminate between that person and others selected from the data set. Rather than using all the data available from the other classes to train the network against an individual, the strategy adopted was to use only negative data that was most similar (using an Euclidean distance metric) to the positive data.

This strategy is based on the assumption that similarity leads to confusion, so the inclusion of this type of negative evidence in the training should improve discrimination. This data would be the hardest to learn to discriminate 'for' and 'against' the individual, since it would be the most ambiguous.

The reduction in the size of the network using the face unit organization plus the use of negative knowledge should allow a more efficient coding of the information. Furthermore, people can be added to the data set of a trained set of networks by the creation of a new 'face unit' network for each new individual to be added without retraining the entire database, as the reorganized scheme is completely modular. In the standard RBF network, a new individual means a complete retraining with the expanded dataset.

1 The Face Unit Network Model

The face unit network is essentially a normal RBF network with two output units, see Figure 1, which produces a positive signal only for the particular person it is trained to recognize. It differs from these RBF networks [1] only in the selection of training data, the data for the face unit network being manipulated to present a many-class problem as a two-class problem: (1) a particular class and (2) all others.

Unlike the standard RBF network, which has positive output signals (one per class) only, the face unit network has two output units, one positive, denoting 'yes' for the current class and, and one negative, ('no') for all other classes. We use the term *pro* to denote hidden units or evidence for the class, and *anti* for that against the class, the negative evidence. For each individual, a face unit RBF network can be trained to discriminate between that person and others selected from the data set, using this pro (supporting) and anti (differentiating) evidence for and against the individual. The ratio between the two can be varied.

Although this approach increases complexity, as more networks need to be trained and the training data needs to be manipulated differently for each face unit, the splitting of the training for individual classes into

Face Unit Radial Basis Function Networks   319

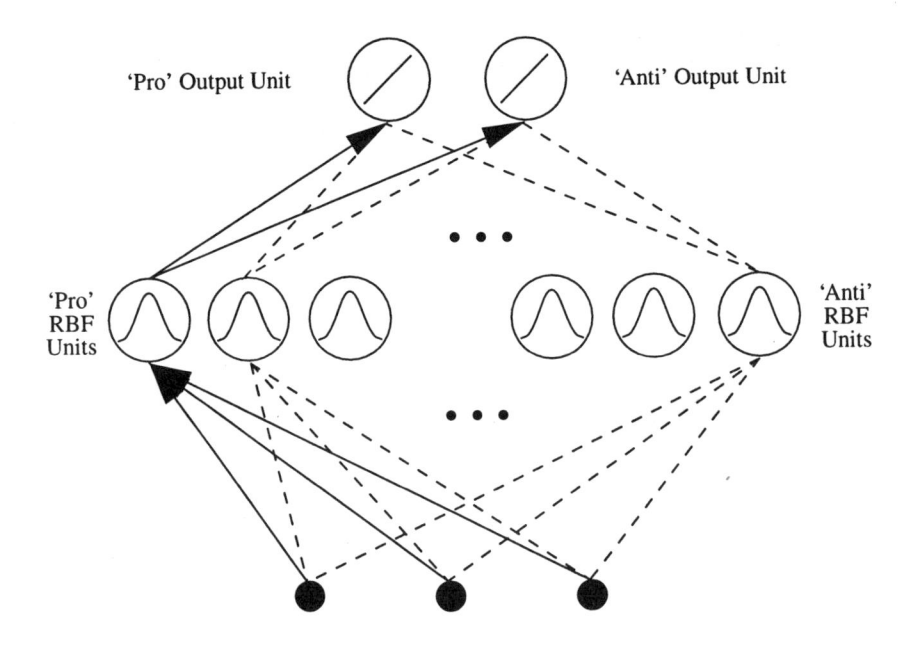

Figure 1. General structure for a 'face unit' RBF network. Although there can be a varying number of and ratio between pro and anti hidden units, there are always two output units (for and against the class learnt by the network). All hidden units are fully connected to both output units.

separate networks gives a modular structure that can potentially support large numbers of classes, since network size and computational load for weight calculations for the 'standard' RBF model may become impractical as the number of classes increases.

1.1 Selection of Negative Evidence

The fundamental process in the face unit network is the splitting of the training data into two halves: class and non-class. The small size of the network is due to the limited amount of non-class data used for training, only those that are seen as hardest to distinguish with the class are included. This selection of negative evidence was based on Euclidean vector distance comparisons of the class face image with images of the same pose angle of non-class faces. In order to make the most efficient arrangement of training examples, the 'anti' evidence was taken from the class that was the closest (in Euclidean distance terms) to the 'pro' class. As the RBF network's hidden units response is based on the same

Euclidean distance comparison, it is important to distinguish the closest non-class examples, as these will be the most 'confusable' for the network, and any other other non-class images further away will then be automatically excluded.

1.2 Types of Face Unit Networks

To investigate the characteristics of the face unit network model, several different network configurations are devised. To assess how varying the pro/anti balance affected performance, two general types of network layout are used:

'Single anti' face unit network This uses equal numbers of pro and anti hidden units.

'Double anti' face unit network This uses two anti hidden units for every one pro.

The double anti face unit network is closer than the single anti arrangement to the full standard RBF model, in that it uses more negative than positive evidence. It is included in the tests to show whether this additional information would give the network better discrimination from the negative classes than the single anti arrangement. This characteristic will be more important as the number of classes in the dataset increases, as the number of negative classes will become proportionately greater.

We can compare the relative sizes of the face unit network and the standard RBF network. The standard RBF network uses cn hidden units, where c is the number of identity classes and n is the number of training examples per class. This gives $10n$ hidden units in total when using the Sussex database, as shown in Table 1. The single anti face unit network has only two classes for training (for and against a single person) and a single anti hidden unit for every pro unit, and therefore has $2n$ hidden units in total (however many identity classes there are). The double anti face unit network uses two anti hidden units for every one pro, and therefore has $3n$ hidden units in all. The outcome of this is that as c, the number of identity classes, increases, the face unit network required for a particular task will becomes much smaller relative to the standard RBF network needed for the same task.

Table 1. Numbers of hidden units used by different RBF networks for same task (when using the Sussex database).

Network	Training Examples per Class	
	1	5
Standard RBF network	10	50
Single anti Face Unit Network	2	10
Double anti Face Unit Network	3	15

Once the number of examples is chosen, we then use two different strategies for the selection of the anti evidence. This gives two further types of network:

'Single best negative' (sbn) face unit networks These use an average of all vector distances between the pro image and all anti images, within each pose angle, averaged over all pose angles to compare whole classes rather than individual images from classes. The lowest overall average value was used to select one anti class, which then represented all negative evidence at all pose angles.

'Multiple best negative' (mbn) face unit networks These use the closest anti image to the pro image for each pose angle, so that several anti classes may be used for a face unit network with more than one training example.

It was anticipated that *sbn* face unit networks would be superior to *mbn* face unit networks, as a more coherent 3-D class boundary would be given by a single negative person-class for all pose angles. On the other hand, the *mbn* approach may utilize local class differences to learn a more efficient solution.

1.3 Face Unit Network Terminology

As the face unit networks are arranged differently to the standard RBF networks, they are labelled slightly differently. The face unit network size is denoted here by '$p + a$', where p is the number of pro hidden units, and a is the number of anti hidden units. Tests were made on a range of network sizes from $1 + 1$ to $6 + 12$ on the standard 100-image

Sussex database (if these networks had been labelled in the standard 'train/test' form, this would correspond to a range between 2/98 and 18/82 networks). To give an optimal spread of the image data for training, fixed selections of pose angle were used for each size of network. For instance, the 5 + 5 and 5 + 10 networks used poses 10°, 30°, 50°, 70° and 90°, where the pose range was 0° (frontal) to 90° (profile).

Figure 2 shows how the images used for training were selected for a 5 + 10 *mbn* face unit network in the experiment. This illustrates not only how several anti classes are used in the *mbn* scheme, but also how they are ranked for the double anti arrangement.

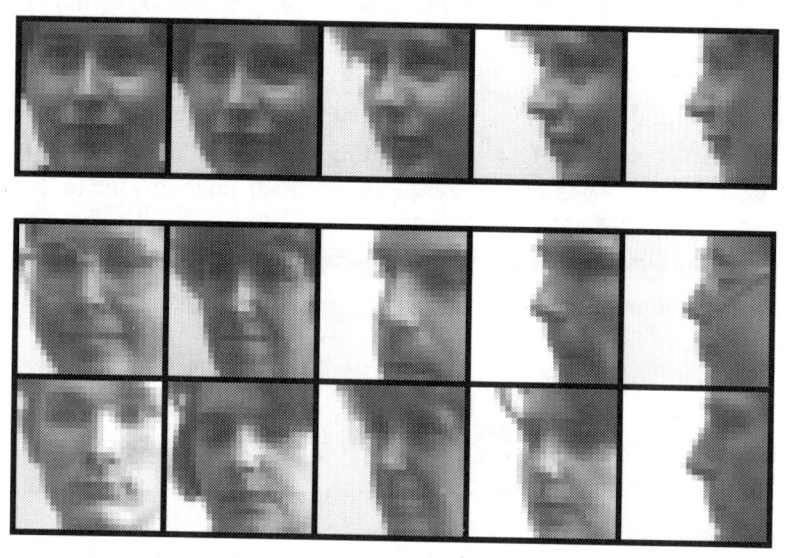

Figure 2. Example of the range of negative classes that can be selected during the training of a 5 + 10 double anti, multiple best negative (*mbn*) face unit RBF network. The top line shows the supporting, 'pro' evidence, the middle and bottom lines the differentiating, 'anti' evidence (middle line is the closest to the pro class, bottom line the second closest).

1.4 Results

For clarity, our tests use two standard preprocessing methods only: the single-scale Difference of Gaussians (DoG) and the Gabor filtering with four scales and three orientations (details in [1]).

Face Unit Radial Basis Function Networks

Figure 3 summarizes the overall results for the various types of face unit networks, with different pro/anti ratios and different strategies for selection of anti images. To simplify the information, these graphs do not show the rates after discard, but these gave a consistent improvement of about 7–15% over rates before discard for all networks.

(a) DoG preprocessing

(i) (ii)

(b) Gabor preprocessing

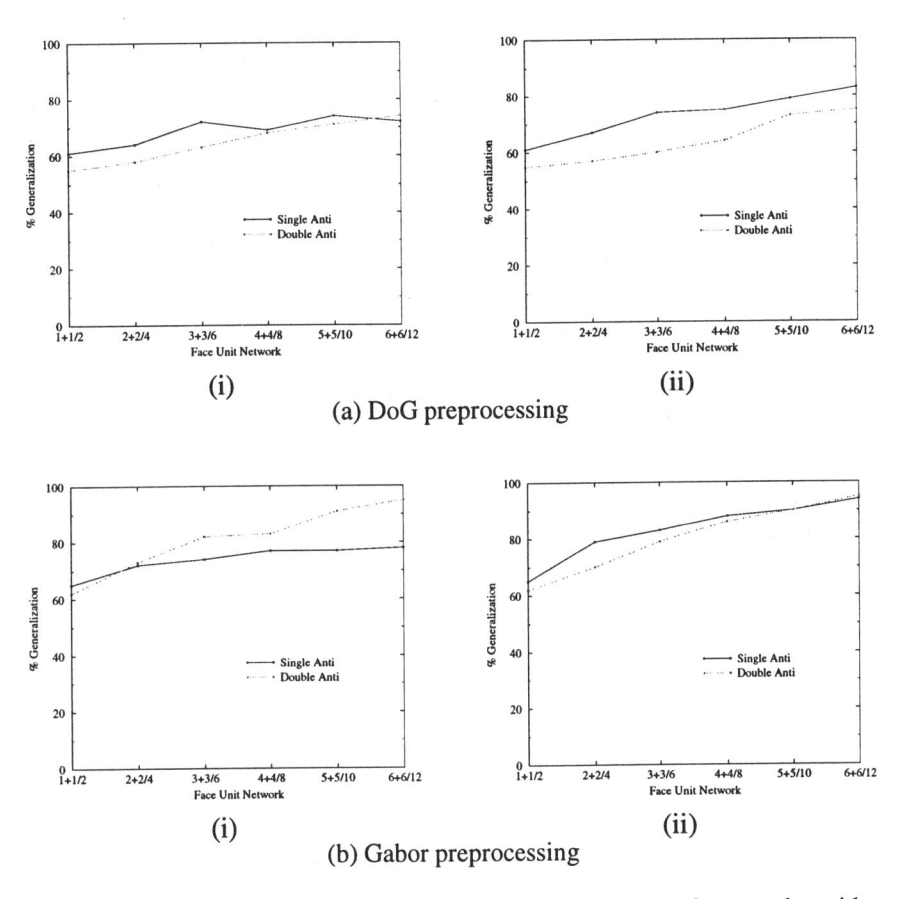

(i) (ii)

Figure 3. Comparing single and double anti training for face unit networks, with average generalization for all face units shown, but no discard results: (i) Single best negative (*sbn*) networks (ii) Multiple best negative (*mbn*) networks.

The face unit networks are essentially working in a two-class classification problem, so a random level of generalization would be 50%. Interestingly, the double anti network arrangement did not appear to give radically better performance than the single anti, except for the 5- and 6-example networks using Gabor preprocessed data. This indicates that the selection of appropriate anti images is efficient enough by itself to create a division in image space between the class and all others without requiring additional negative examples.

Table 2 shows specific generalization rates for the 5-example (5+5 and 5+10) face unit networks before and after discard. It can be seen here that the Gabor preprocessed data allowed the RBF network to perform more efficiently than the DoG preprocessed data, both in lower discard rates and generalization before and after discard.

Table 2. Test generalization for 5-example face unit networks (5+5 and 5+10) using the Sussex database.

(a) Single best negative (*sbn*) networks

Preprocessing	Network	Initial %	% Discarded	% After Discard
DoG	5+5	74	75	87
	5+10	71	49	81
Gabor	5+5	77	50	84
	5+10	91	39	98

(b) Multiple best negative (*mbn*) networks

Preprocessing	Network	Initial %	% Discarded	% After Discard
DoG	5+5	79	73	97
	5+10	73	42	75
Gabor	5+5	90	47	97
	5+10	90	40	99

1.4.1 Summary

The *mbn* strategy for selecting anti evidence seemed slightly better than the *sbn*, indicating that dealing with local (at a pose level) confusions was more efficient that trying to identify one global class with which the main class should be contrasted.

1.5 Shift and Scale-Varying Data

To assess learnt invariance to the shift and scale-varying Sussex data, tests were made using 5-pose-example face unit networks. Single and double anti networks were tested to check which reacted best to the more demanding datasets.

Table 3 shows that the networks were able to learn shift and scale invariance very similarly to the standard RBF network, in that the scale-varying data was learnt more easily than the shift-varying data, and the Gabor preprocessing allowed both higher generalization and lower discard rates than the DoG preprocessing. The double anti networks did not give higher generalization overall, but did give lower discard rates on all tests.

Table 3. Generalization for 5-pose-example multiple best negative (*mbn*) face unit networks (25+25 and 25+50) with shift and scale varying data.

Variation	Preprocessing	Network	Initial %	% Discarded	% After Discard
Shift	DoG	25+25	70	67	82
		25+50	69	46	69
	Gabor	25+25	83	35	92
		25+50	86	27	92
Scale	DoG	25+25	78	50	89
		25+50	79	40	84
	Gabor	25+25	88	29	94
		25+50	90	23	96

1.6 Discussion

From the results, the most useful configuration of face unit RBF network should have:

- more than one training example for both pro and anti data.

- use equal numbers for pro and anti, although exceptions for particular conditions can be seen.

- use the multiple best negative (*mbn*) strategy to identity the most useful anti evidence to match each pro example on a pose-by-pose basis.

This section has shown that the face unit network can operate to a high level of performance when used in isolation. The next section will show how cooperation with a standard RBF network can be accomplished, and assess the usefulness of such combined information.

2 Face Unit Networks as Adjudicators

One potential drawback to using face unit networks is that the processing required to input the test image to every network may become excessive for large number of classes. It would be possible to take advantage of the specialized training characterized by each individual face unit network by using them in cooperation with other networks.

For instance, a single face unit network could be used to confirm or dispute a classification from a standard RBF network trained on all individuals. The initial output from the multi-class network would be used to index into the group of face unit networks to identify which one was needed, and the outputs from the two networks could then be used in conjunction. It is anticipated that this will give a more reliable result.

2.1 Confidence Measures

The standard confidence measure, which has been used in all tests so far for both face unit and standard RBF networks, is based on the difference between the highest and second highest output values. Classifications with a large difference (generally a ratio of 1.8:1 or above) are labelled as high confidence, all the rest as low confidence.

The outputs of face unit networks and standard RBF networks can be combined, using this standard confidence measure for both networks. Several levels of classification confidence are then possible, shown in Table 4. These range from 1, the highest, where both networks have high confidence, to 8, the lowest, where the standard RBF network has low confidence and the face unit network has high confidence *against* the classification.

2.1.1 Confidence Rating Thresholds

A threshold based on these cooperative ratings can be used to control which classifications are thought of as high confidence. All classifications rated above the threshold are accepted, all below are discarded.

Table 4. Possible outcomes given a particular classification from a standard RBF network when used to index into one specific face unit network, based on the outputs of the two networks. These can be combined to give levels of cooperative confidence in accepting the initial classification, ranging from 1 (the highest) to 8 (lowest).

Multi-Class RBF Network Confidence	Face Unit Network Output	Face Unit Network Confidence	Confidence to Accept
High	Yes	High	1
		Low	2
	No	High	6
		Low	5
Low	Yes	High	3
		Low	4
	No	High	8
		Low	7

If the multi-class and face unit network concur, even if both are low confidence, then we might say that it is reasonable evidence for a correct classification. According to how confident we want our networks to be in this agreement, we can set a threshold on the confidence ratings between 1 and 4.

It is harder to decide on heuristics for where the two networks disagree. Thresholds set at levels 5 and 6 might still leave useful classifications undiscarded. Some conflicts could be decided on the basis of accepting the decision of which ever network had the higher confidence.

2.2 Results

The face unit and standard RBF networks were tested together, the face unit chosen to test each image according to the output of the standard RBF network, and the results were arranged according to the cooperating confidence rating thresholds from 1 to 8. Tests were made with both single and double anti networks.

Tables 5 and 6 show that the cooperating networks are able to give a much finer gradation of confidence levels than the normal confidence measure based on the standard RBF network only.

Table 5. Generalization and discard rates for different discard measures: 'Standard RBF Network Only' is the result using a simple discard measure applied to the output of a standard 50/50 multi-class RBF network by itself, the 'Cooperative Threshold' is a threshold value applied to the confidence rating arising from cooperating 50/50 multi-class standard RBF networks and 5+5 single anti multiple best negative (*mbn*) face unit RBF networks.

(a) DoG preprocessing

Discard Measure		Initial %	% Discarded	% after Discard	Ratio after Discard
Standard RBF Network Only		78	52	100	24/24
Cooperative Threshold	1	78	66	100	17/17
	2	78	52	100	24/24
	3	78	46	93	25/27
	4	78	6	79	37/47
	5	78	6	79	37/47
	6	78	6	79	37/47
	7	78	0	78	39/50
	8	78	0	78	39/50

(b) Gabor preprocessing

Discard Measure		Initial %	% Discarded	% after Discard	Ratio after Discard
Standard RBF Network Only		96	20	98	39/40
Cooperative Threshold	1	96	38	97	30/31
	2	96	20	98	39/40
	3	96	20	98	39/40
	4	96	2	96	47/49
	5	96	2	96	47/49
	6	96	2	96	47/49
	7	96	0	96	48/50
	8	96	0	96	48/50

Face Unit Radial Basis Function Networks

Table 6. As Table 5, except the 'Cooperative Threshold' measure uses a 5+10 double anti multiple best negative (*mbn*) face unit RBF network.

(a) DoG preprocessing

Discard Measure		Initial %	% Discarded	% after Discard	Ratio after Discard
Standard RBF Network Only		78	52	100	24/24
Cooperative Threshold	1	78	86	100	7/7
	2	78	58	100	21/21
	3	78	58	100	21/21
	4	78	54	91	21/23
	5	78	48	92	24/26
	6	78	48	92	24/26
	7	78	6	81	38/47
	8	78	0	78	39/50

(b) Gabor preprocessing

Discard Measure		Initial %	% Discarded	% after Discard	Ratio after Discard
Standard RBF Network Only		96	20	98	39/40
Cooperative Threshold	1	96	64	100	18/18
	2	96	20	98	39/40
	3	96	20	98	39/40
	4	96	14	98	42/43
	5	96	14	98	42/43
	6	96	14	98	42/43
	7	96	6	96	45/47
	8	96	0	96	48/50

Although the single anti face unit networks discarded less when in combination with the standard RBF networks than with the double anti networks, their performance was worse on the whole. The double anti networks gave more useful results, giving a good increase in performance compared to no discard at all, and generally equivalent generalization performance to the conventional, one network discard, with lower discard rates.

2.2.1 Summary

Threshold levels of 1 and 6 on the cooperative confidence rating scale were found to be useful in practice. The highest confidence rating threshold discard level, **1** requires both standard and face unit network to have high confidence for the same class. This threshold value can be used to give better, or at least as good, generalization performance, after discard, as that provided by the original confidence measure used with the standard RBF network alone. This superior performance is at the cost of higher proportion of discarded classifications.

The confidence rating threshold level **6**, which ignores low-confidence output from the face unit network, gives roughly equivalent generalization to that given using the normal confidence measure with output from the standard network only, but with much lower discard rates. This could well be the most useful configuration for general use.

Stages 4 and 8 on the confidence rating threshold scale were not found to be useful, but are worth mentioning to clarify the coordinating confidences threshold process. The confidence rating threshold level **4** is equivalent to not using the normal confidence measure for either network, relying on the values of the two 'raw' network classifications. This does not appear to be a useful arrangement, giving no advantage over the use of the standard confidence measure with standard RBF network alone.

A confidence rating threshold of **8** is not useful, as it allows no discard at all. This is because all face unit network output and the confidence rating of the standard RBF network are both effectively ignored. This is demonstrated by the 0% discard levels shown for the threshold set to 8 in Tables 5 and 6.

2.3 Shift and Scale-Varying Data

As in Section 1.5, tests were made to assess learnt invariance to the shift and scale-varying Sussex data, this time using 5-pose-example face unit networks in cooperation with standard RBF networks, as in the previous section.

Face Unit Radial Basis Function Networks

Table 7 shows a similar gradation of performance, controlled by confidence rating threshold, to that found in the previous section.

Table 7. Generalization and discard rates for different discard measures with shift and scale varying data: 'Standard RBF Network Only' is the result using a simple discard measure applied to the output of a standard 50/50 RBF network by itself, the 'Cooperative Threshold' is a threshold value applied to the confidence rating arising from cooperating 250/250 multi-class standard RBF networks and 25+50 double anti multiple best negative (*mbn*) face unit RBF networks, using Gabor preprocessing.

(a) Shift-varying data

Discard Measure		Initial %	% Discarded	% after Discard	Ratio after Discard
Standard RBF Network Only		85	35	98	159/163
Cooperative Threshold	1	85	49	98	126/128
	6	85	13	91	197/217

(b) Scale-varying data

Discard Measure		Initial %	% Discarded	% after Discard	Ratio after Discard
Standard RBF Network Only		90	26	97	178/184
Cooperative Threshold	1	90	42	98	141/144
	6	90	10	94	212/226

2.4 Discussion

The cooperative use of the face unit network with the standard multi-class RBF network shown in this section can be seen as a more subtle approach to assessing classification confidence than the simple, one-network threshold used previously.

Different rating threshold levels can be used with the cooperative scheme to give either *high confidence with high discard* (using a rating threshold of 1), or *moderate confidence with low discard* (rating threshold 6), compared to the *moderate confidence with moderate discard* provided by the original confidence measure using the standard RBF network alone. Intermediate threshold levels (from 2 to 7) provide other combinations of confidence and discard ratios. This ability to vary the system behavior via the threshold level would be useful for real-life applications, as it allows the user to engineer the solution required.

3 Updating Face Units

This section is about how face unit networks can be retrained during use. Face unit networks allow a flexible approach to learning in dynamic environments compared to other neural networks models which have to be completely retrained if the training data is altered in any way.

The face unit network only uses a few of the total number of classes in a problem to train, so operations on any of the other classes not used for training will leave it unaffected. As the number of classes increases, the chance of each face unit network needing retraining due to an operation on another class will become less.

3.1 Adding Face Units

To add a new person-class, vector differences need to be compared for all training images, just as for the initial training. Distance calculations for all classes each time a change is made can be avoided, however, by saving the Euclidean distance information, so that only the values for the new class need to be calculated.

Any face unit where an image from the new class is closer than its existing anti evidence would need to be re-trained. All other face units would not require further training. In the worse case, this would mean the entire system of face unit networks being re-trained, but this is less likely as the number of classes increases.

3.2 Removing Face Units

Removal of a particular face unit is simpler, as it just requires a check for other face units currently using that face class as anti evidence. Only those that did use the removed face unit would require retraining.

To update an old face unit would require two steps, as it would first need to be removed and then the new data added.

3.3 Discussion

This section has tried to address the issue of long-term use of face recognition systems. If our task requirements (such as in Chapter 7) specify a tolerance to middle-term (makeup, facial hair, etc) and long-term (ageing, etc) changes in appearance, potentially suitable systems will need to be flexible enough to update their training data for such changes.

Although our standard RBF network is fairly fast in retraining completely, this might not be the case if many more examples for each class are used (representing x- and z-axis movement, for example).

It is an open issue how an automatic system would determine that a known individual required retraining due to change in their appearance – could the system itself monitor how confident it was recognizing that person and retrain when this fell below some limit, or would it require manual intervention from the user to initiate the process? A system that could be aware of these changes automatically would be more useful than one which simply failed to recognize a previously known person.

4 General Discussion

This chapter has presented experimental work using a novel variant of the RBF network model, the face unit network, which learns to distinguish a single class from a range of other classes. This can be used either in groups, one for each class, or singly in conjunction with a multi-class network to give greater reliability to classification.

The most useful configuration of face unit RBF network overall seems to be the single anti multiple best negative (*mbn*) face unit network, which selects the most useful anti evidence to match each pro example on a pose-by-pose basis.

The standard RBF network will give similar positive and negative information about classes, because of the fully interconnected hidden to output unit layer, but the face unit network, by concentrating only on distinguishing one class at a time, allows the negative influences of such non-class connections to be more specialized, indeed optimized, to give the most effective 'one class against all others' partitioning in image space.

The modular approach presented in this chapter using face unit RBF networks to learn identity is especially attractive for the unconstrained recognition task, as it allows the modification of the learned element of the system during use, and can give a secondary classification decision which can either confirm or dispute the primary RBF network output.

References

[1] Howell, A.J. (1997), *Automatic face recognition using radial basis function networks*, PhD thesis, University of Sussex, U.K.

[2] Hay, D.C. and Young, A. (1982), "The human face," in H.D. Ellis (Ed.), *Normality and Pathology in Cognitive Functions*, Academic Press, San Diego, CA.

[3] Bruce, V. and Young, A. (1986), "Understanding face recognition," *British Journal of Psychology*, 77:305–327.

[4] Edelman, S., Reisfeld, D., and Yeshurun, Y. (1992), "Learning to recognize faces from examples," in G. Sandini (Ed.), *Proceedings of European Conference on Computer Vision, Lecture Notes in Computer Science*, Vol. 588, pp. 787–791, Santa Margherita Ligure, Italy. Springer-Verlag.

Chapter 10:

Face Recognition from Correspondence Maps

FACE RECOGNITION FROM CORRESPONDENCE MAPS

Rolf P. Würtz

Institute für Neuroinformatik
Ruhr-Universität, D-44780 Bochum, Germany
http://www.neuroinformatik.ruhr-uni-bochum.de/ini/PEOPLE/rolf/
Rolf.Wuertz@neuroinformatik.ruhr-uni-bochum.de

A number of systems for the recognition of human faces are presented. They all comprise three steps: feature extraction, solution of the correspondence problem, and the actual comparison with stored faces. Two of them are implemented in the Dynamic Link Architecture and are, therefore, close to neurobiological models. The others are more technical in nature but also have some biological plausibility. Finally, the coherence with the results of psychophysical experiments on human face recognition and the applicability to general object recognition are discussed.

1 Introduction

This chapter presents several algorithms that have proved useful in the task of recognizing human faces from gray-value images without additional information. Their common feature is the extraction of *correspondence maps* between an *image* to be analyzed and several stored views of known objects, which are called *models*. This means that pairs of points in model and image must be found which are images of the same point on the physical face. This is not trivial and has acquired the name *correspondence problem*. Its importance becomes clear immediately if one tries to compare images that are identical copies of each other except for a constant shift in the image plane. If they are compared on a pixel by pixel basis, huge differences will generally occur. This can be overcome by first *aligning* the images, e.g., by finding the maximum of the autocorrelation function. This alignment is the simplest case of determining which points correspond to each other. In more complicated cases such as movements in three-dimensional space, partial occlusion, additional noise or difficult lighting conditions, the correspondence problem must be solved in the general form outlined above.

Its difficulty depends on the choice of features. For example, if gray values of pixels are taken as local features, there is a lot of ambiguity, i.e., many points from very different locations share the same pixel value without being correspondent points. A possible remedy to that consists in combining local patches of pixels, which of course reduces this ambiguity. If this is done too extensively, i.e., if local features are influenced by a large area, the ambiguities disappear if identical images are used, but the features become more and more sensitive to distortions and changes in background. Therefore, methods for establishing correspondences must make use of robust features *and* additional information about their relative locations, which will be discussed in Sections 3 and 4.

2 Correspondence-Based Recognition

Once a mapping is established, the feature similarities of corresponding points are added up (or averaged) over the whole mapping. This yields a similarity value for each stored model, and the highest similarity belongs to the recognized person. A measure for the *reliability* or *significance* of the recognition is derived by a simple statistical analysis of the series of all similarity values: The similarity histogram of one image against all models is analyzed by dividing the distance of the highest similarity value by the standard deviation of the distribution of all similarities *except* the highest one (Figure 1). Alternatively, the distance of the highest similarity to the runner-up can be used. A reliability measure is required because the case of an unknown person (low similarities and low significance) must be distinguished from the case of a highly distorted image of a known person (low similarities but high significance). In combination with a hierarchical matching scheme such as the one described in Section 4.3 the significance measure can be used to stop the refinement once a significant recognition has been achieved. On average, this yields a higher recognition speed and interesting insights into the distribution of recognition-relevant information over spatial scales.

Face Recognition from Correspondence Maps

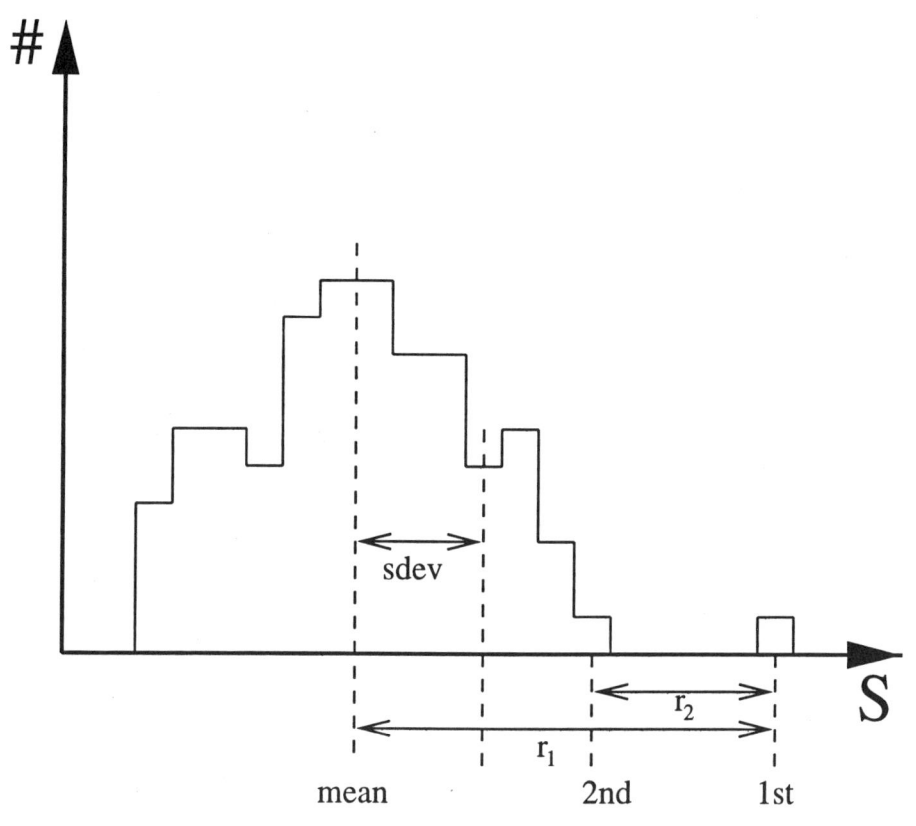

Figure 1. Determining recognition significance from the similarity histogram. For explanations see Section 2.

3 Features

The processing of a retinal gray-level image in simple cells of the primary visual cortex can be modeled by a wavelet transform based on complex-valued Gabor functions (Figure 2) [4], [25]. They are parameterized by their two-dimensional spatial frequency vector. The responses of all spatial frequencies of some fixed length form a *frequency level*, which assigns a small feature vector to all image points on an appropriate sampling grid. These features have turned out to be a good compromise in the dilemma discussed above. Furthermore, the complex numbers invite a splitting into modulus and phase which is very convenient for matching purposes (see Section 4.3). The systems in this chapter differ in the *sampling* of this transform. The ones in [4], [22] need a sampling grid

Figure 2. The form of Gabor wavelets. The left figure shows the Gabor kernel as it is found as receptive field profile in the visual cortex and is used for feature extraction in [25], [26], [29] (real and imaginary part, respectively). On the right the type of kernel used in [4], [23] is shown. The complete wavelet transform consists of convolutions with scaled and rotated versions of these kernels.

which is uniformly dense for all spatial frequencies, the one in [25], [26] uses a pyramidal arrangement.

The appeal of Gabor wavelets in comparison to the immense spectrum of linear filters used in image processing is twofold. First, they model a certain type of nerve cells, which is found in the early visual processing of mammals (simple cells). Second, they have complex values, which makes it natural to represent the transform as amplitude and phase rather than real and imaginary part. This simple nonlinearity will have important (and pleasant) consequences in the algorithms. Again, the amplitudes are models for another type of visual cells (complex cells).

An important advantage of the use of feature *vectors* (called *jets* in [4], [23]) instead of scalar features consists in the fact that the local transformations caused by 3D-movements can be estimated. The recognition performance for faces rotated in depth can be improved significantly by applying this transformation [9].

4 Topology

The only additional information that can be used to overcome the feature ambiguity is the *relative position* of the features. This section describes some of the possible ways to do this, namely *elastic graph matching*, *coarse-to fine pyramid matching* in a technical and a more biologically flavored version, and *learned point distribution models*.

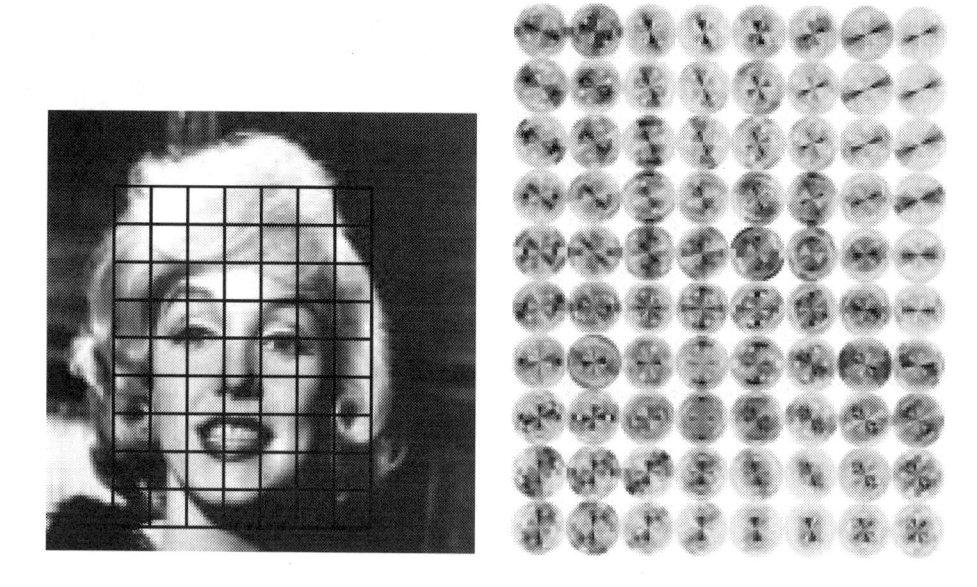

Face Recognition from Correspondence Maps 341

Figure 3. Visualization of Gabor features with frequency-independent sampling. Each point in the grid is assigned a little frequency space, which is visualized on the right-hand side. The gray values of the little segments show the moduli of the Gabor responses as a function of the (two-dimensional) spatial frequency.

4.1 Elastic Graphs

In the system described in [3], [4], [22] the model faces are represented by sparse graphs (see Figure 3) whose vertices are labeled with the vectors of all Gabor features at an image location (jets) and whose edges are labeled with the distance vectors of the connected vertices. Matching is done by first optimizing the similarity of an undistorted copy of the graph in the input image and then optimizing the individual locations of the vertices. This results in a distorted graph whose vertices are at approximately corresponding locations to the ones in the model graph (see Figure 4). The rectangular model graph arrangement has been chosen in [4] and in [22] the vertices have been carefully placed on salient points, thus yielding a larger recognition rate.

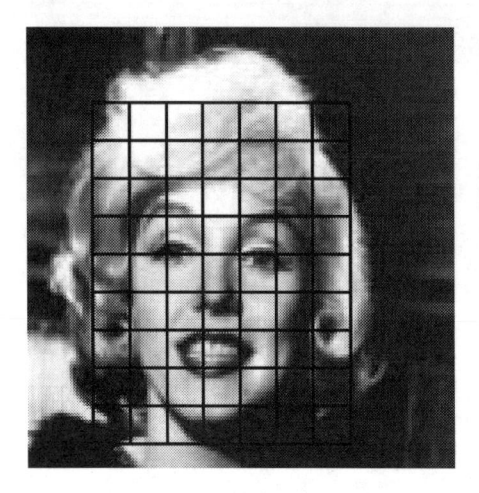

Figure 4. The result of labeled graph matching. The right-hand side shows the graph with maximal similarity to the one on the left.

4.2 Bunch Graphs

The major drawback of correspondence-based recognition systems is that the computationally expensive procedure of creating a correspondence map must be carried out for *each* of the stored models. This has partly been overcome by the concept of *bunch graphs* [20], [23]. The idea is that the database of models is arranged in such a way that corresponding graph nodes are already located at corresponding object points, e.g., a certain node lies on the left eye in all models. For large databases, this reduces the recognition time by orders of magnitude. Another important innovation consists in the possibility of matching the most similar features from *different* persons and thus acquiring information on, e.g., gender, beardedness or the presence of glasses about an unknown person [21].

Although it is not straightforward to apply the bunch graph principle to arbitrary object classes, it has been applied successfully to hand gesture recognition [17], [18].

4.3 Coarse-to-Fine Matching

Another possibility to take the feature arrangement into account is given by coarse-to-fine methods. In that case, the sampling of the Gabor responses depends on the spatial frequency, therefore complete jets can no longer be extracted. The resulting *pyramidal representations* of image and model [25], [26] can be matched using the following modules:

1. The *coarse localization* of the counterpart of the model in the image is done by global template matching of the vectors of Gabor amplitudes on the lowest frequency level. This is not very expensive, because the resolution is low, and yields a first, rough correspondence mapping.

2. Mappings acquired using only the amplitudes of the Gabor responses are not very precise, because the fine geometrical information resides in the phases. On the other hand, the phases or the full complex responses are not suitable for template matching because they depend strongly on the sampling grid. Therefore, *local phase matching* is applied, which enhances the accuracy of amplitude-based mappings. This can be done in parallel at all model points.

3. In order to cope with occlusion problems, it must be possible to *exclude* points from the mapping. This is done on the basis of poor similarity, which is also possible in parallel at all image points.

4. Finally, any mapping can be refined by local template matching with amplitudes from the next higher frequency level. For this, the model is split up into several patches that independently search for correspondences in an area defined by the coarser mapping determined before.

The pyramidal arrangement has the advantage that all responses in the stored model which are influenced by the background can be discarded, but the object (the face without hair in [25], [26]) is still represented well enough for recognition to be possible (see Figure 5 for an illustration). Note that the image to be analyzed consists of a full pyramid, a preceding segmentation is not required. However, the concept lends itself to a combination with segmentation methods, which can partly replace or assist the global template matching. The resulting invariance under changes

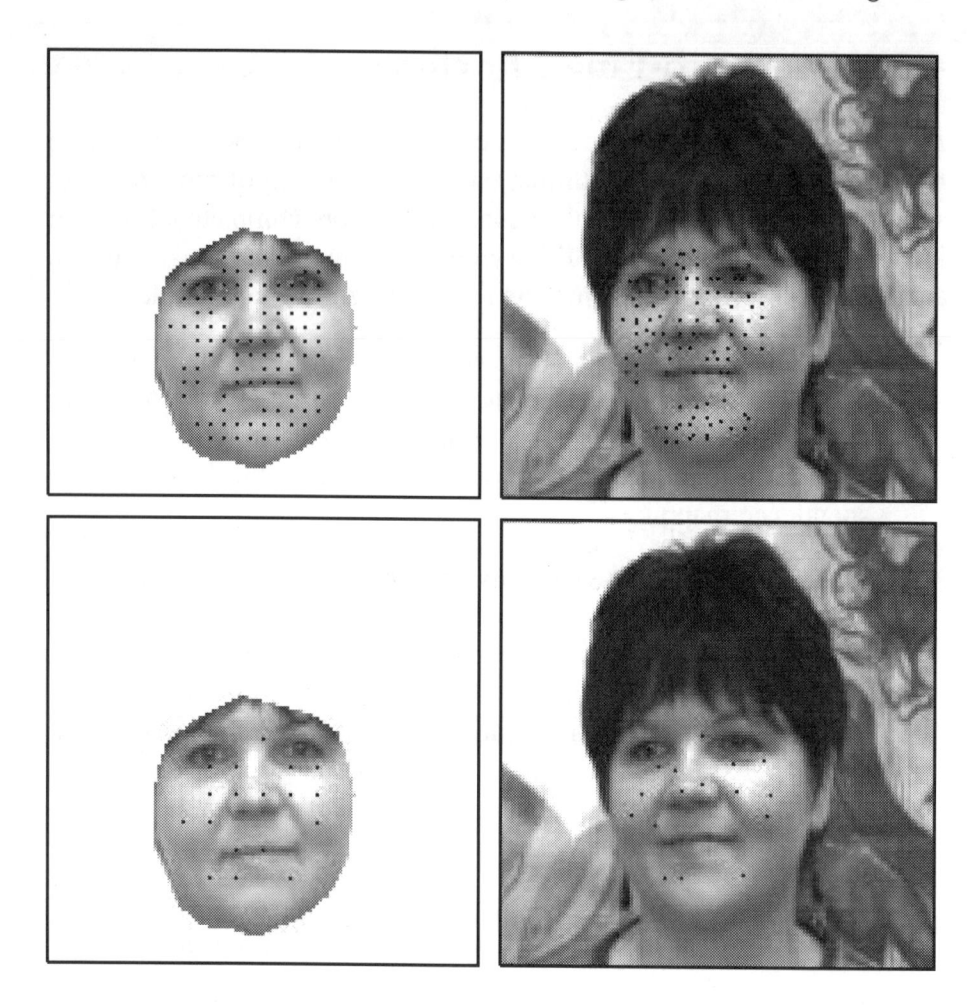

Figure 5. Course-to-fine matching. The left column shows the model representation on the lowest and next higher frequency levels, the right column the matching points in the image. Due to phase matching correspondences are fairly accurate on the lowest level already.

in hairstyle and background constitutes an important advantage of this system as compared with the ones described in [4] and [22].

4.4 Point Distribution Models

Additionally, the system yields fairly dense and, due to phase adjustment, very accurate correspondence maps. These can be used for tracking facial points. In this case, coarse-to-fine matching becomes impractical, and has been replaced by a shape model learned from examples [11]. Such tracking is a prerequisite for the automatic evaluation of emotions. A similar system based on graph matching is described in [10].

5 Neural Models

All the systems discussed here have been inspired by a biologically plausible framework, Von der Malsburg's Dynamic Link Architecture [7], [8]. The basic idea is as follows. Neurons that have physical connections (with a long-term strength) may in addition have a dynamic link between them, i.e. a connection which is modifiable on a very short time scale by a combination of Hebbian learning and competition.

This can be used to solve the correspondence problem as follows. Two layers of neurons that represent the image space in model and image, respectively, are fully interconnected by dynamic links (lower half of Figure 6). They have an internal wiring that supports moving localized blobs of activity, which code for the proximity of links. The link development is supported by feature similarity and synchronous activation of the connected neurons and limited by competition between all links originating at or targeting on the same neuron.

For the start of the process, the blobs on both layers are moving independently of each other, but after a while the similarity-dependent growth rates develop such that corresponding positions become more probable, which in turn yield higher synchronous activity at those positions. Thus, the link dynamics converge to a correspondence mapping. This has been extended by competition between a multitude of model layers to a full-blown neural face recognition system [24].

Such a matching scheme suffers from the inherently sequential component presented by the moving blobs. There are hints that the time complexity will be $O(N^4)$ for sequential, $O(N)$ for fully parallel hardware, if N^2 denotes the number of neurons in a square layer [27], [28]. Therefore, the possibility of speedup by coarse-to-fine matching has been studied [25], [30]. Model and image representations are the same as for pyramid matching (Section 4.3), thus the neural system shows the same independence of the background.

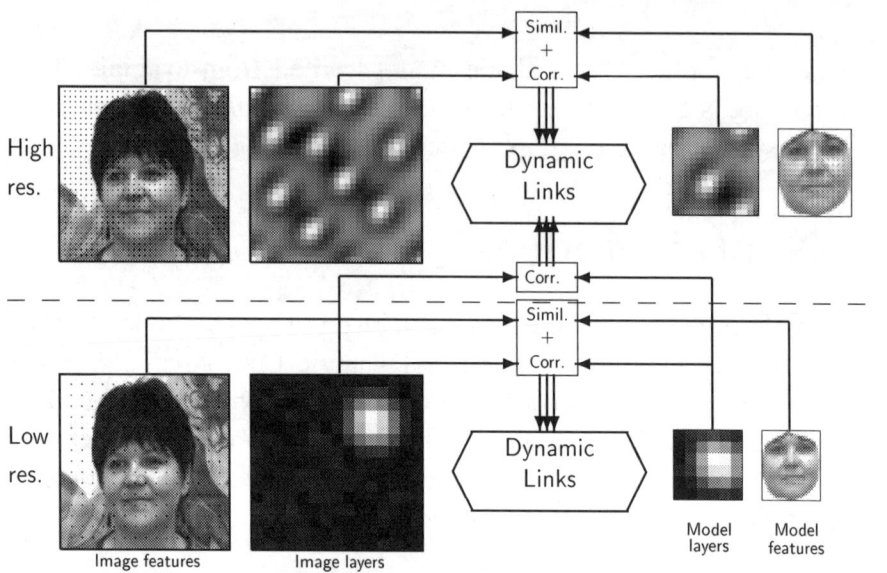

Figure 6. The setup of the matching structure for Hierarchical Dynamic Link Matching.

6 Discussion

The performance of the methods presented is analyzed in detail in the respective publications. An evaluation of all of them on identical datasets has not been attempted and would be of relatively small value because they all have different virtues and shortcomings. Commercially successful derivates such as ZN-Face [3] and PersonSpotter [16] use various refinements and thrive on combinations with segmentation methods and product-specific hardware interfaces. I would therefore like to devote this discussion to comparison with psychological data on human face recognition.

Humans do very poorly in recognizing faces from photographic negatives. One recognition system [4] has no problem with this, because Gabor amplitudes are identical for positive and negative images. The recognition rate of the pyramidal system [25], [26] drops to zero when negatives are presented. This shows that some processing of Gabor phases must occur in face processing. The hierarchical recognition scheme mentioned briefly in Section 2 can also model the fact that a subsampled image represented by few square pixels can be recognized better after low-pass filtering (see [25] for a details and an example).

Face Recognition from Correspondence Maps 347

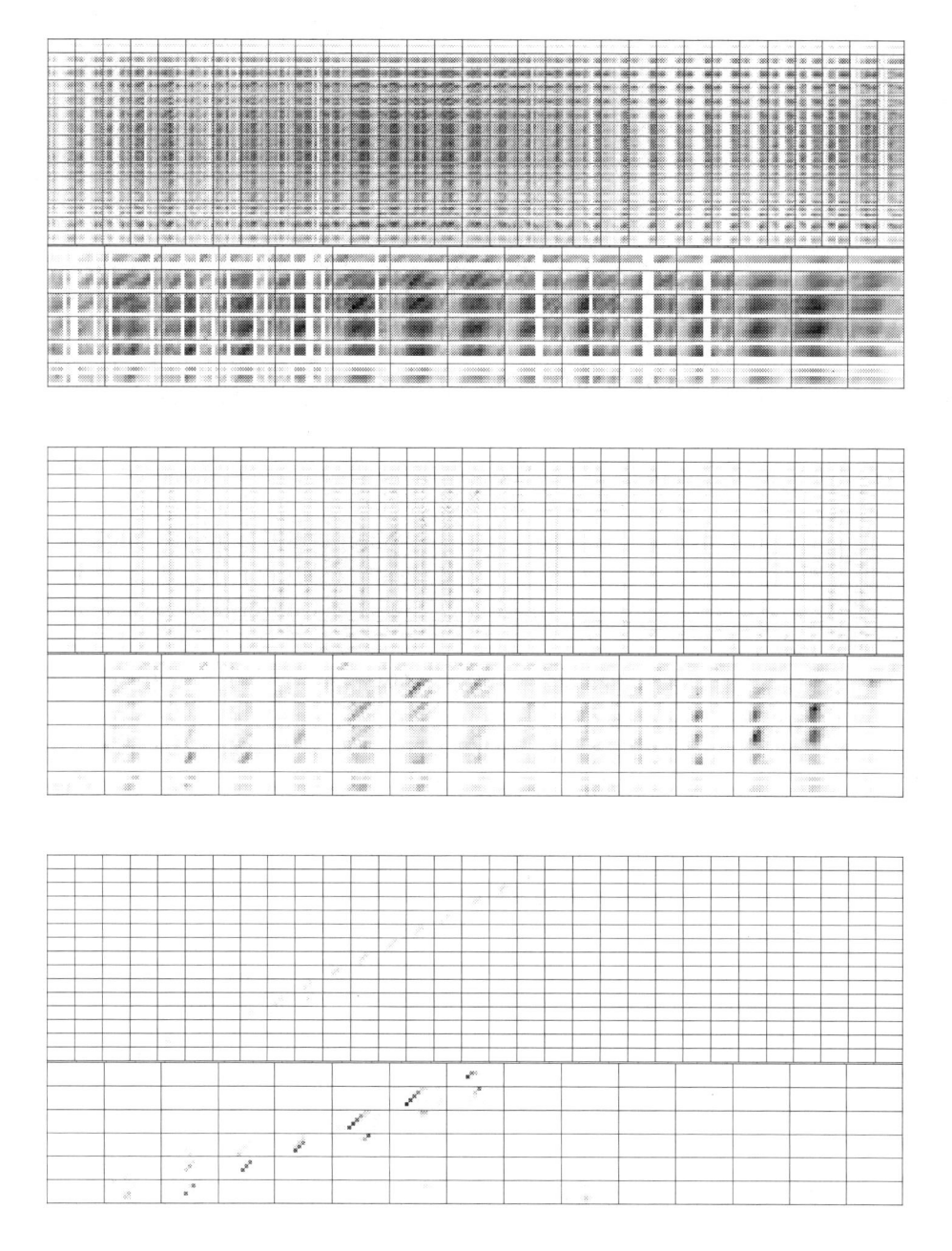

Figure 7. The development of the dynamic links from the highly ambiguous feature similarities ($t = 0$, top) to a clear one-one-mapping ($t = 600$, bottom). The second box shows the intermediate state at $t = 100$.

The major competitor for correspondence-based matching schemes in the realm of automatic face recognition are *eigenface approaches* [12], [15], [19]. They work very well if all models and the image are geometrically normalized and segmented. Achieving normalization of the image is tantamount to constructing a good correspondence map for at least the most significant points. Thus, eigenface techniques are an efficient way of organizing the memory of stored models, but they tend to be very sensitive to poor normalization and background changes not corrected for. From a comprehensive comparison the conclusion can be drawn that a bunch graph matching scheme performs about as well as the best eigenface methods, with a slight advantage in difficult cases such as pictures taken at an interval of more than a year [14].

There is evidence that the nonlinear pixel combinations presented by the Gabor amplitudes are closer to what goes on in the human recognition system. A detailed study [1] has shown that the similarity values found by the system from [22] are correlated with human judgments of similarity both in faces with and without hair. The authors conclude that this system captures essential *face* features better than eigenface approaches, which seem to be superior in capturing essential properties of a single *image*. Comparisons with the system from [25], [26] are currently in progress.

The systems presented in this chapter have been particularly successful in the recognition of human faces. However, they can also be applied to general object recognition; an example is elastic graph matching for object recognition [2] and bunch graph matching for the recognition of hand gestures [17]. Several problems remain with this attempt at generalization. General objects require many more views for even a single object, and for many objects the outlines are more important than the two-dimensionally extended Gabor features. Such features require different matching mechanisms; in [5], [6], an alternative version of graph matching is proposed which relies on outline graphs consisting of gray-value edges and corners rather than elastic graphs. Consequently, the matching algorithm is discrete subgraph matching rather than the stochastic optimization described in Section 4.1. A more general system should incorporate (at least) both types of features and must therefore use a novel matching method. Finally, much work remains to be done on the efficient organization of the model database to avoid recognition times proportional to the number of models, which are too large for realistic models of human perception.

There is now increasing evidence that the human brain employs different modules for the recognition of faces and general objects (see, e.g., [13]). The results obtained so far seem to indicate that correspondence-based recognition is a good model for face recognition but has to be complemented by other mechanisms in order to model general object recognition. The precise range of tasks where such methods are useful and plausible remains to be determined.

Acknowledgments

Most of the work described here has been done by or in close cooperation with other people. The author wishes to thank Christoph von der Malsburg, Jan C. Vorbrüggen, Laurenz Wiskott, Wolfgang Konen, Shaogang Gong, Stephen McKenna, and Tino Lourens for excellent cooperation and Jan C. Vorbrüggen for critical remarks on the manuscript.

References

[1] Hancock, P.J.B., Bruce, V., and Burton, M.A. (1998), "A comparison of two computer-based face identification systems with human perception of faces," *Vision Research*, 38:2277–2288.

[2] Kefalea, E., Rehse, O., and Von der Malsburg, C. (1997), "Object classification based on contours with elastic graph matching," *Proc. 3rd Int. Workshop on Visual Form, Capri, Italy*, pp. 287–297, World Scientific.

[3] Konen, W. and Schulze-Krüger, E. (1995), "ZN-Face: A system for access control using automated face recognition," in M. Bichsel (Ed.), *Proceedings of IWAFGR95, Zürich, June 1995*, pp. 18–23.

[4] Lades, M., Vorbrüggen, J.C., Buhmann, J., Lange, J., Von der Malsburg, C., Würtz, R.P., and Konen, W. (1993), "Distortion invariant

object recognition in the dynamic link architecture," *IEEE Transactions on Computers*, 42(3):300–311.

[5] Lourens, T. (1998), *A Biologically Plausible Model for Corner-based Object Recognition from Color Images*, Shaker, Maastricht.

[6] Lourens, T. and Würtz, R.P. (1998), "Object recognition by matching symbolic edge graphs," in R. Chin and T.-C. Pong (Eds.), *Computer Vision — ACCV'98*, volume 1352 of *Lecture Notes in Computer Science*, pp. II–193 – II–200, Springer Verlag.

[7] Von der Malsburg, C. (1981), "The correlation theory of brain function," technical report, Max-Planck-Institute for Biophysical Chemistry, Postfach 2841, Göttingen, FRG; Reprinted (1994) in Schulten, K. and Van Hemmen, H.J. (Eds.), *Models of Neural Networks*, Vol. 2, Springer.

[8] Von der Malsburg, C. (1985), "Nervous structures with dynamical links," *Ber. Bunsenges. Phys. Chem.*, 89:703–710.

[9] Maurer, T. and Von der Malsburg, C. (1995), "Single-view based recognition of faces rotated in depth," in M. Bichsel (Ed.), *Proceedings of IWAFGR95, Zürich, June 1995*, pp. 248–253.

[10] Maurer, T. and Von der Malsburg, C. (1996), "Tracking and learning graphs and pose on image sequences of faces," *Proceedings of the 2nd International Conference on Automatic Face- and Gesture-Recognition*, pp. 176–181, Killington, Vermont, USA.

[11] McKenna, S.J., Gong, S., Würtz, R.P., Tanner, J., and Banin, D. (1997), "Tracking facial feature points with Gabor wavelets and shape models," in J. Bigün, G. Chollet, and G. Borgefors (Eds.), *Proceedings of the First International Conference on Audio- and Video-based Biometric Person Authentication Crans-Montana, Switzerland, March 1997*, volume 1206 of *LNCS*, pp. 35–42, Springer Verlag.

[12] Moghaddam, B. and Pentland, A. (1997), "Probabilistic visual learning for object representation," *IEEE Trans. PAMI*, 19(7):696–710.

[13] Moscovitch, M., Winocur, G., and Behrmann, M. (1997), "What is special about face recognition?: Nineteen experiments on a person with visual object agnosia and dyslexia but normal face recognition," *J. Cognitive Neuroscience*, 9(5):555–604.

[14] Okada, K., Steffens, J., Maurer, T., Hong, H., Elagin, E., Neven, H., and Von der Malsburg, C. (1998), "The Bochum/USC Face Recognition System And How it Fared in the FERET Phase III test," in H. Wechsler, P.J. Phillips, V. Bruce, F. Fogeman Soulie, and T.S. Huang (Eds.), *Face Recognition: From Theory to Applications*, Springer Verlag.

[15] Sirovich, L. and Kirby, M. (1987), "Low-dimensional procedure for the characterization of human faces," *Journal of the Optical Society of America A*, 4:519–524.

[16] Steffens, J., Elagin, E., and Neven, H. (1998), "PersonSpotter – fast and robust system for human detection, tracking, and recognition," *Proceedings of the 3rd International Conference on Face and Gesture Recognition, Nara, Japan, April 1998*, pp. 516–521.

[17] Triesch, J. and Von der Malsburg, C. (1998), "A gesture interface for robotics," *FG'98, The IEEE Third International Conference on Automatic Face and Gesture Recognition*, pp. 546–551, IEEE Computer Society Press.

[18] Triesch, J. and Von der Malsburg, C. (1996), "Robust classification of hand postures against complex backgrounds," *Proceedings of the Second International Conference on Automatic Face and Gesture Recognition 1996, Killington, Vermont, USA, October 14–16*, pp. 170–175, IEEE Computer Society Press.

[19] Turk, M. and Pentland, A. (1991), "Eigenfaces for Recognition," *Journal of Cognitive Neuroscience*, 3(1):71–86.

[20] Wiskott, L. (1996), *Labeled Graphs and Dynamic Link Matching for Face Recognition and Scene Analysis*, Reihe Physik. Verlag Harri Deutsch, Thun, Frankfurt am Main.

[21] Wiskott, L. (1997), "Phantom faces for face analysis," *Pattern Recognition*, 30(6):837–846.

[22] Wiskott, L., Fellous, J.-M., Krüger, N., and Von der Malsburg, C. (1995), "Face recognition and gender determination," in M. Bichsel (Ed.), *Proceedings of IWAFGR95, Zürich, June 1995*, pp. 92–97.

[23] Wiskott, L., Fellous, J.-M., Krüger, N., and Von der Malsburg, C. (1997), "Face recognition by elastic bunch graph matching," *IEEE Transactions on Pattern Analysis and Machine Intelligence*, 19(7):775–779.

[24] Wiskott, L. and Von der Malsburg, C. (1996), "Face recognition by dynamic link matching," in J. Sirosh, R. Miikkulainen, and Y. Choe (Eds.), *Lateral Interactions in the Cortex: Structure and Function*, The UTCS Neural Networks Research Group, Austin, TX, Electronic book, ISBN 0-9647060-0-8, http://www.cs.utexas.edu/users/nn/web-pubs/htmlbook96.

[25] Würtz, R.P. (1995), *Multilayer Dynamic Link Networks for Establishing Image Point Correspondences and Visual Object Recognition*, volume 41 of *Reihe Physik*, Verlag Harri Deutsch, Thun, Frankfurt am Main.

[26] Würtz, R.P. (1997), "Object recognition robust under translations, deformations and changes in background," *IEEE Transactions on Pattern Analysis and Machine Intelligence*, 19(7):769–775.

[27] Würtz, R.P., Konen, W., and Behrmann, K.-O. (1996), "How fast can neuronal algorithms match patterns?" in C. von der Malsburg, J.C. Vorbrüggen, W. von Seelen, and B. Sendhoff (Eds.), *Artificial Neural Networks – ICANN 96*, volume 1112 of *Lecture Notes in Computer Science*, pp. 145–150, Springer Verlag.

[28] Würtz, R.P., Konen, W., and Behrmann, K.-O. (1998), "On the performance of neuronal matching algorithms," *Neural Networks*, in press.

[29] Würtz, R.P. and Lourens, T. (1997), "Corner detection in color images by multiscale combination of end-stopped cortical cells," in W. Gerstner, A. Germond, M. Hasler, and J.-D. Nicoud (Eds.), *Artificial Neural Networks – ICANN '97*, volume 1327 of *Lecture Notes in Computer Science*, pp. 901–906, Springer Verlag.

[30] Würtz, R.P. and Von der Malsburg, C. (1998), "A hierarchical dynamic link network for correspondence maps between image points," in preparation.

Chapter 11:

Face Recognition by Elastic Bunch Graph Matching

FACE RECOGNITION BY ELASTIC BUNCH GRAPH MATCHING[*][†]

Laurenz Wiskott[1][‡] **Jean-Marc Fellous**[2][§]
Norbert Krüger[1][¶] **and Christoph von der Malsburg**[1,2]

[1] Institute for Neural Computation
Ruhr-University Bochum
D-44780 Bochum, Germany

[2] Computer Science Department
University of Southern California
Los Angeles, CA 90089, USA

We present a system for recognizing human faces from single images out of a large database containing one image per person. The task is difficult because of image variation in terms of position, size, expression, and pose. The system collapses most of this variance by extracting concise face descriptions in the form of *image graphs*. In these, fiducial points on the face (eyes, mouth, etc.) are described by sets of wavelet components (*jets*). Image graph extraction is based on a novel approach, the *bunch graph*, which is constructed from a small set of sample image graphs. Recognition is based on a straightforward comparison of image graphs. We report recognition experiments on the FERET database as well as the Bochum database, including recognition across pose.

---

[*]Supported by grants from the German Federal Ministry for Science and Technology (413-5839-01 IN 101 B9) and from ARPA and the U.S. Army Research Lab (01/93/K-109).

[†]Portions reprinted, with permission, from IEEE Transactions on Pattern Analysis and Machine Intelligence 19(7):775–779, July 1997. ©1997 IEEE.

[‡]Current address: Institute for Advanced Studies, Wallotstrasse 19, D-14193 Berlin, Germany, wiskott@wiko-berlin.de.

[§]Current address: Computational Neurobiology Laboratory, The Salk Institute for Biological Studies, San Diego, CA 92186-5800, fellous@salk.edu.

[¶]Current address: Institute for Computer Science, Christian-Albrecht-University Kiel, Preusserstrasse 1-9, D-24105 Kiel, Germany, nkr@informatik.uni-kiel.de.

1 Introduction

We set ourselves the task of recognizing persons from single images by reference to a gallery, which also contained only one image per person. Our problem was to address image variation due to differences in facial expression, head pose, position, and size (to name only the most important). Our task is thus a typical discrimination-in-the-presence-of-variance problem, where one has to try to collapse the variance and to emphasize discriminating features. This is generally only possible with the help of information about the structure of variations to be expected.

Classification systems differ vastly in terms of the nature and origin of their knowledge about image variations. Systems in Artificial Intelligence and Computer Vision often stress specific designer-provided structures, for instance explicit models of three-dimensional objects or of the image-generation process, whereas Neural Network models tend to stress absorption of structure from examples with the help of statistical estimation techniques. Both of these extremes are expensive in their own way and fall painfully short of the ease with which natural systems pick up essential information from just a few examples. Part of the success of natural systems must be due to general properties and laws on how object images transform under natural conditions.

Our system has an important core of structure which reflects the fact that the images of coherent objects tend to translate, scale, rotate, and deform in the image plane. Our basic object representation is the labeled graph; edges are labeled with distance information and nodes are labeled with wavelet responses locally bundled in *jets*. Stored *model graphs* can be matched to new images to generate *image graphs*, which can then be incorporated into a gallery and become model graphs. Wavelets as we use them are robust to moderate lighting changes and small shifts and deformations. Model graphs can easily be translated, scaled, oriented, or deformed during the matching process, thus compensating for a large part of the variance of the images. Unfortunately, having only one image for each person in the galleries does not provide sufficient information to handle rotation in depth analogously. However, we present results on recognition across different poses.

This general structure is useful for handling any kind of coherent object and may be sufficient for discriminating between structurally different object types. However, for in-class discrimination of objects, of which face recognition is an example, it is necessary to have information specific to the structure common to all objects in the class. This is crucial for the extraction of those structural traits from the image which are important for discrimination ("to know where to look and what to pay attention to"). In our system, class-specific information has the form of *bunch graphs*, one for each pose, which are stacks of a moderate number (70 in our experiments) of different faces, jet-sampled in an appropriate set of fiducial points (placed over eyes, mouth, contour, etc.). Bunch graphs are treated as combinatorial entities in which, for each fiducial point, a jet from a different sample face can be selected, thus creating a highly adaptable model. This model is matched to new facial images to reliably find the fiducial points in the image. Jets at these points and their relative positions are extracted and are combined into an image graph, a representation of the face which has no remaining variation due to size, position (or in-plane orientation, not implemented here).

A bunch graph is created in two stages. Its qualitative structure as a graph (a set of nodes plus edges) as well as the assignment of corresponding labels (jets and distances) for one initial image is designer-provided, whereas the bulk of the bunch graph is extracted semi-automatically from sample images by matching the embryonic bunch graph to them, less and less often intervening to correct incorrectly identified fiducial points. Image graphs are rather robust to small in-depth rotations of the head. Larger rotation angles, i.e., different poses, are handled with the help of bunch graphs with a different graph structure and designer-provided correspondences between nodes in different poses.

After these preparations our system can extract from single images concise invariant face descriptions in the form of image graphs (called model graphs when in a gallery). They contain all information relevant for the face discrimination task. For the purpose of recognition, image graphs can be compared with model graphs at small computing cost by evaluating the mean jet similarity.

In summary, our system is based to a maximum on a general data structure — graphs labeled with wavelet responses — and general transformation properties. These are designer-provided, but due to their generality and simplicity the necessary effort is minimal. As described here our system makes use of hand-crafted object-specific graph structures and a moderately labor-intensive procedure to generate bunch-graphs. We plan to eliminate this need for human intervention and guess-work with the help of statistical estimation methods (cf. [16] and [21]). Our system comes close to the natural model by needing only a small number of examples to handle the complex task of face recognition.

In the discussion we will compare our system to others and to our own previous work. This work has been described in short form in [44], from which portions have been adopted for this text.

2 The System

2.1 Preprocessing with Gabor Wavelets

The representation of local features is based on the Gabor wavelet transform; see Figure 1. Gabor wavelets are biologically motivated convolution kernels in the shape of plane waves restricted by a Gaussian envelope function [7]. The set of convolution coefficients for kernels of different orientations and frequencies at one image pixel is called a jet. In this section we define jets, different similarity functions between jets, and our procedure for precise localization of jets in an image.

2.1.1 Jets

A *jet* describes a small patch of gray values in an image $\mathcal{I}(\vec{x})$ around a given pixel $\vec{x} = (x, y)$. It is based on a wavelet transform, defined as a convolution

$$\mathcal{J}_j(\vec{x}) = \int \mathcal{I}(\vec{x}')\psi_j(\vec{x} - \vec{x}')d^2\vec{x}' \quad (1)$$

with a family of *Gabor kernels*

$$\psi_j(\vec{x}) = \frac{k_j^2}{\sigma^2} \exp\left(-\frac{k_j^2 x^2}{2\sigma^2}\right)\left[\exp(i\vec{k}_j\vec{x}) - \exp\left(-\frac{\sigma^2}{2}\right)\right] \quad (2)$$

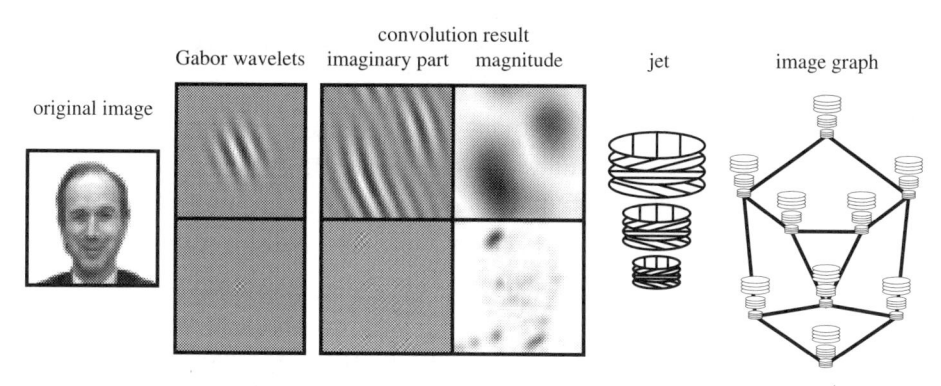

Face Recognition by Elastic Bunch Graph Matching

Figure 1. The graph representation of a face is based on the Gabor wavelet transform, a convolution with a set of wavelet kernels. These have the shape of plane waves restricted by a Gaussian envelope function. We compute 40 coefficients (5 frequencies × 8 orientations). Coefficient phase varies approximately with wavelet frequency (see imaginary part), magnitude varies slowly. The set of 40 coefficients obtained for one image point is referred to as a jet (for clarity, only 3 frequencies and 4 orientations are represented in the figure). A sparse collection of such jets together with some information about their relative location constitutes an image graph, used to represent an object, such as a face.

in the shape of plane waves with wave vector $\vec{k}_j$, restricted by a Gaussian envelope function. We employ a discrete set of 5 different frequencies, index $\nu = 0, ..., 4$, and 8 orientations, index $\mu = 0, ..., 7$,

$$\vec{k}_j = \begin{pmatrix} k_{jx} \\ k_{jy} \end{pmatrix} = \begin{pmatrix} k_\nu \cos \varphi_\mu \\ k_\nu \sin \varphi_\mu \end{pmatrix}, \; k_\nu = 2^{-\frac{\nu+2}{2}} \pi, \; \varphi_\mu = \mu \frac{\pi}{8}, \quad (3)$$

with index $j = \mu + 8\nu$. This sampling evenly covers a band in frequency space. The width σ/k of the Gaussian is controlled by the parameter $\sigma = 2\pi$. The second term in the bracket of Eq. (2) makes the kernels *DC-free*, i.e., the integral $\int \psi_j(\vec{x}) d^2\vec{x}$ vanishes. This is known as a wavelet transform because the family of kernels is self-similar, all kernels being generated from one *mother wavelet* by dilation and rotation.

A jet $\mathcal{J}$ is defined as the set $\{\mathcal{J}_j\}$ of 40 complex coefficients obtained for one image point. It can be written as

$$\mathcal{J}_j = a_j \exp(i\phi_j) \quad (4)$$

with magnitudes $a_j(\vec{x})$, which slowly vary with position, and phases $\phi_j(\vec{x})$, which rotate at a rate approximately determined by the spatial frequency or wave vector $\vec{k}_j$ of the kernels; see Figure 1.

Gabor wavelets were chosen for their robustness as a data format and for their biological relevance. Since they are DC-free, they provide robustness against varying brightness in the image. Robustness against varying contrast can be obtained by normalizing the jets. The limited localization in space and frequency yields a certain amount of robustness against translation, distortion, rotation, and scaling. Only the phase changes drastically with translation. This phase variation can be either ignored, or it can be used for estimating displacement, as will be shown later. A disadvantage of the large kernels is their sensitivity to background variations. It has been shown, however, that if the object contour is known the influence of the background can be suppressed [33]. Finally, the Gabor wavelets have a similar shape as the receptive fields of simple cells found in the visual cortex of vertebrate animals [8], [12], [32] and can be statistically derived from images of natural scenes, at least qualitatively [2], [25].

2.1.2 Comparing Jets

Due to phase rotation, jets taken from image points only a few pixels apart from each other have very different coefficients, although representing almost the same local feature. This can cause severe problems for matching. We therefore either ignore the phase or compensate for its variation explicitly. The similarity function

$$S_a(\mathcal{J}, \mathcal{J}') = \frac{\sum_j a_j a'_j}{\sqrt{\sum_j a_j^2 \sum_j a_j'^2}}, \qquad (5)$$

already used by [18], ignores phase. With a jet $\mathcal{J}$ taken at a fixed image position and jets $\mathcal{J}' = \mathcal{J}'(\vec{x})$ taken at variable position $\vec{x}$, $S_a(\mathcal{J}, \mathcal{J}'(\vec{x}))$ is a smooth function with local optima forming large attractor basins (see Figure 2a), leading to rapid and reliable convergence with simple search methods such as stochastic gradient descent.

Using phase has two potential advantages. Firstly, phase information is required to discriminate between patterns with similar magnitudes, should they occur, and secondly, since phase varies so quickly with location, it provides a means for accurate jet localization in an image. Assuming that two jets $\mathcal{J}$ and $\mathcal{J}'$ refer to object locations with small relative displacement $\vec{d}$, the phase shifts can be approximately compensated for by the terms $\vec{d}\vec{k}_j$, leading to a phase-sensitive similarity function

$$S_\phi(\mathcal{J},\mathcal{J}') = \frac{\sum_j a_j a'_j \cos(\phi_j - \phi'_j - \vec{d}\vec{k}_j)}{\sqrt{\sum_j a_j^2 \sum_j a'^2_j}}. \tag{6}$$

To compute it, the displacement $\vec{d}$ has to be estimated. This can be done by maximizing S_ϕ in its Taylor expansion, as explained in the following section. It is actually a great advantage of this second similarity function that it yields this displacement information. Profiles of similarities and estimated displacements are shown in Figure 2.

2.1.3 Displacement Estimation

To estimate the displacement vector $\vec{d} = (d_x, d_y)$, we have adopted a method used for disparity estimation [9], [36]. The idea is to maximize the similarity S_ϕ in its Taylor expansion:

$$S_\phi(\mathcal{J},\mathcal{J}') \approx \frac{\sum_j a_j a'_j [1 - 0.5(\phi_j - \phi'_j - \vec{d}\vec{k}_j)^2]}{\sqrt{\sum_j a_j^2 \sum_j a'^2_j}}. \tag{7}$$

Setting $\frac{\partial}{\partial d_x} S_\phi = \frac{\partial}{\partial d_y} S_\phi = 0$ and solving for $\vec{d}$ leads to

$$\vec{d}(\mathcal{J},\mathcal{J}') = \begin{pmatrix} d_x \\ d_y \end{pmatrix} = \frac{1}{\Gamma_{xx}\Gamma_{yy} - \Gamma_{xy}\Gamma_{yx}} \times \begin{pmatrix} \Gamma_{yy} & -\Gamma_{yx} \\ -\Gamma_{xy} & \Gamma_{xx} \end{pmatrix} \begin{pmatrix} \Phi_x \\ \Phi_y \end{pmatrix}, \tag{8}$$

if $\Gamma_{xx}\Gamma_{yy} - \Gamma_{xy}\Gamma_{yx} \neq 0$, with

$$\Phi_x = \sum_j a_j a'_j k_{jx}(\phi_j - \phi'_j),$$
$$\Gamma_{xy} = \sum_j a_j a'_j k_{jx} k_{jy},$$

and $\Phi_y, \Gamma_{xx}, \Gamma_{yx}, \Gamma_{yy}$ defined correspondingly. In addition, the phase differences may have to be corrected by $\pm 2\pi$ to put them in the range of $\pm\pi$.

This equation yields a straightforward method for estimating the displacement or disparity between two jets taken from object locations close

similarity without phase (a) -----
similarity with phase (b) ———
estimated displacement / 8 (c) ·····

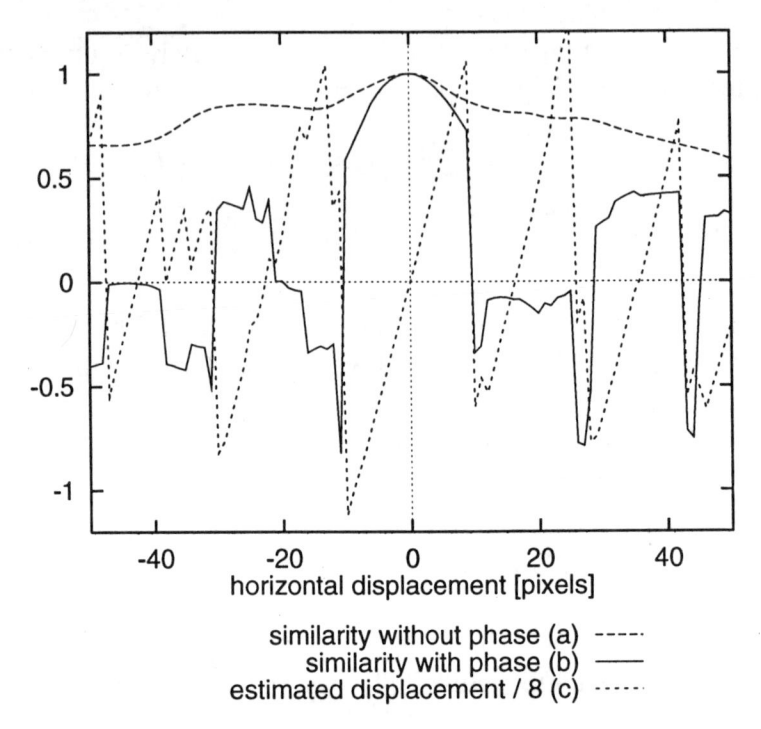

Figure 2. (a) Similarity $\mathcal{S}_a(\mathcal{J}(\vec{x}_1), \mathcal{J}'(\vec{x}_0))$ with jet $\mathcal{J}'$ taken from the left eye of the face shown in Figure 1 and jet $\mathcal{J}$ taken from pixel positions of the same horizontal line, $\vec{x}_1 = \vec{x}_0 + (d_x, 0), d_x = -50, ..., 50$ (The image in Figure 1 has a width of 128 pixels). The similarity potential is smooth and has a large attractor basin. (b) Similarity $\mathcal{S}_\phi(\mathcal{J}(\vec{x}_1), \mathcal{J}'(\vec{x}_0))$ and (c) estimated displacement $\vec{d}(\mathcal{J}(\vec{x}_1), \mathcal{J}'(\vec{x}_0))$ for the same jets as in (a) (using focus 1). The similarity potential has many more local optima. The right eye is 24 pixels away from the left eye, generating a local maximum for both similarity functions close to $d_x = -24$. The estimated displacement is precise around the 0-position and rougher at other local optima, especially at the other eye. (The displacement values are divided by 8 to fit the ordinate range.)

enough that their Gabor kernels are highly overlapping. Without further modifications, this equation can determine displacements up to half the wavelength of the highest frequency kernel, which would be two pixels for $k_0 = \pi/2$. The range can be increased by using low frequency kernels only. For the largest kernels, the estimated displacement may be 8 pixels. One can then proceed with the next higher frequency level and refine the result. When stepping to the next higher frequency, the phases of the higher frequency coefficients have to be corrected by multiples of 2π to match as closely as possible the expected phase differences inferred from the displacement estimated on the lower frequency level. This correction may lead to absolute phase differences larger than π. We refer to the number of frequency levels used for the first displacement estimation as *focus*. A focus of 1 indicates that only the lowest frequency level is used and that the estimated displacement may be up to 8 pixels. A focus of 5 indicates that all five levels are used, and the disparity may only be up to 2 pixels. In any case, all five levels are eventually used in the iterative refinement process described above.

If one has access to the whole image of jets, one can also work iteratively. Assume a jet $\mathcal{J}$ is to be accurately positioned in the neighborhood of point $\vec{x}_0$ in an image. Comparing $\mathcal{J}$ with the jet $\mathcal{J}_0 = \mathcal{J}(\vec{x}_0)$ yields an estimated displacement of $\vec{d}_0 = \vec{d}(\mathcal{J}, \mathcal{J}(\vec{x}_0))$. Then a jet $\mathcal{J}_1$ is taken from position $\vec{x}_1 = \vec{x}_0 + \vec{d}_0$ and the displacement is estimated again. But since the new location is closer to the correct position, the new displacement $\vec{d}_1$ will be smaller and can be estimated more accurately with a higher focus, converging eventually to subpixel accuracy. We have used this iterative scheme in the matching process described in Section 2.3.

2.2 Face Representation

2.2.1 Individual Faces

For faces, we have defined a set of *fiducial points*, e.g., the pupils, the corners of the mouth, the tip of the nose, the top and bottom of the ears, etc. A *labeled graph* $\mathcal{G}$ representing a face consists of N nodes on these fiducial points at positions $\vec{x}_n, n = 1, ..., N$ and E edges between them. The nodes are labeled with jets $\mathcal{J}_n$. The edges are labeled with distances $\Delta\vec{x}_e = \vec{x}_n - \vec{x}_{n'}, e = 1, ..., E$, where edge e connects node n' with n. Hence the edge labels are two-dimensional vectors. (When referring to the geometrical structure of a graph, unlabeled by jets, we call it a *grid*.) This face graph is *object-adapted*, since the nodes are selected from face-specific points (fiducial points); see Figure 4.

Graphs for different head pose differ in geometry and local features. Although the fiducial points refer to corresponding object locations, some may be occluded, and jets as well as distances vary due to rotation in depth. To be able to compare graphs for different poses, we have manually defined pointers to associate corresponding nodes in the different graphs.

2.2.2 Face Bunch Graphs

To find fiducial points in new faces, one needs a general representation rather than models of individual faces. This representation should cover a wide range of possible variations in the appearance of faces, such as differently shaped eyes, mouths, or noses, different types of beards, variations due to sex, age, race, etc. It is obvious that it would be too expensive to cover each feature combination by a separate graph. We instead combine a representative set of individual model graphs into a stack-like structure, called a *face bunch graph* (FBG); see Figure 3. Each model has the same grid structure and the nodes refer to identical fiducial points. A set of jets referring to one fiducial point is called a *bunch*. An eye bunch, for instance, may include jets from closed, open, female, and male eyes, etc., to cover these local variations. During the location of fiducial points in a face not seen before, the procedure described in the next section selects the best fitting jet, called the *local expert*, from the bunch dedicated to each fiducial point. Thus, the full combination of jets in the bunch graph is available, covering a much larger range of facial variation than represented in the constituting model graphs themselves. A similar data structure based on templates has been developed independently by Beymer [3].

Assume for a particular pose that there are M model graphs $\mathcal{G}^{\mathcal{B}m}$ ($m = 1, ..., M$) of identical structure, taken from different model faces. The corresponding FBG $\mathcal{B}$ is then given the same structure, its nodes are labeled with bunches of jets $\mathcal{J}_n^{\mathcal{B}m}$ and its edges are labeled with the averaged distances $\Delta \vec{x}_e^{\mathcal{B}} = \sum_m \Delta \vec{x}_e^{\mathcal{B}m} / M$.

How large should an FBG be and which models should be included? This depends first of all on the variability of faces one wants to represent. If the faces are of many different races, facial expression, age, etc., the FBG must contain many different models to cope with this variability. The required FBG size also increases with the desired matching accuracy for finding the fiducial points in a new face. The accuracy can be estimated

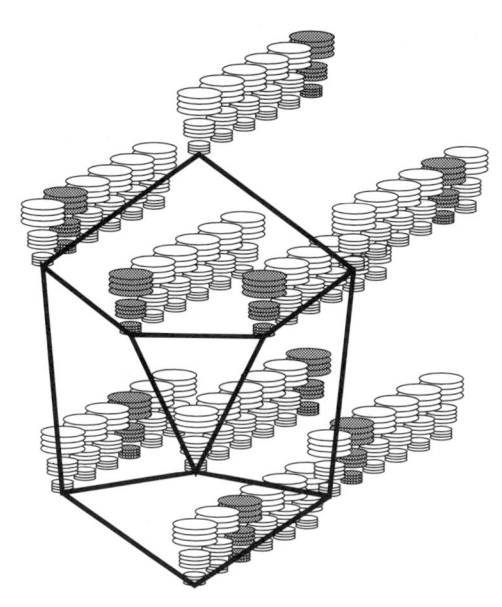

Figure 3. The Face Bunch Graph (FBG) serves as a representation of faces in general. It is designed to cover all possible variations in the appearance of faces. The FBG combines information from a number of face graphs. Its nodes are labeled with sets of jets, called bunches, and its edges are labeled with averages of distance vectors. During comparison to an image, the best fitting jet in each bunch, indicated by gray shading, is selected independently.

by matching the FBG to face images for which the fiducial points have been verified manually; cf. Section 3.2.3. FBG size does not depend on gallery size. In general, the models in the FBG should be as different as possible to reduce redundancy and maximize variability. Here we used FBGs with 30 models for the normalization stage and 70 models for the final graph extraction stage; cf. Section 2.3.4. These sizes seemed to give sufficient matching accuracy and reliability. We selected the models arbitrarily and did not optimize for maximal variability.

2.3 Generating Face Representations by Elastic Bunch Graph Matching

So far we have only described how individual faces and general knowledge about faces are represented by labeled graphs and the FBG, respectively. We are now going to explain how these graphs are generated. The simplest method is to do so manually. We have used this method to generate initial graphs for the system, one graph for each pose, together with pointers to indicate which pairs of nodes in graphs for different poses correspond to each other. Once the system has an FBG (possibly consisting of only one manually defined model), graphs for new images can be generated automatically by Elastic Bunch Graph Matching. Initially, when the FBG contains only a few faces, it is necessary to review and correct the resulting matches, but once the FBG is rich enough (approximately 70 graphs) one can rely on the matching and generate large galleries of model graphs automatically.

2.3.1 Manual Definition of Graphs

Manual definition of graphs is done in three steps. First, we mark a set of fiducial points for a given image. Most of these are positioned at well-defined features which are easy to locate, such as left and right pupil, the corners of the mouth, the tip of the nose, the top and bottom of the ears, the top of the head, and the tip of the chin. These points were selected to make manual positioning easy and reliable. Additional fiducial points are positioned at the center of gravity of certain easy-to-locate fiducial points. This allows automatic selection of fiducial points in regions where well-defined features are missing, e.g., at the cheeks or the forehead. Then, edges are drawn between fiducial points and edge labels are automatically computed as the differences between node positions. Finally, the Gabor wavelet transform provides the jets for the nodes.

In general, the set of fiducial points should cover the face evenly. But depending on the task, it may be appropriate to emphasize certain regions by additional nodes. For face finding, for example, we place more nodes on the outline, because with homogeneous background the contour is a good cue for finding faces. For face recognition, on the other hand, we place more nodes in the interior of the faces, because of its importance for recognition. A more systematic way of selecting nodes from a dense set is presented in [15] and [17]. More nodes tend to yield better results, because more information is used, but this effect saturates if the nodes are too close and the corresponding Gabor coefficients become highly correlated due to overlap between the kernels. On the other hand, the computational effort increases linearly with the number of nodes. The optimal number of nodes will therefore be a compromise between recognition performance and speed.

2.3.2 The Graph Similarity Function

A key role in Elastic Bunch Graph Matching is played by a function evaluating the *graph similarity* between an image graph and the FBG of identical pose. It depends on the jet similarities and the distortion of the image grid relative to the FBG grid. For an image graph $\mathcal{G}^\mathcal{I}$ with nodes $n = 1, ..., N$ and edges $e = 1, ..., E$ and an FBG $\mathcal{B}$ with model graphs $m = 1, ..., M$ the similarity is defined as

$$S_\mathcal{B}(\mathcal{G}^\mathcal{I}, \mathcal{B}) = \frac{1}{N} \sum_n \max_m \left(S_\phi(\mathcal{J}_n^\mathcal{I}, \mathcal{J}_n^{\mathcal{B}m})\right) - \frac{\lambda}{E} \sum_e \frac{(\Delta \vec{x}_e^\mathcal{I} - \Delta \vec{x}_e^\mathcal{B})^2}{(\Delta \vec{x}_e^\mathcal{B})^2}, \quad (9)$$

where λ determines the relative importance of jets and metric structure. $\mathcal{J}_n$ are the jets at node n, and $\Delta \vec{x}_e$ are the distance vectors used as labels at edges e. Since the FBG provides several jets for each fiducial point, the best one is selected and used for comparison. These best fitting jets serve as *local experts* for the image face.

2.3.3 Matching Procedure

The goal of Elastic Bunch Graph Matching on a probe image is to find the fiducial points and thus to extract from the image a graph which maximizes the similarity with the FBG as defined in Eq. (9). In practice, one has to apply a heuristic algorithm to come close to the optimum within a reasonable time. We use a coarse to fine approach in which we introduce the degrees of freedom of the FBG progressively: translation, scale, aspect ratio, and finally local distortions. We similarly introduce phase information and increase the focus of displacement estimation: no phase, phase with focus 1, and then phase with focus 1 up to 5. The matching schedule described here assumes faces of known pose and approximately standard size, so that only one FBG is required. The more general case of varying size is sketched in the next section.

Step 1: Find approximate face position. Condense the FGB into an *average graph* by taking the average magnitudes of the jets in each bunch of the FBG (or, alternatively, select one arbitrary graph as a representative). Use this as a rigid model ($\lambda = \infty$) and evaluate its similarity at each location of a square lattice with a spacing of 4 pixels. At this step the similarity function $\mathcal{S}_a$ without phase is used instead of $\mathcal{S}_\phi$. Repeat the scanning around the best fitting position with a spacing of 1 pixel. The best fitting position finally serves as the starting point for the next step.

Step 2: Refine position and size. Now the FBG is used without averaging, varying it in position and size. Check the four different positions ($\pm 3, \pm 3$) pixels displaced from the position found in Step 1, and at each position check two different sizes which have the same center position, a factor of 1.18 smaller or larger than the FBG average size. This is without effect on the metric similarity, since the vectors $\vec{x}_e^B$ are transformed accordingly. We still keep $\lambda = \infty$. For each of these eight variations, the best fitting jet for each node is selected and its displacement according to Eq. (8) is computed. This is done with a focus of 1, i.e., the displacements may be of a magnitude up to eight pixels. The grids are then rescaled and repositioned to minimize the square sum over the displacements. Keep the best of the eight variations as the starting point for the next step.

Step 3: Refine size and find aspect ratio. A similar relaxation process as described for Step 2 is applied, but relaxing the x- and y-dimensions independently. In addition, the focus is increased successively from 1 to 5.

Step 4: Local distortion. In a pseudo-random sequence the position of each individual image node is varied to further increase the similarity to the FBG. Now the metric similarity is taken into account by setting $\lambda = 2$ and using the vectors $\vec{x}_e^B$ as obtained in Step 3. In this step only those positions are considered for which the estimated displacement vector is small ($d < 1$, see Eq. (8)). For this local distortion the focus again increases from 1 to 5.

The resulting graph is called the *image graph* and is stored as a representation of the individual face of the image.

Face Recognition by Elastic Bunch Graph Matching

grids for face finding

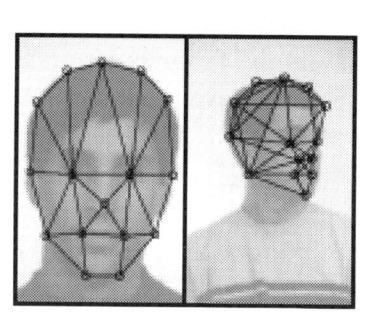

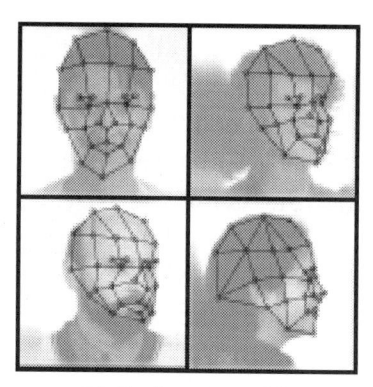

grids for face recognition

Figure 4. Object-adapted grids for different poses. The nodes are positioned automatically by elastic graph matching against the corresponding face bunch graphs. The two images on the left show originals with widely differing size and grids, as used for the normalization stage with many nodes on the outline for reliable face finding. The images on the right are already rescaled to normal size. Their grids have more nodes on the face, which is more appropriate for recognition (The grids used in Section 3.2 had about 14 additional nodes, which, for simplicity, are not shown here). One can see that, in general, the matching finds the fiducial points quite accurately. But mismatches occurred, for example, for the bearded man. The chin was not found accurately; the leftmost node and the node below it should be at the top and the bottom of the ear respectively.

2.3.4 Schedule of Graph Extraction

To minimize computing effort and to optimize reliability, we extract a face representation in two stages, each of which uses a matching procedure as described in the previous section. The first stage, called the *normalization stage* and described in greater detail in [17], has the purpose of estimating the position and size of the face in the original image, so that the image can be scaled and cut to standard size. The second stage takes this image as input and extracts a precise image graph appropriate for face recognition purposes. The two stages differ in emphasis. The first one has to deal with greater uncertainty about size and position of the head and has to optimize the reliability with which it finds the face, but there is no need to find fiducial points with any precision or extract data important for face recognition. The second stage can start with little uncertainty about position and size of the head, but has to extract a detailed face graph with high precision.

In the experiments on the FERET database described below, original images had a format of 256×384 pixels, and the faces varied in size by a factor of three; see Figure 4. The poses were given and did not need to be determined. The normalization stage used three FBGs of appropriate pose which differed in face size. We somewhat arbitrarily picked approximately 30 images to form each FBG. More careful selection of images to cover a wider range of variations can only improve system performance. The grids used in the construction of the FBGs put little emphasis, i.e., few nodes, on the interior of the face and have fewer nodes than those used for the second stage; see Figure 4 for two examples. The smaller number of nodes speeds up the process of face finding. Using a matching scheme similar to the one described in Section 2.3.3, we match each of the three FBGs to the input image. We select the graph that matches best, cut a frame of appropriate size around it from the image and resize it to 128×128 pixels. The poses could be determined analogously [17], but here the poses are assumed to be known. In our experiments, normalization took approximately 20 seconds on a SPARCstation 10-512 with a 50 MHz processor and identified face position and scale correctly in approximately 99% of the images.

The second stage uses the matching procedure exactly as described in Section 2.3.3, starting the match at standard size and position. The face bunch graphs used in this stage have more nodes, which we have placed in positions we believe are important for person identification, emphasizing the interior of the face. Each of the three principal poses (frontal, half-profile, and profile; left-facing poses are flipped to right-facing poses) is matched with a different grid structure and with a different FBG, formed by using 70 arbitrarily chosen images. This stage took approximately 10 seconds.

2.4 Recognition

After having extracted model graphs from the gallery images and image graphs from the probe images, recognition is possible with relatively little computational effort by comparing an image graph to all model graphs and picking the one with the highest similarity value. A comparison against a gallery of 250 individuals took slightly less than a second. The similarity function we use here for comparing graphs is an average over the similarities between pairs of corresponding jets. For image and model graphs referring to different pose, we compare jets according to the manually provided correspondences. If $\mathcal{G}^\mathcal{I}$ is the image graph, $\mathcal{G}^\mathcal{M}$ is the model graph, and node $n_{n'}$ in the model graph corresponds to node n' in the image graph, we define *graph similarity* as:

$$\mathcal{S}_\mathcal{G}(\mathcal{G}^\mathcal{I},\mathcal{G}^\mathcal{M}) = \frac{1}{N'} \sum_{n'} \mathcal{S}_a(\mathcal{J}^\mathcal{I}_{n'}, \mathcal{J}^\mathcal{M}_{n_{n'}}), \qquad (10)$$

where the sum runs only over the N' nodes in the image graph with a corresponding node in the model graph. We use the jet similarity function without phase here. It turned out to be more discriminative, possibly because it is more robust with respect to change in facial expression and other variations. Here we ignore the jet distortions created by rotation in depth, but will take up the subject in the discussion.

This graph similarity induces a ranking of the model graphs relative to an image graph. A person is recognized correctly if the correct model yields the highest graph similarity, i.e., if it is of rank one. A confidence criterion on how reliably a person is recognized can easily be derived from the statistics of the ranking [18]. However, we have restricted our results to unqualified recognition rates, which already give a good impression of system performance.

3 Experiments

3.1 Databases

For the experiments we used image galleries taken from two different databases. Both of them explicitly distinguish different poses, and images are labeled with pose identity.

The first one is the ARPA/ARL FERET database provided by the US Army Research Laboratory. The poses are: frontal, quarter view, half-profile right or left (rotated by about 40-70°), and profile right or left; see Figure 5 for examples. We disregarded quarter view images, because there were only a few of them. For most faces there are two frontal views with different facial expression. Apart from a few exceptions, there are no disguises, variations in hair-style or in clothing. The background is always a homogeneous light or gray, except for smoothly varying shadows. The size of the faces varies by about a factor of three (but is constant for each individual, information which we could have used to improve recognition rates, but didn't). The format of the original images is 256×384 pixels.

frontal views A and B   half-profiles left and right   profiles left and right

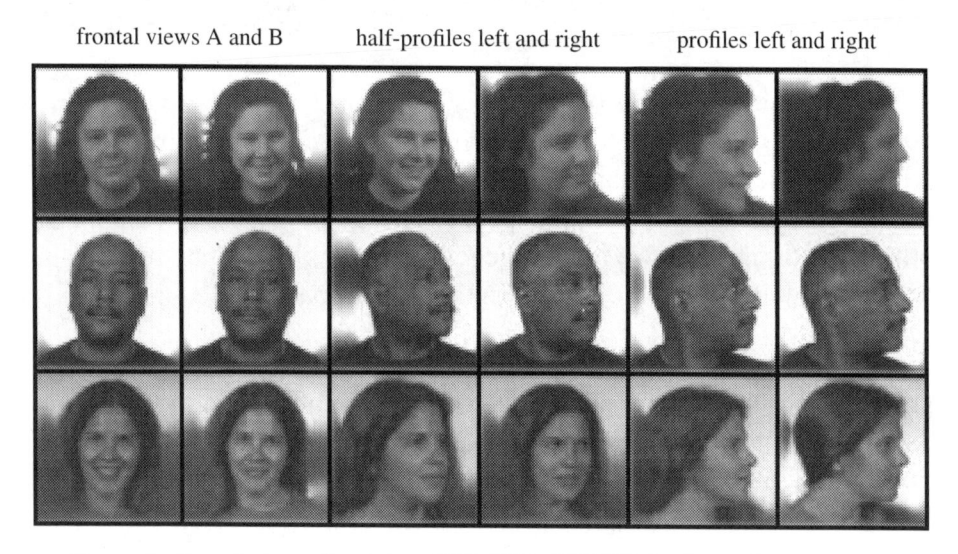

Figure 5. Sample faces from the ARPA/ARL FERET database: frontal views, half-profiles, and profiles. Pictures for left-facing poses are flipped around a vertical axis, and all images have been rescaled to standard size by our normalization stage (Section 2.3.4). Notice the large variation in the rotation angle of half-profiles and that some faces have no variation in facial expression in the two frontal views.

The second database has been collected at the Institute for Neural Computation in Bochum and has been partly described in [18]. The poses are: frontal, 11° and 22° rotated. The 11° and 22° angles have been estimated from the distance between the eyes (THOMAS MAURER, personal communication, 1996). People were actually told to orient towards 15° and 30° marks on the wall, but these angles hold only for the gaze; the heads are usually less rotated. The 11° rotated faces are referred to as 15° rotated in [18]. For all faces there are two frontal views, one neutral and one with a different facial expression. The latter includes a few cases where half of the face is occluded by hair or a hand. Two frontal views in this database generally differ more than those in the FERET database. All images were taken with the same set-up, so that faces varied in size only within the natural range. Our tests on this database allow direct comparison with the preceding system [18]. The description of our algorithm in Section 2 referred to the FERET database. For the Bochum database we did not use the normalization stage, because faces varied only a little in size. For matching, the FBGs were comprised of all the available images except for the image onto which the FBG was matched.

3.2 Results

3.2.1 FERET Database

We used various model and probe galleries with faces of different pose. Each model gallery contained 250 faces with just one image per person. We relied on the explicitly labeled pose identity instead of using our own pose recognition capability. Recognition results are shown in Table 1 (from [44]).

Table 1. Recognition results for cross-runs between different galleries (f: frontal views; a, b: expression a and b; h: half-profiles; p: profiles; l, r: left and right). Each gallery contained only one image per person; the different compositions in the four bottom rows are due to the fact that not all poses were available for all people. The table shows how often the correct model was identified as rank one and how often it was among the first 10 (4%).

Model gallery	Probe images	First rank #	First rank %	First 10 ranks #	First 10 ranks %
250 fa	250 fb	245	98	248	99
250 hr	181 hl	103	57	147	81
250 pr	250 pl	210	84	236	94
249 fa + 1 fb	171 hl + 79 hr	44	18	111	44
171 hl + 79 hr	249 fa + 1 fb	42	17	95	38
170 hl + 80 hr	217 pl + 33 pr	22	9	67	27
217 pl + 33 pr	170 hl + 80 hr	31	12	80	32

The recognition rate is very high for frontal against frontal images (first row). This is mainly due to the fact that in this database two frontal views show little variation, and any face recognition system should perform well under these circumstances, cf. Table 3. See the results on the Bochum database for a more challenging test.

Before comparing left against right poses, we flipped all left pose images over. Since human heads are to some degree bilaterally symmetric and since our present system performs poorly on such large rotations in depth (see below), we proceeded under the expectation that it would be easier to deal with differences due to facial asymmetry than with differences caused by substantial head rotation. This assumption is borne out at least by the high recognition rate of 84% for right profile against left profile (third row). The sharply reduced recognition rate of 57% (second row) when comparing left and right half-profiles could be due to inherent facial asymmetry, but the more likely reason is the poor control in rotation angle in the database — inspection of images shows that right and left rotation angles differ by up to 30°, cf. Figure 5.

When comparing half-profiles with either frontal views or profiles, another reduction in recognition rate is observed (although even a correct recognition rate of 10% out of a gallery of 250 is still high above chance level, which would be 0.4%!). The results are asymmetrical, performance being better if frontal or profile images serve as model gallery than if half-profiles are used. This is due to the fact that both frontal and profile poses are much more standardized than half-profiles, for which the angle varies approximately between 40° and 70°. We interpret this as being due to the fact that similarity is more sensitive to depth-rotation than to inter-individual face differences. Thus, when comparing frontal probe images to a half-profile gallery, a 40° half-profile gallery image of a wrong person is often favored over the correct gallery image if in the latter the head is rotated by a larger angle. A large number of such false positives degrades the correct-recognition rate considerably. In these experiments we also flipped all left pose images over, so that, to a large extent, the recognition was not only done across pose but also across mirror reflection.

3.2.2 Bochum Database

We used 108 neutral frontal views as a model gallery and the other images as probe galleries. For comparison we also give recognition rates obtained with the preceding system [18]. Recognition results are shown in Table 2 (from [44]).

Face Recognition by Elastic Bunch Graph Matching

Table 2. Recognition results for cross-runs between different galleries (f: frontal views; a: neutral; b: different facial expression; 11°, 22°: rotated faces). Each gallery contained only one image per person. For matching, only face bunch graphs of frontal pose and 22° pose were used. Matching on 11° pose images was done with the frontal pose face bunch graph. The table shows how often the correct model was identified as rank one and how often it was among the first 4 (4%). The results for the preceding system have been obtained with the original software of [18]. All results are from [44].

Model gallery	Probe images	Preceding system First rank #	Preceding system First rank %	This system First rank #	This system First rank %	This system First 4 ranks #	This system First 4 ranks %
108 fa	108 fb	99	92	98	91	102	94
108 fa	108 11°	105	97	101	94	105	97
108 fa	108 22°	92	85	95	88	103	95

On this database, recognition rates for frontal views are lower than on the FERET database. This is due to the fact that the frontal views in the Bochum gallery differ in facial expression more than those in the FERET database. This is consistent with the results for 11°-rotated but neutral faces, which are higher than those for the frontal views. This also indicates that the variation due to facial expression is relatively large.

Comparison with the preceding system [18] shows that both systems perform equally well on the Bochum gallery. In comparing the results, two differences in the algorithms should be taken into account; see Section 4.1.1 for a discussion of the algorithmic differences between the two systems. Firstly, the preceding system used 70 nodes while our system used only 30 nodes for the Bochum galleries. We have noticed that taking more nodes leads to higher recognition rates. Secondly, the preceding system did not scale model grids to compensate for size variations. Since all images were taken from the same distance, faces had natural size variation relative to the fixed grid and size could then implicitly contribute to recognition. Though our system did not resize the images themselves, the grids were averaged in size to generate the face bunch graph and matched to the size of each face individually with no cost to the similarity function (although, the jets were not scaled). Thus our system yields the same recognition performance with less information.

The preceding system [18] was implemented on a system with 23 transputers. The Gabor wavelet transform required less than 7 seconds. Comparing one image with a gallery of 87 models required about 25 seconds. On a SPARCstation 10-512, a single comparison of an image graph with a model graph would probably require about 0.2 seconds. The system presented here requires less than 30 seconds to extract the image graph once and can then compare with about 300 model graphs per second. Thus, there is a significant increase in speed of our system over the preceding system for large galleries.

3.2.3 Matching Accuracy

We have introduced phase information to improve matching accuracy. Thomas Maurer [personal communication, 1996] has tested the accuracy on the Bochum database by matching a face bunch graph onto images for which all fiducial points were controlled manually. He always left the person on the image out of the face bunch graph, so that no information about that particular person could be used for matching. He ran the same algorithm with phase information and without phase information, i.e., all phases set to zero. Matching accuracy was calculated as the mean Euclidean distance between matching positions and manually controlled reference positions. It was 1.6 with and 5.2 pixels without phase, and the histograms had their maximum at 1 and 4 pixels distance, respectively. The images had a size of 128×128 pixels. Notice that since the reference positions were set manually, one cannot expect a matching accuracy much better than one pixel. This is because manual positioning focuses more on local high frequency information, while the matching system takes into account low frequencies as well. In addition, the manual positioning may be inaccurate itself. One can get an impression of the matching accuracy of the preceding system from Figure 6 in [18]. A typical effect without phase is that a node is positioned at the wrong side of an edge, e.g., fifth node from the top in the rightmost column.

To investigate the importance of matching accuracy for recognition performance, Maurer has performed three different cross-runs of the 22° probe images against the neutral frontal view gallery. In the first run he used manually controlled node positions for the probe images, in the second run positions were obtained by matching with phase information,

Face Recognition by Elastic Bunch Graph Matching

and in the third run without phase information. The frontal gallery was always the same. Numbers of correctly recognized faces in these three cases were 96 (89%) for manual positioning, 95 (88%) with phase, and 72 (67%) without phase. This shows that the matching accuracy with phase is sufficient, while using no phase would cause a significant degradation in recognition performance.

Notice that the preceding system [18] achieves high recognition rates without using phase information for the matching. Two reasons for this may be the larger number of nodes and the advantage of using different grid sizes if faces are of different but reliable size, as discussed above. Another reason is that phase information becomes more important as more degrees of freedom are introduced. Apart from local distortions, the preceding system only varied the location of individual graphs while our system also varied grid size, aspect ratio, and the identity of the local experts during the matching. With these additional degrees of freedom, the matching is more likely to fail without phase information, while the preceding algorithm was still robust.

Another study on face recognition was also based on face bunch graphs (including the correct face) and Gabor jets, but the matching algorithm was much simpler and constrained to a sparse grid of points in the image [42]. Matching accuracy was therefore limited by the spacing of the grid points, which was 8 pixels in images of 128×128 pixels. Tests on the same 108 images of the Bochum database as used here confirmed that matching with phase (and recognition without phase) yields the highest recognition rates. However, it was surprising that for frontal views with different mimic expression (fb) and for 11° rotated faces such a simple matching algorithm achieved recognition rates of 92% and 94%, respectively, which is comparable to the performance of our system. It was only for the 22° rotated faces that the more sophisticated method presented here performed significantly better (88%) than the simple matching algorithm (81%).

4 Discussion

The system presented here is general and flexible. It is designed for an *in-class recognition* task, i.e., for recognizing members of a known class of objects. We have applied it to face recognition, but the system is in no way specialized in faces, and we assume that it can be directly applied to other in-class recognition tasks, such as recognizing individuals of a given animal species, given the same level of standardization of the images. In contrast to many neural network systems, no extensive training for new faces or new object classes is required. Only a moderate number of typical examples have to be inspected to build up a bunch graph, and individuals can then be recognized after storing a single image.

We tested the system with respect to rotation in depth and differences in facial expression. We did not investigate robustness to other variations, such as illumination changes or structured background. The performance is high on faces of the same pose. We also showed robustness against rotation in depth up to about 22°. For large rotation angles the performance degrades significantly.

4.1 Comparison to Other Systems

There is a considerable literature on face recognition, and many different techniques have been applied to the task; see [5], [34], [38] for reviews. Here we relate our system to those of others in regard to conceptual and performance aspects.

4.1.1 Comparison to the Preceding System

We developed the system presented here based on [18], with several major modifications. We now utilize wavelet phase information for accurate node localization. Previously, node localization was rather imprecise. We have introduced the potential to specialize the system for specific object types and to handle different poses with the help of object-adapted grids. The face bunch graph is able to represent a wide variety of faces, which allows matching on face images of previously unseen individuals. These improvements make it possible to extract an image graph from

Face Recognition by Elastic Bunch Graph Matching

a new face image in one matching process. Even if the person of the new image is not included in the FBG, the image graph reliably refers to the fiducial points. This considerably accelerates recognition from large databases, since for each probe image, correct node positions need to be searched only once instead of in each attempted match to a gallery image, as was previously necessary. The ability of the new system to refer to object-fixed fiducial points irrespective of pose represents an advantage in itself and is essential for some interesting graph operations; cf. Section 4.2. Computational efficiency, the ability to deal with different poses explicitly, and greater potential for further developments are the major advantages of the new system compared to the preceding one. We did not expect and experiments do not show an immediate improvement of recognition performance on faces of similar orientation.

4.1.2 Recognizing Faces of the Same View

Some face recognition systems are based on user-defined face-specific features. Yuille [47], for example, represented eyes by a circle within an almond-shape and defined an energy function to optimize a total of 9 model parameters for matching it to an image. The drawback of these systems is that the features as well as the procedures to extract them must be newly defined and programmed by the user for each object class, and the system has no means to adapt to samples for which the feature models fail. For example, the eye models mentioned above may fail for faces with sunglasses or have problems if the eyes are closed. In these cases the user has to design new features and new algorithms to extract them. With this type of approach, the system can never be weaned from designer intervention. Our system, in contrast, can be taught exceptional cases, such as sunglasses or beards, or entirely new object classes, by the presentation of examples and incorporation into bunch graphs.

An approach to face recognition also avoiding user-defined features is based on principal component analysis (PCA) [13], [26], [35], [37]. In this approach, faces are first aligned with each other and then treated as high-dimensional pixel vectors from which eigenvectors, so-called eigenfaces, are computed, together with the corresponding eigenvalues. A probe face is decomposed with respect to these eigenvectors and is efficiently represented by a small number, say 30, of expansion coefficients.

(The necessary image alignment can be done automatically within the PCA framework [23], [37]). PCA is optimal with respect to data compression and is successful for recognition purposes.

In its original, simple form, PCA treats an entire face as one vector, which causes two major problems. Firstly, because PCA is linear in the image space, it cannot deal well with variations in geometry. Consider, for example, two faces which have the mouth at a different height. Any linear combination of these two images can only generate a mouth at either height or two superimposed mouths but never a natural-looking mouth at an intermediate height. Linear combinations of images do not interpolate geometry. As a consequence, face images have to be aligned carefully before applying PCA or computing the expansion coefficients. Usually, face images are at least scaled, rotated, and shifted to align the eyes; sometimes also the aspect ratio is changed to also align the mouth. But then other facial features may still be misaligned. A solution to this problem is to factorize geometry and texture completely by warping face images to an average geometry. PCA is then applied to the warped face and to the geometrical features separately. Craw *et al.* [6] have shown that this technique is advantageous for recognition. They used manually defined fiducial points for warping. Lanitis *et al.* [19] apply a graph-matching algorithm similar to ours to find the fiducial points. Even more extreme in this sense are the systems in [4], [39], and [40]. They use an image-flow algorithm to match each pixel of a face image to a pixel in a different image. The warping is correspondingly accurate. These latter systems, however, require carefully taken images of high quality and are less robust against perturbations, such as occlusions or glasses.

A second problem of the original PCA approach is its sensitivity to occlusions or other localized perturbations, such as variations in hair style or facial hair. In a more localized feature representation, some regions can be explicitly treated as occluded, yielding good recognition results despite large occlusions [45]. In the holistic representation of PCA, any local image perturbation will have an effect on all expansion coefficients and cannot be easily disregarded. One way this problem has been dealt with was by treating small image regions centered on fiducial points (eyes, nose, mouth) as additional pixel vectors from which to extract more features by PCA [23]. A more systematic approach has been developed by Penev and Atick [27]. They explore spatial correlations within the set of eigenvectors found by PCA and generate a redundant set of localized kernels, one for each pixel location. Applying these kernels is called local feature analysis. A few of the local kernel responses are selected to generate a sparse representation.

The PCA approach with both extensions, a matching and warping stage at the beginning and PCA on localized regions or local feature analysis, becomes quite similar to our approach. A graph of fiducial points, labeled with example-derived feature vectors is matched to the image, and the optimally matching "grid" is used to extract a structural description from the image in the form of small sets of expansion coefficients in the PCA approach or jets in our approach. Recognition is then based on this. A remaining difference between the approaches lies in the nature of the underlying feature types: principal components statistically derived from a specific set of images of an object class, or Gabor-wavelets, which can be statistically derived from a more general set of natural images, at least qualitatively [2], [25]. It remains to be seen which of the approaches has the greater potential for development.

It is worth considering the system by Lanitis *et al.* [19] in more detail because of its close relationship to our system. Both systems apply a graph-matching process for finding fiducial points and extract local features for recognition. The system by Lanitis *et al.* [19] in addition warps the face to average geometry and applies PCA to it. There are two differences we want to point out. Firstly, the matching process differs in the way in which distortions are treated. Our system assumes a simple spring model, which introduces a large number of degrees of freedom and also includes distortions which are unrealistic, cf. mismatches in Figure 4. Lanitis *et al.* [19], on the other hand, use PCA also to analyze distortion patterns of sample face images, the first eigenvectors providing a relatively small set of plausible geometrical distortions. Allowing the matching using only these distortions significantly reduces the number of degrees of freedom and leads to more reliable matchings in this respect. In addition, information about pose and shape can be inferred more easily and used for recognition. Secondly, the local features used in their system are relatively simple compared to our bunches of jets. The features are local gray value profiles along a line described by a few parameters. We assume that a combination of these two systems, using few geometrical distortion patterns and bunches of jets as local features, would improve matching performance compared to either system.

4.1.3 Performance Comparison on the FERET Database

To obtain a meaningful performance comparison between different face recognition systems, the Army Research Laboratory has established a database of face images (ARPA/ARL FERET database) and compared our and several other systems in a blind test. Official results have been published in [29] and [31]. Here we summarize results which other groups have reported for their systems tested on the FERET database. The recognition rates are given in Table 3. It should be noted that we could not find results for the face recognition system developed by Atick et al. [1] and Penev and Atick [27]. Their system performed well in one of the official tests [31].

Gordon [10] has developed a system which automatically selects regions around left eye, right eye, nose, and mouth for frontal views and a region covering the profile for profile views. The faces are then normalized for scale and rotation. The recognition is based on normalized cross-correlation of these five regions compared to reference models. Results are given for the fully automatic system, also for frontal views only, and for a system in which the normalization points, i.e., pupil centers, nose tip, and chin tip, are selected by hand. For the combined gallery (fa + pl), there is a great difference between the performance of the fully automatic system and that with manually located normalization points. This indicates that the automatic location of the normalization points is the main weakness of this system.

Gutta et al. [11] have collected the images for the FERET database. They have tested the performance of a standard RBF (radial basis function) network and a system based on geometrical relationships between facial features, such as eyes, nose, mouth, etc. The performance of the latter was very low and is not summarized in Table 3.

Moghaddam and Pentland [22] have presented results based on the PCA approach discussed in the previous section. A front-end system normal-

Table 3. Methods and performances of the different systems discussed. For our system we repeat results from two different publications for comparison. For some systems it was not reported whether fa or fb was used for the model gallery; we consistently indicate the frontal model gallery by fa. When comparing the results, notice that the first rank recognition rates depend on gallery size. Only [22] have reported results on half-profiles and on profiles; none of the groups has reported results across different poses, such as half-profile probe images against profile gallery.

Reference	Method	Model gallery	Probe images	First rank %
Gordon [10]	normalized cross-correlation on different regions in a face - manually located normalization points - fully automatic system	202 fa + pl 194 fa + pl 194 fa	202 fb + pr 194 fb + pr 194 fb	96 72 62
Gutta et al. [11]	radial basis function network on automatically segmented face images	100 fa	100 fb	83
Moghaddam and Pentland [22]	principal component analysis on the whole face - fully automatic system	150 fa 150 hr 150 pr	150 fb 150 hl 150 pl	99 38 32
Phillips and Vardi [30] Phillips [28]	trained matching pursuit filters for different regions in a face - manually located feature points - fully automatic system - manually located feature points - fully automatic system	 172 fa 172 fa 311 fa 311 fa	 172 fb 172 fb 311 fb 311 fb	 98 97 95 95
Wiskott et al. (1995) [43] (1997) [44]	Gabor wavelets, labeled graphs, elastic bunch graph matching - fully automatic system - fully automatic system	300 fa 250 fa 250 hr 250 pr	300 fb 250 fb 181 hl 250 pl	97 98 57 84

ized the faces with respect to translation, scale, lighting, contrast, and slight rotations in the image plane. The face images were then decomposed with respect to the first eigenvectors, and the corresponding coefficients were used for face representation and comparison. The performance on frontal views, which are highly standardized, was high and comparable to that of our system, but the performance on half-profiles and profiles was relatively low. That indicates that the global PCA-approach is more sensitive to rotation in depth.

Phillips and Vardi [30] and Phillips [28] have trained two sets of matching pursuit filters for the tasks of face location and identification. The filters focus on different regions: the interior of the face, the eyes, and the nose for location; tip of the nose, bridge of the nose, left eye, right eye, and interior of the face for identification. The performance is high and comparable to that of our system. The small performance difference between the fully automatic system and the identification module indicates that the location module works reliably.

For none of the systems were results across different poses reported. In the next section we will therefore summarize systems which have been tested for rotation in depth on different databases.

4.1.4 Recognizing Faces Rotated in Depth

While there is a considerable literature on face recognition in the same pose, there are few systems which deal with large rotation in depth. It is difficult to compare these systems in terms of performance, because they have been tested on different galleries. Furthermore, the recognition rates are a result of complete systems and do not necessarily reflect the usefulness of a particular method to compensate for rotation in depth. However, we think it may still be useful to give an overview and to briefly discuss the different approaches. Results are given in Table 4.

The system by Moghaddam and Pentland [23] simply applies several recognition subsystems in parallel, each of which is specialized to one view and is based on the PCA approach described above for recognition of the same views. The subsystem which is specialized for a view closest to the view of the probe image is usually best suited to explain the image data in terms of its eigenvectors. It therefore can be selected automatically to perform the recognition. This system has been tested on galleries of 21 persons in different views. The results listed in Table 4 are averages over several different combinations of training and testing views. The recognition rates are an example of how well a system can

Table 4. Methods and performances of the different systems discussed. Our results are repeated for comparison. When comparing the results, notice that the first rank recognition rates depend on gallery size as well as on the quality of the databases.

Ref.	Database and Method	Model gallery #	Model gallery angle(s)	Probe images #	Probe images angle(s)	First rank %
[23]	PCA approach separately for different views, no specific transformation					
	interpolation performance ±23°	21	±90°, ±45°, 0°	21	±68°, ±23°	90
	extrapolation performance ±23°	21	e.g., −90°, ..., +45°	21	e.g., +68°	83
	extrapolation performance ±45°	21	e.g., −90°, ..., +45°	21	e.g., +90°	50
[44]	Gabor jets, no specific transformation					
	Bochum database	108	0°	108	11°	94
		108	0°	108	22°	88
	FERET database	250	0°	250	45°	18
		250	45°	250	0°	17
		250	45°	250	90°	9
		250	90°	250	45°	12
[20]	Gabor jets, learned normal vectors for geometrical rotation transformation					
	Bochum database, no transformation	110	0°	110	22°	88
	transforming 22° to 0°	110	0°	110	22°	96
	FERET database, no transformation	90	0°	90	45°	36
	transforming 45° to 0°	90	0°	90	45°	50
	transforming 0° to 45°	90	0°	90	45°	53
[4]	well-controlled gallery, little hair information					
	warping between different views	62	(±)20°	620	range ±40°	82
	linear decomposition, no shape info.	62	(±)20°	620	range ±40°	70
[39]	images rendered from 3D face data, no hair information					
	mapping onto 3D-model	100	0°	100	24°	100
	linear decomposition and synthesis	100	0°	100	24°	100
	linear decomposition on 4 subregions	100	0°	100	24°	100

perform if it does not compensate for effects of rotation in depth but relies only on the robustness of the subsystem which is closest to the view of the probe image. Our basic system compensates for rotation in depth only in that matching is done with a bunch graph of the new view and correspondences are defined between fiducial points of the new view and fiducial points of the standard view for which the model graphs are available. Thus, corresponding jets are compared across different views, but the jets are not modified in any way to compensate for the effects of rotation in depth.

There are at least three different approaches to compensating for the effects of rotation in depth more explicitly: transforming feature vectors, warping images of faces, and linear decomposition and synthesis of faces in different views. Let us first consider transforming feature vectors. As an extension to our system, Maurer and Von der Malsburg [20] have applied linear vector transformations to the jets to compensate for the effect of rotation in depth. The assumption was that faces can be locally treated as plane surfaces and that the texture transforms accordingly. Since the total rotation of the faces is known, only the normal of the surface at each node has to be estimated, which is done on a training set of faces available in both views. This results in a significant improvement. Notice that transformations of feature vectors can only be an approximation to the true transformations of images. This is due to the fixed and limited support of the kernels which are used to extract the features. For instance, a circular region on a plane becomes an ellipse if the plane is tilted. Feature vectors based on kernels with circular support can only represent a circular region. If this circular region needs to be transformed into an elliptic region, some information is lost or incorrect information is added to obtain a circular region again. An advantage of this method is that transformations can be performed without reference to the original image.

More accurate results can be obtained if the gray-value distributions of faces are warped from one view to another view directly on a pixel level. Vetter [39] has done this by means of a 3-D model onto which the texture of a face is projected and from which it is then back-projected onto the image plane in a different view. Beymer and Poggio [4] have used a warping transformation derived from sequences of rotating sample faces. Another interesting approach is the concept of linear object classes [40]. It is assumed that objects in one view can be linearly decomposed with respect to images of a set of prototype objects of the same view. When images of these prototypes are available in another view, the object can be linearly synthesized in that view with the same coefficients as used for the decomposition. Vetter [39] tested this method on images rendered from 3D face data. Shape and texture were processed separately, the texture being processed as a whole or broken down into four local regions. Recognition was based on a simple similarity measure, e.g., Euclidean distance, applied directly to the image gray values. The recognition rates of 100% for this and the warping method described above are remarkable. However, it has to be taken into account that the images were rendered from 3-D face representations and that the galleries were correspondingly perfect. Beymer and Poggio [4] also used this method, but they did the decomposition with respect to eigenfaces and did not use shape information. Their model gallery included 20° rotated faces plus the mirror-reflected images, and they tested on probe faces randomly drawn from a range of approximately ±40° rotation angle. They also considered rotation around horizontal axes. This system as well as the one by Vetter [39] used an image-flow algorithm to find correspondences between different faces. We have tested a similar method for our system on the FERET database. For a half-profile face image we used a half-profile face bunch graph to generate a phantom face [41], which was then transformed into a frontal pose by using corresponding jets of the same fiducial points and individuals from a frontal face bunch graph. The idea was that if, for instance, the noses of two persons look similar in one pose, they would look similar in another pose as well. The results were disappointing and are not reported in Table 4. It is surprising that these three systems, which are based on similar ideas, perform so differently. A possible reason might be the different quality of the databases, which was perfect for the system by Vetter [39] and worst for ours.

Each of these three approaches (transforming feature vectors, warping images of faces, and the concept of linear object classes) has its own advantages and drawbacks, and none is clearly superior to the others. Warping can potentially deal well with new types of faces not seen before, e.g., of new race or age, but it cannot be applied to transform between other variations, e.g., in illumination. Linear decomposition, on the other hand, can be applied to different types of variations, but it does probably not extrapolate well to new types of faces. Transforming features may deal well with different kinds of variations as well as new types of faces, but in its current formulation it is limited because of the fixed and limited kernels. The decision for one of the approaches will also depend on how well it integrates into a particular recognition system.

In the following section we discuss some methods which can potentially show or have been shown to further improve the performance of our system; see also [24].

4.2 Further Developments

The current system can be improved in many respects. In Section 4.1, we already argued that the simple spring model used here for the grid has too many degrees of freedom, which could be considerably reduced by using only a small number of typical distortions found by PCA on manually controlled grids [19]. This would probably improve matching accuracy further and would provide more precise geometrical information which could be used to increase recognition performance. However, the matching precision achieved in our system, as compared to the preceding version [18], is already sufficient to apply specific methods which require reliable fiducial points, for instance when the issue is learning about local object properties. One such local property is differential degree of robustness against disturbances. In an extension of the basic system presented here, Krüger [15] and Krüger et al. [17] have developed a method for learning weights emphasizing the more discriminative nodes. On model galleries of size 130–150 and probe images of different pose, the first rank recognition rates have been improved by an average of 6%, from 25% without to 31% with weights. As mentioned in Section 4.1.4, another extension of our system also requiring reliable fiducial points has been developed by Maurer and Von der Malsburg [20] to compensate for rotation in depth.

In [41], the bunch graph technique has been used to fairly reliably determine facial attributes from single images, such as sex or the presence of glasses or a beard. If this technique was developed to extract independent and stable personal attributes, such as age, race, or sex, recognition from large databases could be improved and speeded up considerably by preselecting corresponding sectors of the database.

We did some preliminary experiments on images with structured background and got encouraging results. In this case more nodes in the interior of the faces were used. However, a more principled method should be employed. One could use only Gabor kernels lying within the face [46] or, alternatively, transform jets from contour nodes such that the influence of background structure is suppressed [33]. Robustness with respect to illumination variations also has to be investigated.

The manual definition of appropriate grid structures and the semi-autonomous process of bunch graph acquisition will have to be replaced by a fully autonomous process. Automatic reduction of an FBG to its essential jets in the bunches has been demonstrated in [17]. The creation of a new bunch graph is most easily based on image sequences, which contain many cues for grouping, segmentation, and detecting correspondences. This has been demonstrated for individual graphs in [16] and [21].

In the current system, one recognition against a gallery of several hundred models takes approximately 30 seconds on a SPARCstation 10-512. This is too slow for most applications. However, there are many possibilities to optimize the system with respect to speed: reducing the number of Gabor-kernels, reducing the number of steps in the matching process, using fewer nodes, etc. The system described in [18], which is computationally even more expensive, has been optimized in this way and has been successfully turned into a commercial product for access control [14], which runs on a standard PC. Our group is also currently working on a real-time face-tracking system based on the matching process presented here. For this project a high performance parallel processor system will be employed.

Acknowledgments

We wish to thank Irving Biederman, Ladan Shams, Michael Lyons, and Thomas Maurer for very fruitful discussions and their help in evaluating the performance of the system on the FERET database. Many thanks go to Thomas Maurer also for reviewing and optimizing the code and performing tests on the Bochum gallery. Jan Vorbrüggen performed the tests for the preceding system, which we used for comparison. We acknowledge helpful comments on the manuscript by Jonathon Phillips and Marni Stewart Bartlett. For the experiments we have used the FERET database of facial images collected under the ARPA/ARL FERET program and the Bochum gallery collected at the Institute for Neural Computation, Ruhr-University Bochum.

References

[1] Atick, J.J., Griffin, P., and Redlich, A.N. (1995), "Face-recognition from live video for real-world applications — now," *Advanced Imaging*, 10(5):58–62.

[2] Bell, A.J. and Sejnowski, T.J. (1997), "The independent components of natural scenes are edge filters," *Vision Research*, 37(23):3327–3338.

[3] Beymer, D. (1994), "Face recognition under varying pose," *Proc. IEEE Computer Vision and Pattern Recognition*, pp. 756–761, Seattle, WA.

[4] Beymer, D. and Poggio, T. (1995), "Face recognition from one example view," *Proc. 5th Int'l Conf. on Computer Vision*, pp. 500–507, Cambridge, MA. IEEE Comput. Soc. Press.

[5] Chellappa, R., Wilson, C.L., and Sirohey, S. (1995), "Human and machine recognition of faces: A survey," *Proc. of the IEEE*, 83(5):705–740.

[6] Craw, I., Costen, N., Kato, T., Robertson, G., and Akamatsu, S. (1995), "Automatic face recognition: Combining configuration and texture," in Bichsel, M. (Ed.), *Proc. Int'l Workshop on Automatic Face- and Gesture-Recognition, IWAFGR'95, Zurich*, pp. 53–58. MultiMedia Laboratory, University of Zurich.

[7] Daugman, J.G. (1988), "Complete discrete 2-D Gabor transform by neural networks for image analysis and compression," *IEEE Trans. on Acoustics, Speech and Signal Processing*, 36(7):1169–1179.

[8] DeValois, R.L. and DeValois, K.K. (1988), *Spatial Vision*, Oxford Press.

[9] Fleet, D.J. and Jepson, A.D. (1990), "Computation of component image velocity from local phase information," *Int'l J. of Computer Vision*, 5(1):77–104.

[10] Gordon, G.G. (1995), "Face recognition from frontal and profile views," in Bichsel, M. (Ed.), *Proc. Int'l Workshop on Automatic*

Face- and Gesture-Recognition, IWAFGR'95, Zurich, pp. 47–52. MultiMedia Laboratory, University of Zurich.

[11] Gutta, S., Huang, J., Singh, D., Shah, I., Takacs, B., and Wechsler, H. (1995), "Benchmark studies on face recognition," in Bichsel, M. (Ed.), *Proc. Int'l Workshop on Automatic Face- and Gesture-Recognition, IWAFGR'95, Zurich*, pp. 227–231. MultiMedia Laboratory, University of Zurich.

[12] Jones, J.P. and Palmer, L.A. (1987), "An evaluation of the two-dimensional Gabor filter model of simple receptive fields in cat striate cortex," *J. of Neurophysiology*, 58:1233–1258.

[13] Kirby, M. and Sirovich, L. (1990), "Application of the Karhunen-Loève procedure for the characterization of human faces," *IEEE Trans. on Pattern Analysis and Machine Intelligence*, 12(1):103–108.

[14] Konen, W. and Schulze-Krüger, E. (1995), "ZN-Face: A system for access control using automated face recognition," in Bichsel, M. (Ed.), *Proc. Int'l Workshop on Automatic Face- and Gesture-Recognition, IWAFGR'95, Zurich*, pp. 18–23. MultiMedia Laboratory, University of Zurich.

[15] Krüger, N. (1997), "An algorithm for the learning of weights in discrimination functions using a priori constraints," *IEEE Trans. on Pattern Analysis and Machine Intelligence*, 19(7):764–768.

[16] Krüger, N., Maël, E., Pagel, M., and Von der Malsburg, C. (1998), "Autonomous learning of object representations utilizing self-controlled movements," *Proc. of Neural Networks in Applications, NN'98, Magdeburg, Germany*, pp. 25–29.

[17] Krüger, N., Pötzsch, M., and Von der Malsburg, C. (1997), "Determination of face position and pose with a learned representation based on labelled graphs," *Image and Vision Computing*, 15:665–673.

[18] Lades, M., Vorbrüggen, J.C., Buhmann, J., Lange, J., Von der Malsburg, C., Würtz, R.P., and Konen, W. (1993), "Distortion invariant

object recognition in the dynamic link architecture," *IEEE Trans. on Computers*, 42(3):300–311.

[19] Lanitis, A., Taylor, C.J., and Cootes, T.F. (1995), "An automatic face identification system using flexible appearance models," *Image and Vision Computing*, 13(5):393–401.

[20] Maurer, T. and Von der Malsburg, C. (1995), "Linear feature transformations to recognize faces rotated in depth," *Proc. Int'l Conf. on Artificial Neural Networks, ICANN'95, Paris*, pp. 353–358, Paris. EC2 & Cie.

[21] Maurer, T. and Von der Malsburg, C. (1996), "Tracking and learning graphs and pose on image sequences of faces," *Proc. 2nd Int'l Conf. on Automatic Face- and Gesture-Recognition*, pp. 176–181, Los Alamitos, CA. IEEE Comp. Soc. Press.

[22] Moghaddam, B. and Pentland, A. (1994), "Face recognition using view-based and modular eigenspaces," *Proc. SPIE Conf. on Automatic Systems for the Identification and Inspection of Humans*, volume SPIE 2277, pp. 12–21.

[23] Moghaddam, B. and Pentland, A.P. (1997), "Probabilistic visual learning for object representation," *IEEE Trans. on Pattern Analysis and Machine Intelligence*, 19(7):696–710.

[24] Okada, K., Steffens, J., Maurer, T., Hong, H., Elagin, E., Neven, H., and Von der Malsburg, C. (1998), "The Bochum/USC face recognition system and how it fared in the FERET phase III test," in *Face Recognition: From Theory to Applications*, Springer-Verlag. (to appear).

[25] Olshausen, B.A. and Field, D.J. (1996), "Emergence of simple-cell receptive field properties by learning a sparse code for natural images," *Nature*, 381:607–609.

[26] O'Toole, A.J., Abdi, H., Deffenbacher, K.A., and Valentin, D. (1993), "Low-dimensional representation of faces in higher dimensions of the face space," *J. of the Optical Society of America A*, 10(3):405–411.

[27] Penev, P.S. and Atick, J.J. (1996), "Local feature analysis: A general statistical theory for object representation," *Network: Computation in Neural Systems*, 7(3):477–500.

[28] Phillips, P.J. (1996), *Representation and Registration in Face Recognition and Medical Imaging*, PhD thesis, RUTCOR, Rutgers University.

[29] Phillips, P.J. and Rauss, P.J. (1997), "Face recognition technology (FERET program)," *Proc. Office of National Drug Control Policy*, (in press).

[30] Phillips, P.J. and Vardi, Y. (1995), "Data driven methods in face recognition," in Bichsel, M. (Ed.), *Proc. Int'l Workshop on Automatic Face- and Gesture-Recognition, IWAFGR'95, Zurich*, pp. 65–70. MultiMedia Laboratory, University of Zurich.

[31] Phillips, P.J., Wechsler, H., Huang, J., and Rauss, P. (1998), "The FERET database and evaluation procedure for face-recognition algorithms," *Image and Vision Computing*, 16(5):295–306.

[32] Pollen, D.A. and Ronner, S.F. (1981), "Phase relationship between adjacent simple cells in the visual cortex," *Science*, 212:1409–1411.

[33] Pötzsch, M., Krüger, N., and Von der Malsburg, C. (1996), "Improving object recognition by transforming Gabor filter responses," *Network: Computation in Neural Systems*, 7(2):341–347.

[34] Samal, A. and Iyengar, P.A. (1992), "Automatic recognition and analysis of human faces and facial expressions: A survey," *Pattern Recognition*, 25(1):65–77.

[35] Sirovich, L. and Kirby, M. (1987), "Low-dimensional procedure for the characterization of human faces," *J. of the Optical Society of America A*, 4(3):519–524.

[36] Theimer, W.M. and Mallot, H.A. (1994), "Phase-based binocular vergence control and depth reconstruction using active vision," *CVGIP: Image Understanding*, 60(3):343–358.

[37] Turk, M. and Pentland, A. (1991), "Eigenfaces for recognition," *J. of Cognitive Neuroscience*, 3(1):71–86.

[38] Valentin, D., Abdi, H., O'Toole, A.J., and Cottrell, G.W. (1994), "Connectionist models of face processing: A survey," *Pattern Recognition*, 27(9):1209–1230.

[39] Vetter, T. (1998), "Synthesis of novel views from a single face image," *Int'l J. of Computer Vision*, 28(2):103–116.

[40] Vetter, T. and Poggio, T. (1997), "Linear object classes and image synthesis from a single example image," *IEEE Trans. on Pattern Analysis and Machine Intelligence*, 19(7):733–741.

[41] Wiskott, L. (1997), "Phantom faces for face analysis," *Pattern Recognition*, 30(6):837–846.

[42] Wiskott, L. (1999), "The role of topographical constraints in face recognition," *Pattern Recognition Letters*. (To appear).

[43] Wiskott, L., Fellous, J.-M., Krüger, N., and Von der Malsburg, C. (1995), "Face recognition and gender determination," in Bichsel, M. (Ed.), *Proc. Int'l Workshop on Automatic Face- and Gesture-Recognition, IWAFGR'95, Zurich*, pp. 92–97. MultiMedia Laboratory, University of Zurich.

[44] Wiskott, L., Fellous, J.-M., Krüger, N., and Von der Malsburg, C. (1997), "Face recognition by elastic bunch graph matching," *IEEE Trans. on Pattern Analysis and Machine Intelligence*, 19(7):775–779.

[45] Wiskott, L. and Von der Malsburg, C. (1993), "A neural system for the recognition of partially occluded objects in cluttered scenes," *Int. J. of Pattern Recognition and Artificial Intelligence*, 7(4):935–948.

[46] Würtz, R.P. (1997), "Object recognition robust under translations, deformations, and changes in background," *IEEE Trans. on Pattern Analysis and Machine Intelligence*, 19(7):769–775.

[47] Yuille, A.L. (1991), "Deformable templates for face recognition," *J. of Cognitive Neuroscience*, 3(1):59–70.

Chapter 12:

Facial Expression Synthesis Using Radial Basis Function Networks

FACIAL EXPRESSION SYNTHESIS USING RADIAL BASIS FUNCTION NETWORKS

I. King and **X.Q. Li**
Department of Computer Science & Engineering
The Chinese University of Hong Kong
Shatin, New Territories, Hong Kong

Many multimedia applications require the synthesis of facial expressions. In this chapter, we present a technique that synthesizes 2-D gray-scale facial expressions using Radial Basis Function (RBF) neural networks. There are three types of networks that we have constructed to synthesize facial expressions. The first type is a RBF network which generates the displacement of Facial Characteristic Points (FCPs) of a single expression with different degrees, e.g., from Expressionless to Happy. The second type is also a RBF network that generates a mixture of two expressions, e.g., Happy and Surprised. The third type of network is a hybrid network that combines the first two types of RBF networks to generate FCP displacements for mixture of expressions in various degrees. Once the network has approximated the displacement vector, this information is then fed into a digital warping algorithm which turns a normal expressionless face image into the synthesized image of desired facial expression. We describe our system along with its components and show our experimental results.

1 Introduction

1.1 Application Background

Facial expressions play an important role in non-verbal communication. We use facial gestures to convey our moods and express our feelings. Facial Expression Synthesis (FES) techniques try to simulate people's expressions artificially using computers. With these techniques, we can make an intelligent, friendly, and effective machine-human interface for a variety of applications. Some important FES applications are listed as follows:

- **Communications:** We may use FES to reduce transmission bandwidth by model-based facial expression coding when transmitting facial images in teleconferencing and video-phone applications.

- **Film Making:** The most well known application, as far as the general public is concerned, is FES for films and videos. Many of the latest movies make use of face animation, such as The Abyss, Terminator 2, and Robocop 2.

- **Medical Research:** FES can be used to gain more insight into the contribution of visual information of facial movement in relationship to speech perception permitting a better controlled and more systematic analysis of the auditory perceptual process.

- **Teaching and Speech Aids:** Teaching people with hearing disorders to lip-read or deaf to speak could be made easier with a FES system. By making the learning process seem like a game, the student's attention is held and they can learn more than they would using traditional methods.

1.2 Facial Expression Recognition vs. Facial Expression Synthesis

In the general framework of face computing, Facial Expression Synthesis (FES) and Facial Expression Recognition (FER) can be seen as two complementary processes of each other. Here we will first introduce FER before detailing FES.

In [6] and [7], facial expressions are categorized into six basic groups: Happy, Sad, Angry, Fear, Surprised and Disgusted. In order to recognize such facial expressions, Kobayashi *et al.* used Facial Characteristic Points (FCPs) as landmarks. These FCPs are feature points that represent significant movements during the generation of an expression. During the FER process, they used the FCP displacements between an expressionless and an expressive face as the input to a Backpropagation (BP) neural network. The BP network performs a basic classification task to recognize facial expressions through supervised learning. The FER system has six output units and the value of each unit represents the degree of a particular expression. Furthermore, the six outputs are combined to recognize mixed facial expressions in [7].

Now, let us consider Facial Expression Synthesis (FES). Suppose we are given a 2-D gray-scale expressionless face image of a person, what do we do to generate a new 2-D gray-scale image with a particular expression, e.g., "smile" of the person? One of the ways is to generate the approximate displacements of FCPs between the expressionless and the expressive face. This problem is similar to multivariate approximation since we cannot describe accurately the underlying multidimensional function of the FCP displacements. Hence, a typical FES system will require algorithms to map from expression labels to FCP displacements. In our work, these algorithms are realized in neural networks.

1.3 2-D Expression Synthesis

For 2-D expression synthesis, Arad [1] used image warping techniques to construct the image that is determined by the mapping of a small number of anchor points using Radial Basis Functions (RBFs) for interpolation. However, his system does not provide a mechanism to determine the intermediate points of the anchors of each particular facial expression. In other words, Arad used the RBFs for interpolation instead of approximation. The difficulty of Arad's system is how to generate accurate facial landmark displacement information from inaccurate input data.

The causes of inaccurate input data are:

1. It is hard to generate a set of standardized expressions, e.g., each person may smile differently.

2. It is hard to produce the precise degree of a particular expression, e.g., how to generate a "20% smile" ?

3. It is difficult to mix various facial expressions, e.g., how to make a gesture of combined Happy and Sad face?

For better approximation from inaccurate data, we choose the RBF networks which estimate the mapping between the displacements of FCPs and the six universal facial expressions described in [6] and [7]. Using this method, we can obtain a set of FCP displacements which have greater potentials in revealing changes in displaying a particular facial expression. The FCP displacements for the facial expressions are used to generate control points in the 2-D image warping procedure for generating facial expression image.

The next section will describe in more detail of the FER process which is crucial in understanding the FES process. We will then formulate the RBF networks and demonstrate the FES results of our system in Section 3. We will conclude with some discussions in Section 4.

2 Facial Expression Recognition (FER)

2.1 Feature Preprocessing

In [2], Ekman and Friesen described a set of Action Units (AU) that showed basic movement of face muscles. Among the total of 44 AUs, except for a few AUs corresponding to movement of cheek, chin, and wrinkles, 30 AUs are directly associated with movement of eyes, eyebrows, and mouth. These 30 AUs represent movement of significant landmarks which are desirable for machine recognition of facial expressions. They are equivalent to the FCPs described in [6] and [7]. These FCPs are illustrated in Figure 1. Since the extracted FCPs from different images may be varied, several pre-processing steps are performed.

To compensate for the variations in the size, orientation, and position of the face in images as well as the variations in the size of the face components, i.e., eyes, eyebrows, and mouth, we transform the coordinates of the FCPs so that they are comparable across the set of faces. Hence, the coordinates of these FCPs undergo four transformations: *translation, rotation, normalization,* and *displacement calculation.* They are described as follows:

1. **Translation:** While the origin of the image coordinate system is located at the top left corner of the image, the origin of FCP co-

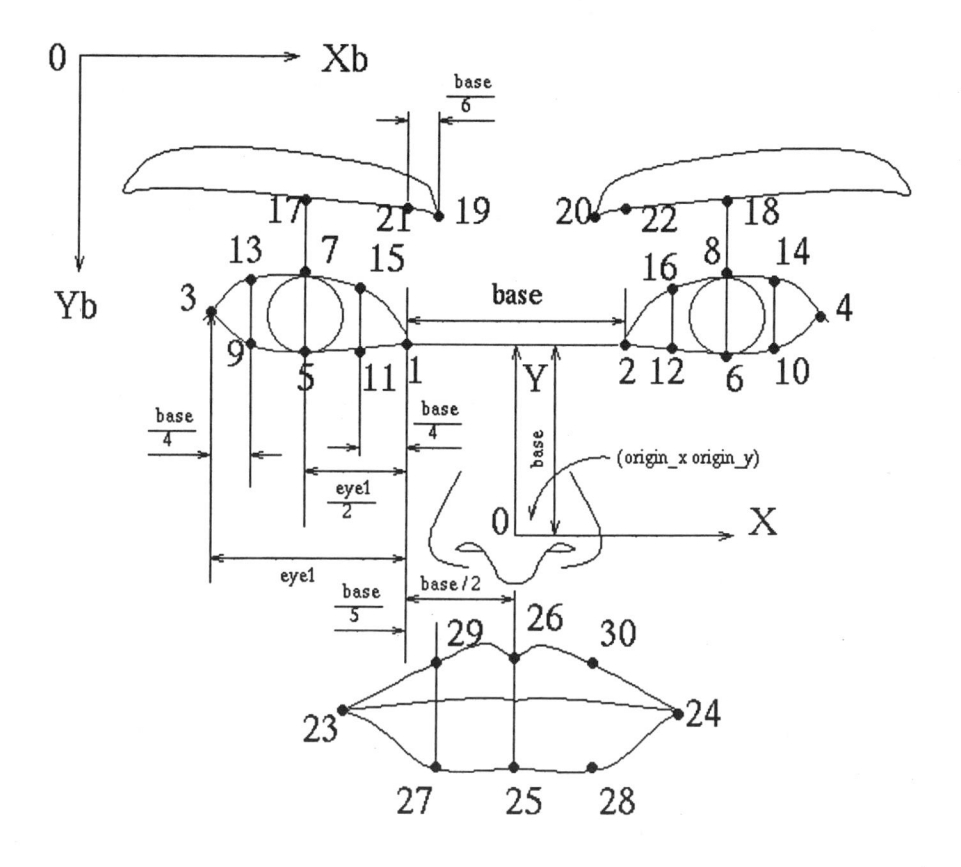

Facial Expression Synthesis Using Radial Basis Function Networks

Figure 1. The 30 facial characteristic points.

ordinate system should be shifted to the tip of the nose of each individual. A quantity, which should remain fairly constant for different facial expressions of each individual, called *base* is defined as

$$base = \sqrt{(xb_2 - xb_1)^2 + (yb_2 - yb_1)^2} \tag{1}$$

where (xb_1, yb_1) and (xb_2, yb_2) are the pixel position of the inner corners of left eye and right eye respectively. The mid-point (x_0, y_0) between (xb_1, yb_1) and (xb_2, yb_2) is also calculated using the mid-point formula, $x_0 = (xb_1 + xb_2)/2$ and $y_0 = (yb_1 + yb_2)/2$.

2. **Rotation:** It is used to correct the rotation of the face so that the coordinates are expressed with respect to the z-axis (camera-to-face) in the FCP coordinate system. The rotation angle of the face

from the horizontal axis, θ, is defined as

$$\theta = \tan^{-1} \frac{(yb_2 - yb_1)}{(xb_2 - xb_1)} \quad (2)$$

Each FCPs in the image is rotated with $-\theta$ accordingly with respect to the origin.

3. **Normalization:** It is introduced to compensate for the distance effect between the camera and the face. The FCPs of the expressionless and expressive face after rotation are divided by the value, $base$, as $(x_i)_{norm} = x_i/base$, $(x_i)_{norm} = x_i/base$. These normalized values of the expressive face are subtracted from those of the normal ones.

4. **Displacement Calculation:** The FCP displacements are obtained from subtracting the FCPs of the expressionless face from the FCPs of an expressive face.

After performing the above preprocessing steps, the processed data are ready as the input feature vector for facial expression classification.

2.2 Facial Expression Recognition Using BP Neural Networks

In the work of Kobayashi and Hara [6], [7], they designed and implemented a human facial expression recognition system using a Backpropagation (BP) neural network [8]. This network can recognize six basic facial expressions: Happy, Sad, Angry, Fear, Surprised and Disgusted. As shown in Figure 2, the network contains sixty input units which correspond to the thirty pairs of FCP displacements information. It has six units at the output layer with one hidden layer of 100 units.

The BP network has a feedforward neural network architecture as shown in Figure 2. In this architecture, nodes are partitioned into layers. The lower-most layer is the input layer and the top-most layer is the output layer. There is at least one hidden layer in the network. The feedforward process involves presenting an input pattern to the input layer neurons that pass the input values onto the first hidden layer. Each node in the hidden layer computes a weighted sum of its inputs before passing the sum through its activation function and presenting the result to the output layer. Its learning algorithm is a generalization of the least-mean-square (LMS) algorithm which modifies network weights to minimize the mean-square error between the desired and actual outputs of the network. This is a form of supervised learning in which the network is trained using data for which inputs as well as desired outputs are known.

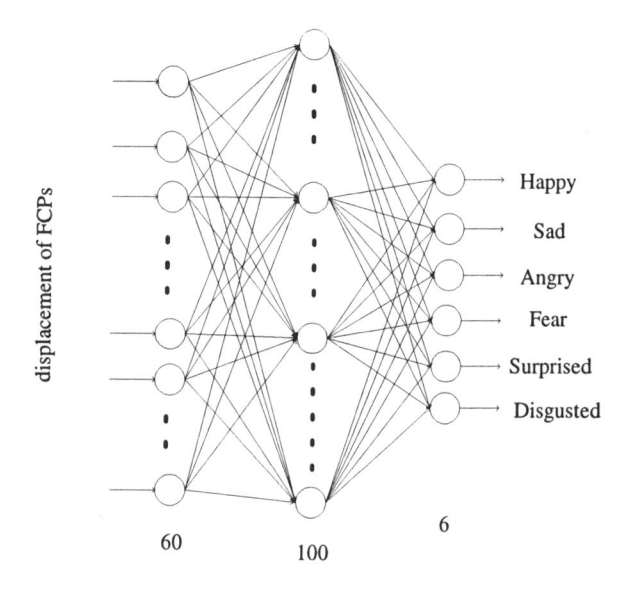

Figure 2. The neural network for FER and FES.

The BP training algorithm uses the gradient descent method to seek out the minima of the error function in the weight space. After choosing the initial weights of the network randomly, the BP training algorithm computes the necessary error corrections and updates its weights accordingly. The algorithm is iterative and can be decomposed in the following four steps: (1) Feed-forward computation for output, (2) Backpropagation of error to the output layer, (3) Backpropagation of error to the hidden layer, and (4) Weight updates. The learning is terminated when the value of the error function has become less than a predefined constant or the iterative number reaches a predefined value. Once trained, the network weights remain unchanged and can be used to compute output values for new input samples.

3 Facial Expression Synthesis (FES)

In this section, we examine the basic principle of FES. The first step is to find out the necessary relative spatial FCP displacements for a particular facial expression. The second step is to use the calculated displacements for generating the facial expression using the image warping technique.

3.1 The Radial Basis Function (RBF) Network

To approximate the displacements of FCPs, we should find a function F which satisfies

$$\vec{o} = F(\vec{d}), \qquad (3)$$

where $\vec{d} \in \Re^6$ denotes the six facial expressions and $\vec{o} \in \Re^{60}$ denotes the spatial displacement of the FCPs.

The above function can be estimated using the interpolative method with a set of training data. This is similar to a high dimensional approximation problem. However, in FES application this approximation is difficult because the function which maps a low dimension vector (six facial expressions) to a high dimension vector (thirty FCP displacements) without prior knowledge is ill-defined. The RBF networks are suitable to solve the FES problem since they can generate smooth outputs based on missing and inaccurate inputs.

RBF networks are multilayer feedforward networks which use a set of RBF-nodes. A function is radially symmetric (or is a RBF) if its output depends on the distance of the input sample (vector) from another stored vector. Neural networks whose node functions are radially symmetric functions are referred to as RBF-nodes [5].

Typically, the RBF can be considered in the form of

$$F_k(\vec{d_i}) = \sum_{j=1}^{c} w_{jk}\, g(\|\vec{d} - \vec{\mu_j}\|), \quad \vec{x} \in \Re^n, k = 1, \cdots, n' \qquad (4)$$

where $\|\cdot\|$ denotes the usual Euclidean norm on $\Re^n$ and $\vec{\mu_j} \in \Re^n$, $j = 1, 2, \cdots, c$ denote the "centers" of the radial basis functions which are given as the known data points. Figure 3 shows the schematic architecture of the RBF neural networks.

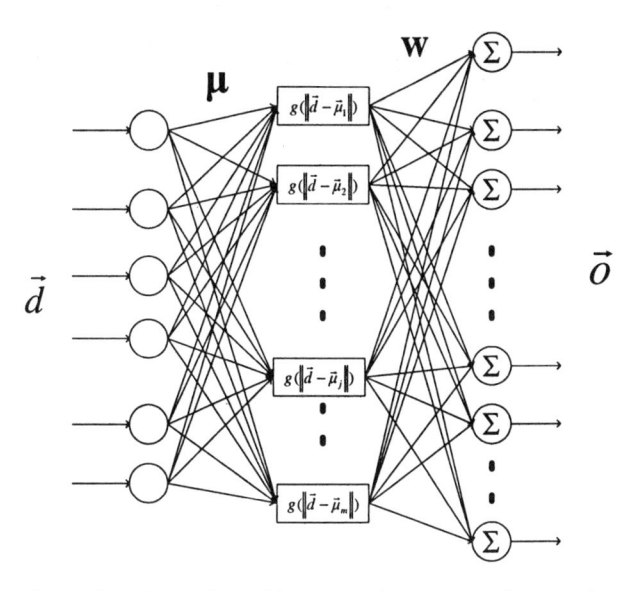

Figure 3. Schematic architecture of RBF neural networks.

In RBF networks, each commonly used radial basis function g is a non-increasing function of a distance measure u which is its only argument, with $g(u_1) \geq g(u_2)$ whenever $u_1 < u_2$. Function g is applied to the Euclidean distance $u = \|\vec{d} - \vec{\mu}\|$, between the "center" or stored vector $\vec{\mu}$ and the input vector $\vec{d}$. The Gaussian function described by the equation

$$g(\vec{x}) \propto e^{-(\vec{x}/\sigma)^2} \quad (5)$$

is the most widely used radially symmetric function. In this chapter, the g is the normalized Gaussian activation function defined as

$$g(\|\vec{d} - \vec{\mu}\|) = \frac{\exp[-(\vec{d} - \vec{\mu})^2 / 2\sigma_j^2]}{\sum_{k=1}^{c} \exp[-(\vec{d} - \vec{\mu})^2 / 2\sigma_k^2]} \quad (6)$$

where $\vec{d}$ is the input vector, $\vec{\mu}$ is a set of weights, and σ is the width of the RBF. The outputs of a RBF network can be written as

$$\begin{pmatrix} o_1 \\ \vdots \\ o_{n'} \end{pmatrix} = \begin{pmatrix} w_{11} & \cdots & w_{1c} \\ \vdots & \ddots & \vdots \\ w_{n'1} & \cdots & w_{n'c} \end{pmatrix} \cdot \begin{pmatrix} z_1 \\ \vdots \\ z_c \end{pmatrix} \quad (7)$$

where z_i is the output of $g(\|\vec{d} - \vec{\mu_i}\|)$.

We now present the training of RBF networks [4]. Consider the training set of m labeled pairs $(\vec{d_i}, \vec{o_i})$ that represents the association of a given mapping or samples of a continuous multivariate function. Furthermore, consider the sum of square error (SSE) criterion function as an error function, E, to be minimized over the given training set. In other words, we seek to develop a training method that minimizes E by adaptively updating the free parameters of the RBF network. These parameters are the receptive field centers ($\vec{\mu_j}$, means of the hidden layer Gaussian units), the receptive field widths (standard deviations, σ_j), and the output layer weights, w_{ij}.

One of the training methods for RBF is a fully supervised gradient descent method over E. In particular, $\vec{u_j}$, σ_j, and w_{ij} are updated by the respective increments defined as follow:

$$\Delta \vec{\mu_j} = -\rho_{\vec{\mu}} \nabla_{\vec{\mu_j}} E, \quad j = 1, 2, \cdots, c \tag{8}$$

$$\Delta \sigma_j = -\rho_\sigma \frac{\partial E}{\partial \sigma_j}, \quad j = 1, 2, \cdots, c \tag{9}$$

$$\Delta w_{ij} = -\rho_w \frac{\partial E}{\partial w_{ij}}, \quad i = 1, 2, \cdots, n', \ j = 1, 2, \cdots, c \tag{10}$$

where $\rho_{\vec{\mu}}$, ρ_σ, and ρ_w are small positive constants as learning rates. In our experiment, we use a single global fixed value σ for all σ_j values in the RBF network for FES. σ determines the generalization ability of the RBFs. The variable σ should be large enough to allow the overlapping of the input regions of radial basis functions. This makes the network function smoother and results in better generalization for new input vectors occurring between input vectors. However, σ should not be too large so that each neuron responds in essentially the same manner, i.e., any information presented to the network becomes lost. Our σ is picked manually by trial and error within maximum and minimum range of the input vectors. The training process is very similar to the process for BP network introduced in Section 2. After training, two sets of weights, $\vec{\mu}$ and $\mathbf{w}$, are obtained. They will be used to produce the spatial FCP displacements.

In the FES applications, it is hard to find a single RBF network that is able to generate mixed expressions with different degrees. The reason is that each expression can be seen as occupying an axis in a multidimensional space whose origin is the normal expression as shown in Figure 4

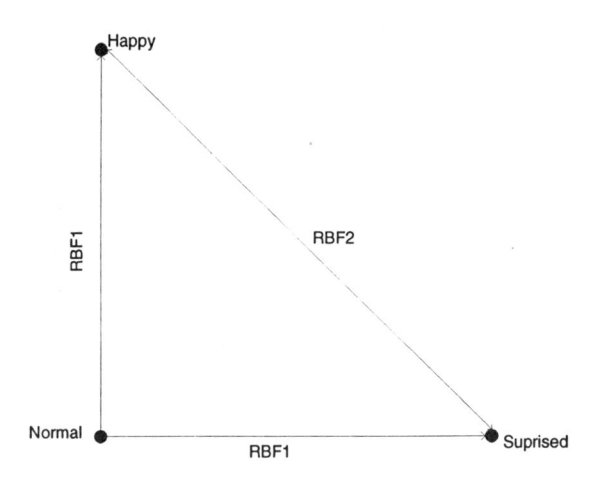

Figure 4. Different use of RBF1 and RBF2.

on axes with the "RBF1" label. Moreover, each training pattern is just a final single expression which can be seen as an end-point on an axis as illustrated in Figure 4, the line with the "RBF2" label. Hence, we use two RBF networks, RBF1 and RBF2, to approximate these two cases. One to approximate along the axis for different degrees of a particular expression and another to approximate between two different expressions. For RBF1, the network can generate five different degrees for each of the six expressions. For RBF2, we can mix two different expressions, but we cannot have different degrees for the mixed expression. The final result of FES is a hybrid network that combines the outputs from RBF1 and RBF2 to obtain different mixtures of expressions of various degrees.

3.2 Extracting FCPs from the Expressionless Face

Before using the RBF networks for FES, we have to obtain the FCPs from the expressionless face. All FCPs lie on five salient regions in our face, namely the two eyebrows, the two eyes and the mouth. In the first step, we use a semi-automatic method to extract these regions from the expressionless face. For accurate extraction, we have the following requirements:

1. There has to be a frontal face without wearing glasses in the image.

2. There is no or slight rotation of head about the vertical axis.

3. The rotation angle of the face from the horizontal axis of the image is within 5 degrees.

4. The width of the face is about 1/2 of the width of the image.

5. The illumination on the face has to be good.

6. The important facial regions, such as eyebrows and eyes, cannot be hidden by any obstacle. For example, the hair cannot cover the eyebrow.

The algorithm for regions extraction requires knowing the nose position of the face by interactive method. We choose this position because the nose is the central and prominent region in a face. Here, all the coordinates are relative to the top left corner of the image. Once the nose point has been located, the search for other features can take advantage of the knowledge of bilateral symmetry and anthropometric measure of facial features.

The input of our algorithm is a 256-level gray-scale image. To obtain a good threshold result, we perform histogram equalization first to enhance the contrast of these regions before converting the grayscale image into a binary image using a hard thresholding method. The method of searching the desired facial regions is based on template matching by means of a correlation coefficient. The extracted results are shown in Figure 5 which illustrates the bounded rectangles with FCPs for different regions. Figure 6 shows the flowchart of our feature extraction algorithm. The size of the templates used in template matching and that of the input image are tabulated in Table 1. Figures 7a to 7d show the different shapes of the facial templates.

Table 1. Size of Image and Templates used in the Project.

Image / Template	Height (in pixels)	Width (in pixels)
Input Image	512	512
Eyes Template	50	240
Single Eye Template	30	30
Mouth Template	60	120
Eyebrow Template	40	120

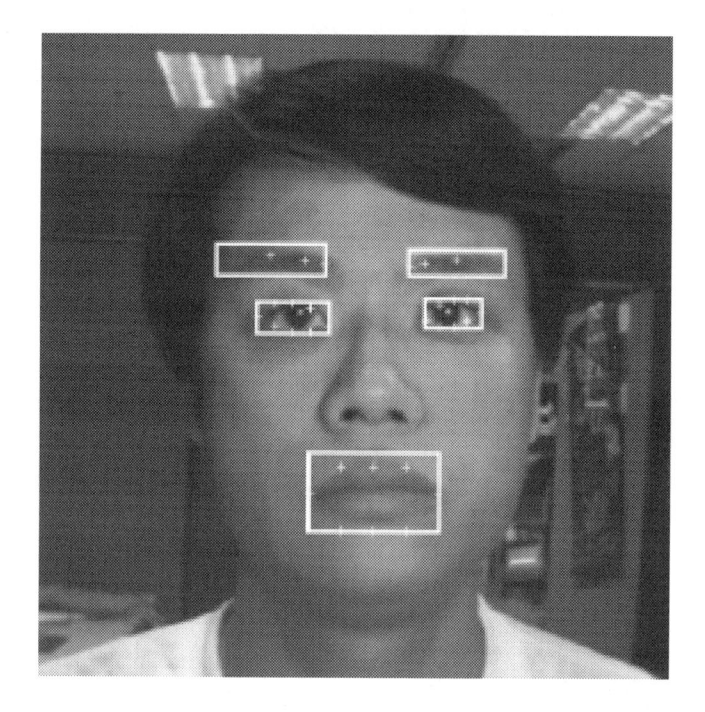

Figure 5. Salient Regions with FCPs for different features.

After we obtained these five face component rectangles, we then finetune our search in these regions to extract FCPs. These regions are described with their top left corner coordinate (l, t) and their width and height (w, h). These five bounded rectangles are denoted as le (left eye), re (right eye), lb (left eyebrow), rb (right eyebrow), and mo (mouth). We acquire the FCPs using the set of predefined positions within the rectangles. Table 2 shows the predefined positions within different regions.

3.3 Image Warping in FES

We employ a digital image warping technique for mapping an expressionless face image to a new expressive face image using FCPs as control points. The set of FCPs of a specific expression is the summation of the expressionless FCPs and the FCP displacements calculated by the RBF network:

$$\begin{cases} x_i = \Delta x_i + (x_i)_{norm} \\ y_i = \Delta y_i + (y_i)_{norm} \end{cases} \quad (11)$$

Flow Chart

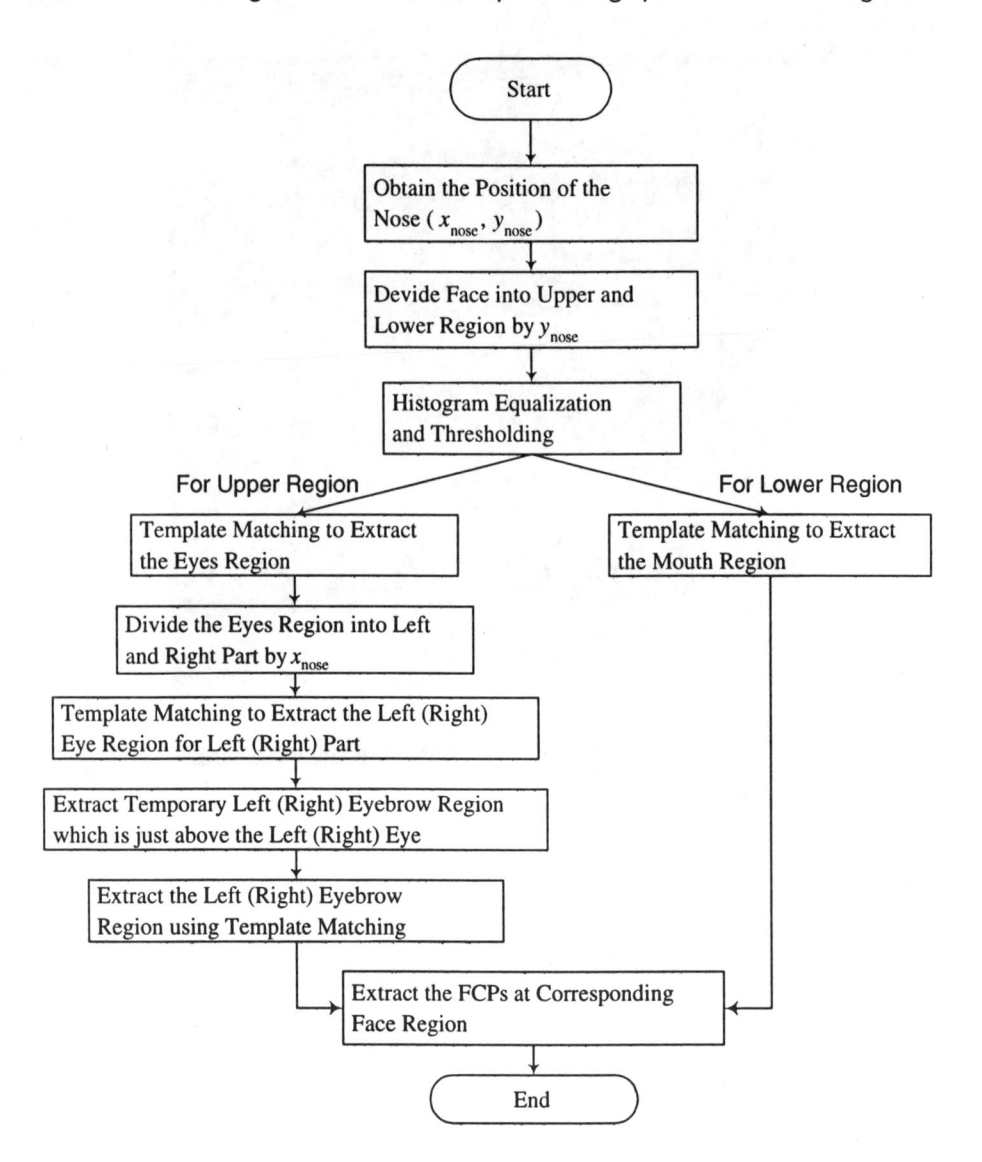

Start → Obtain the Position of the Nose (x_{nose}, y_{nose}) → Devide Face into Upper and Lower Region by y_{nose} → Histogram Equalization and Thresholding

For Upper Region:
- Template Matching to Extract the Eyes Region
- Divide the Eyes Region into Left and Right Part by x_{nose}
- Template Matching to Extract the Left (Right) Eye Region for Left (Right) Part
- Extract Temporary Left (Right) Eyebrow Region which is just above the Left (Right) Eye
- Extract the Left (Right) Eyebrow Region using Template Matching

For Lower Region:
- Template Matching to Extract the Mouth Region

→ Extract the FCPs at Corresponding Face Region → End

Figure 6. Flow Chart for the Feature Extraction Algorithm.

The new FCPs should be de-normalized, rotated, and transformed back before feeding into the image warping algorithm. These steps are the reversal of the preprocesses steps discussed in Section 2.1.

Facial Expression Synthesis Using Radial Basis Function Networks

(a) The Eyes Template for locating eyes

(b) The Single Eye Template for locating left (right) eye

(c) The Eyebrow Template for locating left (right) eyebrow

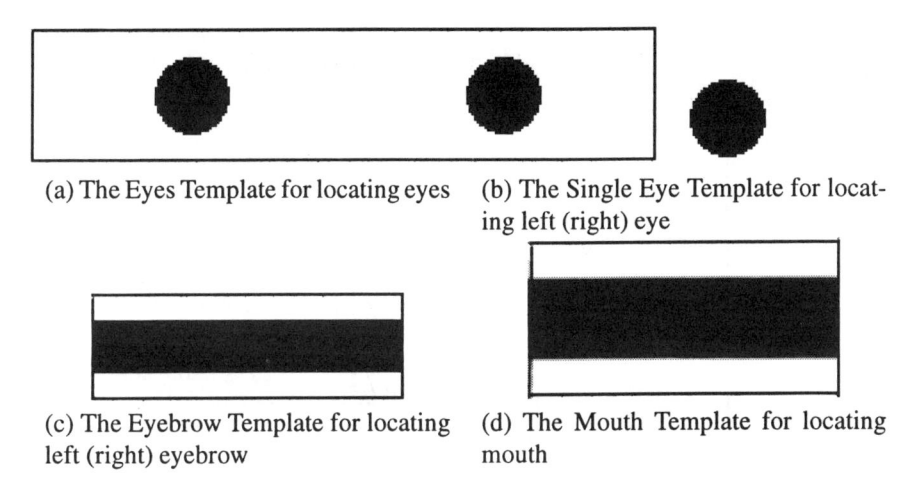

(d) The Mouth Template for locating mouth

Figure 7. Templates for matching different facial regions.

The i-th FCP (x_i, y_i) is de-normalized by the $base$ value at first:

$$\begin{cases} (x_i)_{rot} := x_i * base \\ (y_i)_{rot} := y_i * base \end{cases} \quad (12)$$

and then rotated and translated back:

$$\begin{cases} xb_i = (x_i)_{rot} * \cos\theta - (y_i)_{rot} * \sin\theta + x_{origin} \\ yb_i = (x_i)_{rot} * \sin\theta + (y_i)_{rot} * \cos\theta + y_{origin} \end{cases} \quad (13)$$

where (xb_i, yb_i) is a coordinate pair representing the new position of the i-th FCP for image warping. In the image warping process, the input image is first subdivided into triangular patches as shown in Figure 8.

The 2-D digital image warping is a branch of image processing that deals with the geometric transformation of digital images. A geometric transformation is an operation that redefines the spatial relationship between points in an image. The basis of geometric transformations is the mapping of one coordinate system onto another. This is defined by means of a spatial transformation, a mapping function that establishes a spatial correspondence between all points in the input and output images. Given a spatial transformation, each point in the output assumes the value of its corresponding point in the input image. The correspondence is found by using the spatial transformation mapping function to project the output point onto the input image. Depending on the application, spatial transformation mapping functions may take on many different forms [9]. In

Table 2. FCPs within the Bounded Rectangles.

region	FCP number	x coordinate	y coordinate
left eye	1	$l_{le} + w_{le}$	$t_{le} + h_{le} \times 4/5$
	3	l_{le}	$t_{le} + h_{le}/2$
	5	$l_{le} + w_{le}/2$	$t_{le} + h_{le}$
	7	$l_{le} + w_{le}/2$	t_{le}
	9	$l_{le} + w_{le}/4$	$t_{le} + h_{le} \times 4/5$
	11	$l_{le} + w_{le} \times 3/4$	$t_{le} + h_{le}$
	13	$l_{le} + w_{le}/4$	t_{le}
	15	$l_{le} + w_{le} \times 3/4$	$t_{le} + h_{le}/5$
right eye	2	l_{re}	$t_{re} + h_{re}/2$
	4	$l_{re} + w_{re}$	$t_{re} + h_{re} \times 4/5$
	6	$l_{re} + w_{re}/2$	$t_{re} + h_{re}$
	8	$l_{re} + w_{re}/2$	t_{re}
	10	$l_{re} + w_{re} \times 3/4$	$t_{re} + h_{re} \times 4/5$
	12	$l_{re} + w_{re}/4$	$t_{re} + h_{re}$
	14	$l_{re} + w_{re} \times 3/4$	t_{re}
	16	$l_{re} + w_{re}/4$	$t_{re} + h_{re}/5$
left eyebrow	17	$l_{lb} + w_{lb}/2$	$t_{lb} + h_{lb}/3$
	19	$l_{lb} + w_{lb}$	$t_{lb} + h_{lb} \times 2/3$
	21	$l_{lb} + w_{lb} \times 5/6$	$t_{lb} + h_{lb} \times 4/5$
right eyebrow	18	$l_{rb} + w_{rb}/2$	$t_{rb} + h_{rb}/3$
	20	l_{rb}	$t_{rb} + h_{rb} \times 2/3$
	22	$l_{rb} + w_{rb}/6$	$t_{rb} + h_{rb} \times 4/5$
mouth	23	l_{mo}	$t_{mo} + h_{mo}/2$
	24	$l_{mo} + w_{mo}$	$t_{mo} + t_{mo}/2$
	25	$l_{mo} + w_{mo}/2$	$t_{mo} + h_{mo}$
	26	$l_{mo} + w_{mo}/2$	$t_{mo} + h_{mo}/6$
	27	$l_{mo} + w_{mo}/4$	$t_{mo} + h_{mo}$
	28	$l_{mo} + w_{mo} \times 3/4$	$t_{mo} + h_{mo}$
	29	$l_{mo} + w_{mo}/4$	$t_{mo} + h_{mo}/6$
	30	$l_{mo} + w_{mo} \times 3/4$	$t_{mo} + h_{mo}/6$

our system, we use an inferring affine transformation method which maps each pixel on the source triangles to the target triangles. The general representation of an affine transformation between the target coordinates, uv-, and the source, xy-coordinate system is defined as:

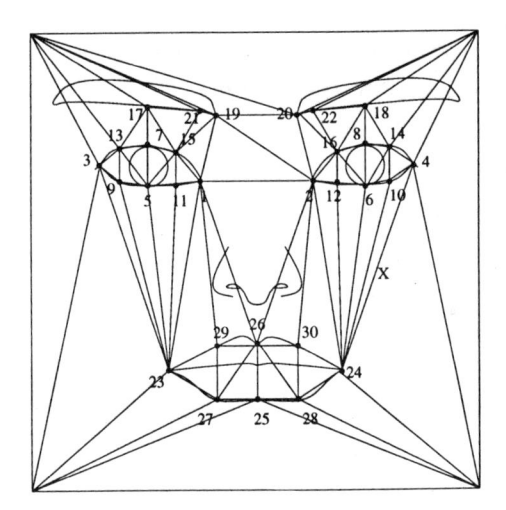

Facial Expression Synthesis Using Radial Basis Function Networks

Figure 8. The FCPs are connected to form patches.

$$[x, y, 1] = [u, v, 1] \begin{bmatrix} a_{11} & a_{12} & 0 \\ a_{21} & a_{22} & 0 \\ a_{31} & a_{32} & 1 \end{bmatrix}, \quad (14)$$

$$\text{where } \mathbf{A} = \begin{bmatrix} a_{11} & a_{12} & 0 \\ a_{21} & a_{22} & 0 \\ a_{31} & a_{32} & 1 \end{bmatrix}$$

is a transformation matrix. Consequently, an affine mapping is characterized by $\mathbf{A}$ whose last column is equal to $[0, 0, 1]^T$. This corresponds to an orthographic or parallel plane projection form the target uv-plane onto the source xy-plane. As a result, affine mappings preserve parallel lines which avoid foreshortened axes when performing 2-D projections. Furthermore, equi-spaced points are preserved, although the actual spacing in the two coordinate systems may differ. For affine transformations, the forward mapping functions are

$$\begin{cases} x = a_{11}u + a_{21}v + a_{31} \\ y = a_{12}u + a_{22}v + a_{32} \end{cases} \quad (15)$$

This equation accommodates translations, rotations, scale, and shear transformations. Since the product of affine transformations is also affine, they can be used to perform a general orientation of a set of points relative to an arbitrary coordinate system while still maintaining a unity value for the homogeneous coordinate. This is necessary for generating composite transformations. An affine transformation has six degrees of freedom, relating directly to coefficients a_{11}, a_{21}, a_{31}, a_{12}, a_{22}, and a_{32}. If we know the vertex positions of a triangle both on the source plane and the destination plane, these six unknown coefficients of the affine mapping can be inferred by solving the system of six linear equations:

$$\begin{bmatrix} x_0 & y_0 & 0 \\ x_1 & y_1 & 0 \\ x_2 & y_2 & 1 \end{bmatrix} = \begin{bmatrix} u_0 & v_0 & 0 \\ u_1 & v_1 & 0 \\ u_2 & v_2 & 1 \end{bmatrix} \begin{bmatrix} a_{11} & a_{12} & 0 \\ a_{21} & a_{22} & 0 \\ a_{31} & a_{32} & 1 \end{bmatrix} \quad (16)$$

Let the system of equations given above be denoted as $\mathbf{X} = \mathbf{UA}$. In order to determine the coefficients, we isolate $\mathbf{A}$ by multiplying both sides with $\mathbf{U}^{-1}$. $\mathbf{U}^{-1} = \mathbf{U}^*/\det(\mathbf{U})$ where $\mathbf{U}^*$ is the adjoint of $\mathbf{U}$ and $\det(\mathbf{U})$ is the determinant. Although the adjoint is always computable, an inverse exists only when the determinant is nonzero. Fortunately, in our application the vertices of the triangular patches are noncollinear; hence, this condition serves to ensure that $\mathbf{U}$ is nonsingular, i.e., $\det(\mathbf{U}) \neq 0$. Consequently, the inverse $\mathbf{U}^{-1}$ is guaranteed to exist and we have

$$\mathbf{A} = \mathbf{U}^{-1}\mathbf{X}, \quad (17)$$

or equivalently,

$$\begin{bmatrix} a_{11} & a_{12} & 0 \\ a_{21} & a_{22} & 0 \\ a_{31} & a_{32} & 1 \end{bmatrix} = \frac{1}{\det(\mathbf{U})} \times \\ \begin{bmatrix} v_1 - v_2 & v_2 - v_0 & v_0 - v_1 \\ u_2 - u_1 & u_0 - u_2 & u_1 - u_0 \\ u_1 v_2 - u_2 v_1 & u_2 v_0 - u_0 v_2 & u_0 v_1 - u_1 v_0 \end{bmatrix} \begin{bmatrix} x_0 & y_0 & 0 \\ x_1 & y_1 & 0 \\ x_2 & y_2 & 1 \end{bmatrix} \quad (18)$$

where $\det(\mathbf{U}) = u_0(v_1 - v_2) - v_0(u_1 - u_2) + (u_1 v_2 - u_2 v_1)$.

After computing the coefficient matrix of each triangle patch, we can map the pixels from the source to the target image. For each pixel in the source image, we first find the triangle it belongs to, then we map the pixel to the target image according to the relevant coefficient matrix.

3.4 Results of FES Using Two RBF Networks

In our experiment, we used a SGI Indigo workstation running C++ with a set of 128×128 grayscale images. The total time for facial feature ex-

Facial Expression Synthesis Using Radial Basis Function Networks 417

traction, pre-processing, neural network calculation, and image warping took less than 15 seconds. Figure 9a shows an expressionless face image in the following experiments and Figure 9b shows the interface of our system.

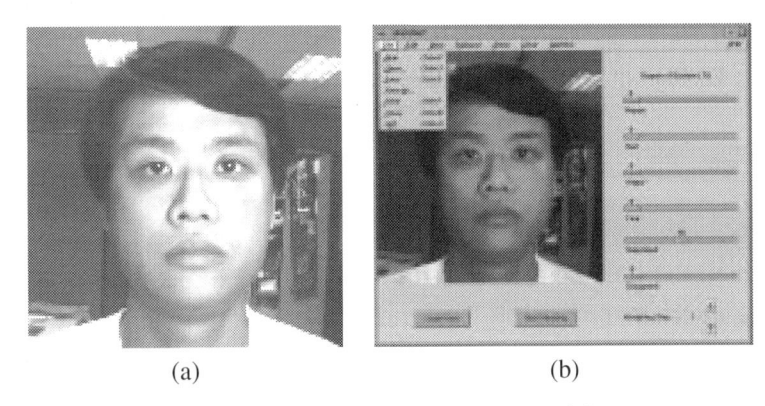

Figure 9. (a) Normal face; (b) The interface of the system.

In order to provide a better approximation, we manually arranged the FCPs to generate more interpolated input training patterns. For each expression, we added four sets of FCP patterns along with the emotional expression FCPs in the order of increasing strength and assigned to each of them the value 0.6, 0.7, 0.8, 0.9, and 1.0 accordingly. Thus the input vector would be in the form (0.6, 0, 0, 0, 0, 0) for the weakest degree of happiness among the five training patterns.

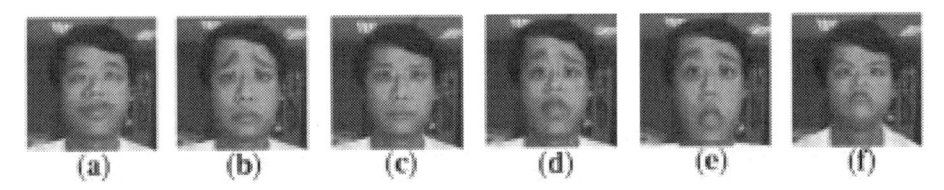

Figure 10. The six universal facial expressions generated by RBF1: (a) Happy, (b) Sad, (c) Angry, (d) Fear, (e) Surprised and (f) Disgusted.

After training, the RBF1 network was able to generate the six universal facial expressions and different degrees of variation for each facial expression. Figure 10 illustrates the resulting images. Note that the neural network is able to capture the features of the facial expressions: for the happy face, the upward movement of mouth corners is captured; for sad face, the inner eyebrows are raised and the mouth corners shift downward; for surprise, the whole eyebrow is raised and the mouth opens

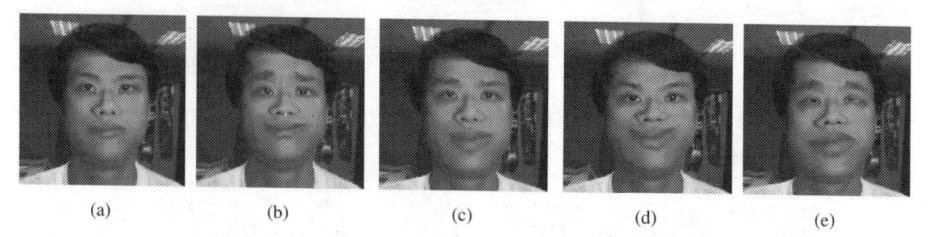

418   Intelligent Biometric Techniques in Fingerprint and Face Recognition

(a)   (b)   (c)   (d)   (e)

Figure 11. Various degrees of the Happy expression increasing from left to right.

widely, etc. Figure 11 displays the five generated different degrees of a happy face.

We trained the RBF2 network with two of the six universal expressions. For instance, we may select Sad and Surprised and use their FCP displacements as one of the training patterns. However, this network is unable to generate different degrees of mixed expression, e.g., 0.5 of Sad and 0.7 of Surprised mixed expression. Figure 12 shows the results obtained from RBF2 network. For generating mixed expressions with different degrees, we must combine the outputs of RBF1 and RBF2 networks. Here, we use the simple averaging method to calculate the final FCP displacements. For example, to generate 0.5 of Sad and 0.7 of Surprised mixed expression we averaged the output from 0.5 of Sad from RBF1, 0.7 of Surprised from RBF1, and Sad-Surprised from RBF2. Figure 13 illustrates some results of our FES system in a grid which shows the Sad expression (at the top center of the figure) mixed with the Happy expression (at the lower right of the figure) and the Surprised expression (lower left of the figure) with different degrees.

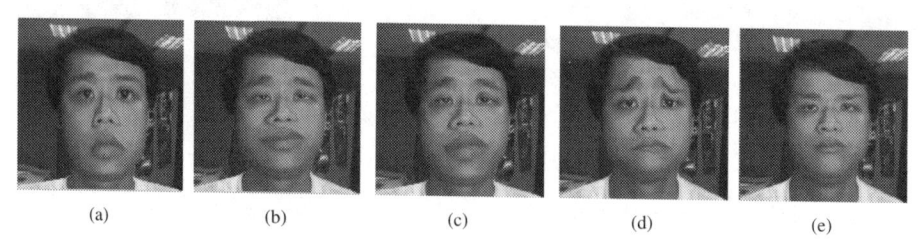

(a)   (b)   (c)   (d)   (e)

Figure 12. A mixture of various expressions: (a) Fear-Surprised, (b) Happy-Sad, (c) Happy-Surprised, (d) Sad-Angry and (e) Sad-Disgusted.

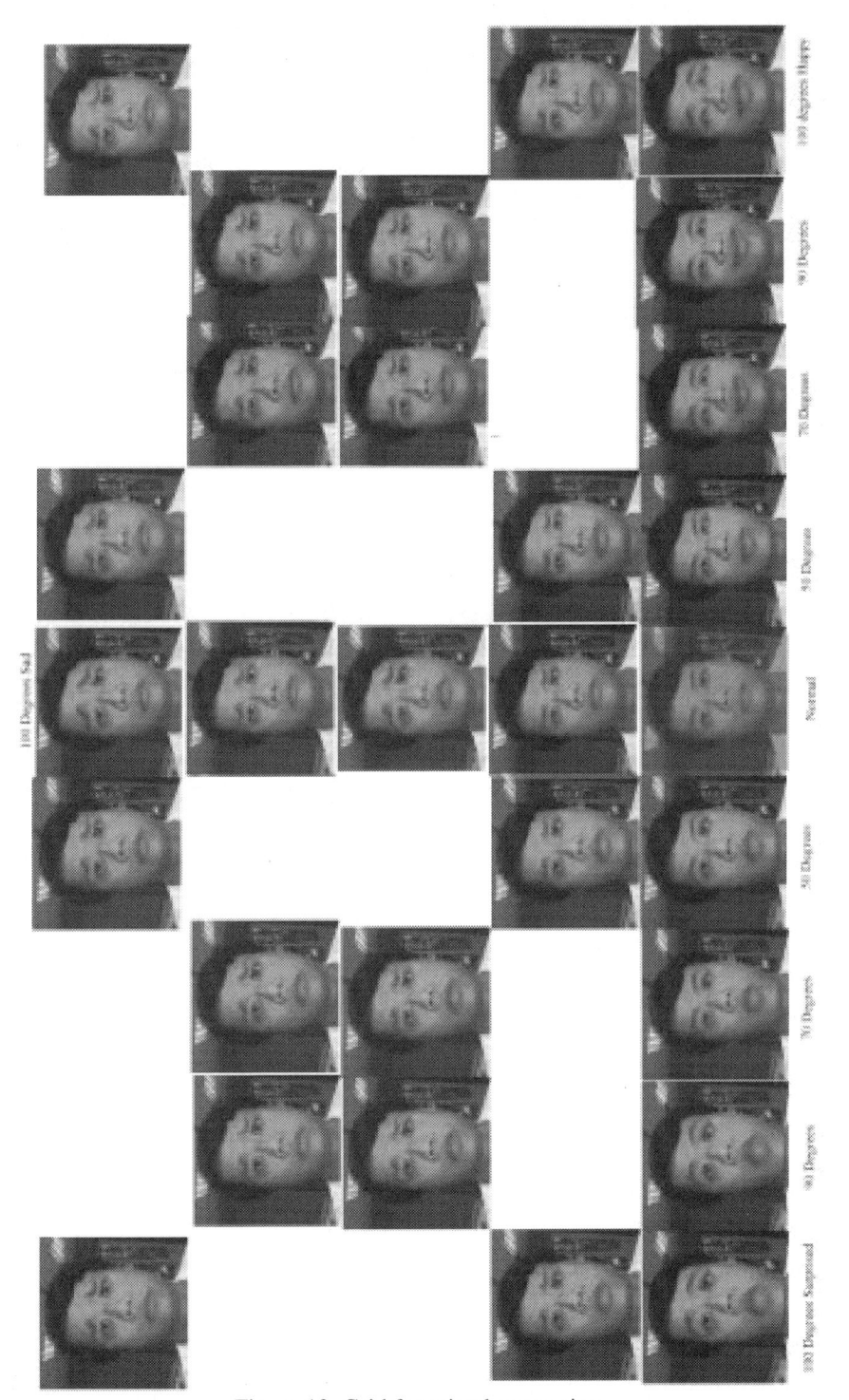

Facial Expression Synthesis Using Radial Basis Function Networks

Figure 13. Grid for mixed expression.

4 Discussion and Conclusion

As a first step towards an automated FES system, our experimental results are encouraging. One main problem is that we are unable to generate transitional expressions with different degrees and mixtures smoothly. Some transitional expressions obviously are unlike the expressions made by people and some are even bizarre. In light of this, we highlight some possible solutions for future improvements:

1. **Hard to obtain accurate FCPs for training**: FES essentially is a function approximation process which seeks a mapping from low-dimension to high-dimension. The precision of the approximation largely depends on the number and accuracy of the training data. We did not provide very accurate nor enough training samples. In fact, it is difficult to ask someone to gesture an expression accurately and objectively. To overcome this, we may need to sample a larger database of expressions from many people. Moreover, a video facility to record continuous movement of transitional expressions will be useful in obtaining intermediate expressions.

2. **Cannot generate artificial features**: Apart from the FCP displacements generated, our system cannot generate movements for subtle facial components such as naso-labial folds or crow's-feet wrinkles on the face. Special technique such as texture mapping may be adopted to add these features to the final image.

3. **Combination of different RBF networks**: Instead of using the averaging weight method in our hybrid network, we may use other methods to combine different RBF networks in order to improve the generation of mixed expressions of various degrees.

In this chapter, we presented some basic concepts on synthesis of facial expressions based on 2-D gray-scale images using Radial Basis Function (RBF) neural networks and image warping. We detailed how RBF networks approximate the mapping function of the displacement of Facial Characteristic Points (FCPs). Specifically, we constructed two different RBF networks that calculate the FCP displacements of (1) the different degrees of an expression and (2) a mixture of two expressions respectively. To generate a mixture of two expressions with various degrees, we averaged the output of the two RBF networks to obtain the result. Finally, we introduced and demonstrated the 2-D image warping technique in generating experimental images of various facial expressions.

References

[1] Arad, N., Dyn, N., Reisfeld, D., and Yeshurun, Y. (1994), "Image warping by radial basis functions-application to facial expressions," *CVGIP-Graphical Models and Image Processing*, 56(2):161–172, March.

[2] Ekman, P. and Friesen, W. (1978), *The facial action coding system: A technique for the measurement of facial movement*, Consulting Psychologists Press, Palo Alto, CA.

[3] Ekman, P. and Friesen, W.V. (1975), *Unmasking the face: A guide to recognizing emotions from facial clues*, Prentice Hall, New Jersey.

[4] Hassoun, M.H. (1995), *Fundamentals of Artificial Neural Networks*, The MIT Press, Cambridge, Massachusetts, London, England.

[5] Mehrotra, K., Mohan, C.K., and Ranka, S. (1997), *Elements of Artificial Neural Networks*, (Complex Adaptive Systems), The MIT Press, Cambridge, Massachusetts, London, England.

[6] Kobayashi, H. and Hara, F. (1992), "Recognition of six basic facial expressions and their strength by neural network," *IEEE International Workshop on Robot and Human Communication*.

[7] Kobayashi, H. and Hara, F. (1991), "The recognition of basic facial expressions by neural network," *Proceedings to the IEEE International Joint Conference on Neural Networks*, Vol. 1, pp. 460–466, Singapore, November.

[8] Werbos, P. (1974), *Beyond regression: New tools for prediction and analysis in the behavioral sciences*, PhD thesis, Harvard University.

[9] Wolberg, G. (Ed.) (1992), *Digital Image Warping*, Image processing – Digital techniques, IEEE Computer Society Press, Los Alamitos, California.

Chapter 13:

Recognition of Facial Expressions and Its Application to Human Computer Interaction

RECOGNITION OF FACIAL EXPRESSIONS AND ITS APPLICATION TO HUMAN COMPUTER INTERACTION

T. Onisawa
Institute of Engineering Mechanics, University of Tsukuba
1-1-1, Tennodai, Tsukuba, 305-8573, Japan

S. Kitazaki
Doctoral Program in Engineering, Onisawa Lab.,
Institute of Engineering Mechanics, University of Tsukuba
1-1-1, Tennodai, Tsukuba, 305-8573, Japan

This chapter describes a model that recognizes emotions from not only facial expressions but also human situations. We use an artificial neural network model and questionnaire data in our approach. Information on situations is inputted to the model indirectly, which is transformed into 6 basic emotions; happiness, sadness, fear, surprise, anger and disgust. The transformation is performed by the use of questionnaire data. The recognition model of emotions is obtained by the backpropagation algorithm and the data obtained by the questionnaire that asks the degree of emotions of facial expressions under presented situations. Seven kinds of models are obtained, which give the degree of each emotion.

Next, we combine the recognition model of emotions and the route decision system to model emotion. This serves as an example of human-computer interaction using facial expressions. The route decision system reaches an instructed destination or sometimes takes a wrong way, where instructions are given by natural language. The route decision system also shows its emotions through facial expressions.

1 Introduction

Human face-to-face communication is a good model of human-computer interaction. In human face-to-face communication, human employs not only verbal information but also non-verbal information [1], e.g., facial expressions, gesture, voice pitch, which is difficult to process using conventional information processing technology. Among these pieces of non-verbal information, facial expressions play an important role in human face-to-face communication since they reflect human emotions and feelings well. Almost all people can see through other people by looking at their faces. If a computer can perceive human facial expressions, then the human-computer interaction is hoped to be smooth. In this sense, the study of human facial expressions becomes important in human-computer interaction [2]. The face recognition is important in areas such as security systems, identification of criminals [3] as well as human-computer interaction. Considering face recognition from the viewpoint of the understanding of the relationship between human facial expressions and human emotions, there have been many approaches such as Fourier descriptors, template matching as a pattern recognition technique [3], a neural network approach [3]-[9], a fuzzy measure and fuzzy integral approach [10], an approach from the psychological point of view [11], [12], and other approaches [13]-[16]. In most of these approaches, only the relationship between facial expressions and emotions is considered. In practice, however, this relationship is often dependent on human situations. Psychologists still argue on the relative importance of expressive information and context information [17], [18].

First, we describe a recognition model of human emotions which considers not only facial expressions but also human situations using an artificial neural network model and questionnaire data. The performance of our model is verified by the use of two types of questionnaires. Next, we combine the recognition model of emotions and the route decision system [19] as an application of the recognition model of emotions. The route decision system reaches an instructed destination, or sometimes loses its way, where instructions of a route to a destination are given with natural language. The route decision system also expresses its emotions through facial expressions according to its situation. The combination is also considered as an example of human-computer interaction with facial expressions, situations and emotions.

The conventional computer science deals with intelligent information that is represented logically. This kind of information has objectivity, uniqueness and universality. This chapter, however, deals with Kansei information (a kind of emotional information) that has subjectivity, vagueness, ambiguity and context dependency. This kind of information is necessary in a human centered system [20].

2 Recognition Model of Emotions

2.1 Model Structure

This chapter employs an artificial neural network model as the recognition model of emotions. In implementing the recognition model, information on situations as well as facial expressions is necessary. With respect to facial expressions, eight parameters, i.e., the size of eyes, the slant of eyes, the shape of eyes, the slant of eyebrows, the shape of eyebrows, the size of a mouth, the width of a mouth and the shape of a mouth are considered. With respect to situations, limited kinds of situations are considered since it is impossible to consider every situation as the input of the model. And information on situations is inputted into the model not directly but indirectly. That is, information on situations is transformed into emotions that are happiness, sadness, fear, surprise, anger and disgust [21]. For example, let us consider the situation where *I am suddenly barked at by a dog*. If I have a feeling of surprise, fear and disgust for the situation, the situation is transformed into surprise, fear and disgust.

The artificial neural network models to recognize emotions from facial expressions and situations have an input layer, an output layer and some hidden layers. The model has 8 input nodes about facial expressions and 6 input nodes about emotions obtained from situations. The model has an output node, that is, one of 6 kinds of emotions (happiness, surprise, anger, disgust, sadness and fear) or unnaturalness. The unnaturalness means the feeling of unnaturalness for the combination of facial expressions and situations. The recognition model is composed of seven neural network models. Each neural network has 14 input nodes about facial expressions and emotions that are transformations of situation information, and one output node about emotions under the situation or unnaturalness as shown in Figure 1.

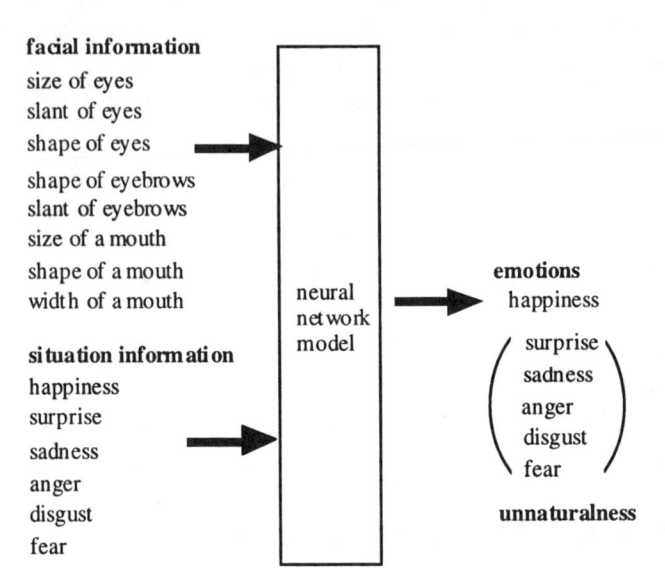

Figure 1. Structure of recognition model from facial expressions and situations.

2.2 Questionnaire about Situations

As mentioned before, a situation is transformed into emotions that are felt under the situation. In order to make the relationship between situations and emotions, the questionnaire about situations is performed on 10 subjects. Subjects are asked to mark the questionnaire sheet as shown in Figure 2 according to the degree of each emotion under the presented situation with one of three scale evaluations, i.e., a weak feeling, a moderate feeling or a strong feeling of happiness, surprise, anger, disgust, sadness and fear. Subjects are undergraduate or graduate students, and 8 students are males and 2 females. In the questionnaire, 18 kinds of situations are prepared as shown in Table 1.

2.3 Questionnaire about Facial Expressions and Situations

In this questionnaire, faces by line drawing as shown in Figure 3 are used since they are controlled more easily than real expressions such as face photographs. Subjects are asked to compare the facial expression at the right side of the arrow with that at the left side as shown in Figure 3, and to mark the questionnaire sheet according to the degree of each emotion felt under the presented situation with one of three scale evaluations, i.e., a weak feeling, a moderate feeling or a strong feeling of happiness, surprise, anger, disgust, sadness and fear. The left side of expression is a standard expression, which is assumed that each subject does not feel emotions.

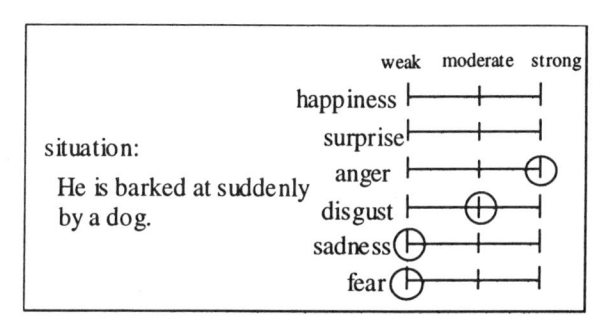

Figure 2. Example of questionnaire about situation.

Table 1. Situations.

no.	situations
1	He passes an exam unexpectedly.
2	He is going to see a pass announcement.
3	He is scoled.
4	He is scoled unfairly.
5	He is praised.
6	He is separated familier friend.
7	He gets bad marks in a exam.
8	He passes an exam.
9	He gets bad marks in an exam unexpectedly.
10	He loses his wallet.
11	He is barked at by a big dog.
12	He is going to have a picnic.
13	He is barked at suddenly by a dog.
14	He is preached.
15	He is forced the work on.
16	He is going to give a talk in large company.
17	He is forced to choose one between the two.
18	He is ill spoken of.

If subjects feel that the combination of a situation and facial expressions is unnatural, they are asked to mark **unnatural** in the sheet. In the questionnaire, 38 kinds of facial expressions and 12 kinds of situations are combined, where situations are selected out of 18 situations considered in Section 2.2. In order to avoid an influence of the former combination of facial expressions and a situation in the questionnaire, only one combination is presented in one questionnaire sheet.

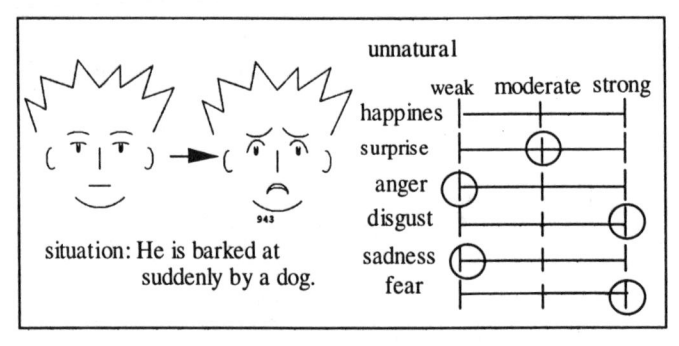

Figure 3. Example of questionnaire about facial expressions and situations.

2.4 Preprocessing of Questionnaire Data and Neural Network Model

Eight parameters about facial expressions, i.e., the size of eyes, the slant of eyes, the shape of eyes, the slant of eyebrows, the shape of eyebrows, the size of a mouth, the width of a mouth and the shape of a mouth, have numerical values in [0, 1] depending on the location of the part in facial expressions as shown in Figure 4. For example, the slant of eyebrows is described by normalizing the difference between the right end of eyebrow and the left end of eyebrow. Figure 4 depicts the parameter value range according to the location of each feature in a face. Each parameter of the standard facial expression is assigned 0.5 except the size of a mouth. With respect to emotions in the questionnaire, natural language expressions correspond to numerical values in [0, 1] as follows:

$$\begin{cases} No\,mark & :0.00 \\ Weak & :0.33 \\ Moderate & :0.67 \\ Strong & :1.00 \end{cases} \quad (1)$$

Recognition of Facial Expressions and its Application to Human Computer Interaction

Value Range	the slant of eyes	the shape of eyes	the size of eyes	the shape of eyebrows	the slant of eyebrows
1 • E 0					

Value Range	the shape of a mouth	the size of a mouth	the width of a mouth
1 • E 0			

Figure 4. Parameter value range according to location of each feature in a face.

If unnatural in the questionnaire sheet is marked, the numerical value about the unnaturalness is assigned 1. Otherwise, the value is 0.

Seven neural network models are obtained by the backpropagation algorithm. Each model has an input layer, an output layer and two hidden layers. The model has 20 nodes in each hidden layer as shown in Figure 5. Outputs of the neural network models are expressed by natural language as shown in Figure 6 and Table 2.

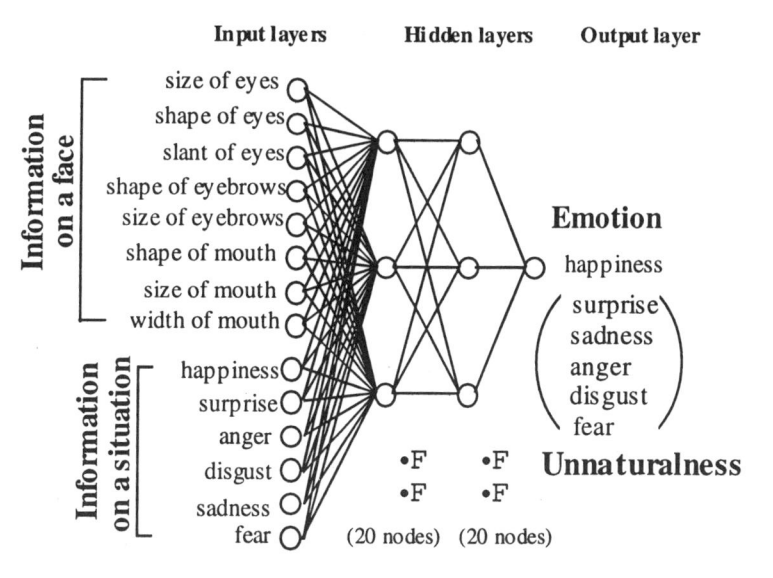

Figure 5. Neural network of recognition model.

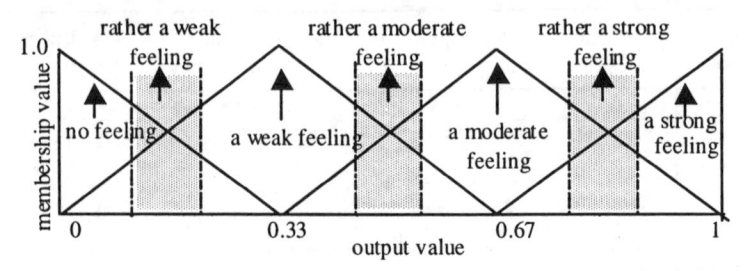

Figure 6. Natural language expressions and their fuzzy sets (1).

Table 2. Natural language expressions.

Expressions	Output Value Ranges
no feeling	[0.00, 0.11)
rather a weak feeling	[0.11, 0.22)
a weak feeling	[0.22, 0.44)
rather a moderate feeling	[0.44, 0.56)
a moderate feeling	[0.56, 0.78)
rather a strong feeling	[0.78, 0.89)
a strong feeling	[0.89, 1.00]

If three or more unpleasant emotions out of the four (anger, disgust, sadness and fear) are felt even a little, the natural language expression *a feeling of unpleasant* is used in order to express those unpleasant emotions together, where the degree of unpleasantness is the largest output degree in those unpleasant emotions. If one of unpleasant emotions is felt much more strongly than the other unpleasant emotions, the degree of those unpleasant emotions is represented separately by natural language. Natural language expressions about unpleasantness are shown in Figure 7 and Table 3. If the output value of unpleasantness is in [0, 0.33], **natural** is not expressed, and natural language expressions of emotions are used instead.

2.5 Performance of Obtained Model

Figure 8 shows examples of recognition results when three kinds of pairs (facial expressions, situations), which have the same facial expressions and different situations, are given. Although the same facial expressions are given, a variety of recognition results are obtained according to the situation.

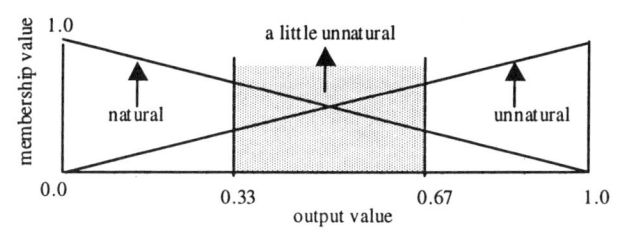

Figure 7. Natural language expressions and their fuzzy sets (2).

Table 3. Natural language expressions about unnaturalness.

Expressions	Output Value Ranges
(natural)	[0.00, 0.33]
a little unnatural	(0.33, 0.67]
unnatural	(0.67, 1.00]

facial expressions			
situations	He is seperated from familier friend.	He gets bad marks in an exam.	He is going to see a pass announcement.
results	He has a strong feeling of sadness.	He has a feeling of unpleasantness.	He has a strong feeling of fear.

Figure 8. Examples of recognition results.

In order to verify the model performance, two kinds of questionnaires are carried out. In the first, subjects evaluate the model performance by the questionnaire about model recognition results. In the second, the model performance is evaluated by comparing its recognition results with subjects ones. In the first kind of the questionnaire, subjects are asked to evaluate model recognition results by the use of the questionnaire sheet as shown in Figure 9. In this questionnaire, 57 combinations of facial expressions and situations are used. Some of them are not used in the questionnaire described in Sections 2.3. Each model gets the point according to subjects' evaluations as shown in Table 4. The average point of the evaluation among 57 combinations is 0.90. The best average evaluation point is 2.00 and the worst average point is –0.43 as shown in Figure 10.

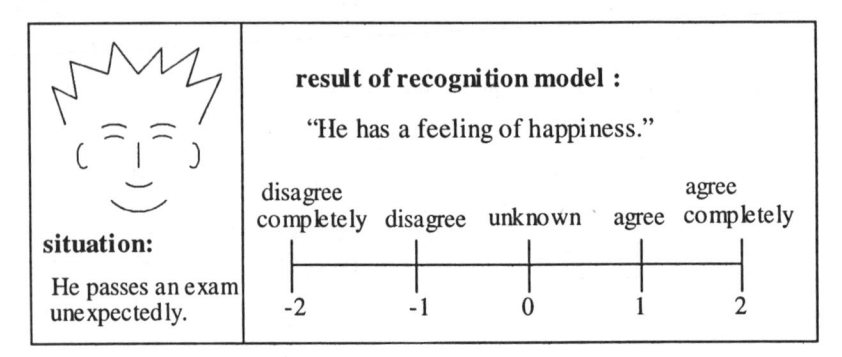

Figure 9. Example of questionnaire about model results.

Figure 10. Evaluation point of model.

Table 4. Evaluation points.

Evaluation	Points
The subject does not agree with the model recognition result at all.	-2
The subject does not agree with the model recognition result.	-1
The subject does not conclude whether he/she agree with the model recognition result or not.	0
The subject agrees with the model recognition result.	+1
The subject agrees with the model recognition result completely.	+2

In the second kind of the questionnaire, 10 subjects are asked to mark the same questionnaire sheet as shown in Figure 3. In this questionnaire, 22 kinds of pairs of facial expressions and situations are selected, where some of them are not used in the questionnaire in Section 2.3. The degree of emotions recognized by the model is transformed into one of four scales in the questionnaire, i.e., no feeling, weak, moderate or strong. Model recognition results and questionnaire results are compared with. Let the difference of the degree between, for example, weak and moderate, be one scale. Figure 11 shows examples of different scales of the degree per kind of emotion in one facial expression.

The average difference scale of the degree per kind of emotion in one facial expression is 0.63. In order to evaluate the value 0.63, the numbers 0, 1, 2 and 3 are generated at random. It is assumed that the numbers 0, 1, 2 and 3 correspond to no feeling, a weak feeling, a moderate feeling and a strong feeling of emotions respectively. Considering that the average of the difference between the random number and model recognition result is 1.3 per kind of emotion in one facial expression, it is found that the obtained model has a high evaluation point.

From the above evaluations of models, it is found that the models reflect the recognition of emotions by the subjects very well when facial expressions and situations are given.

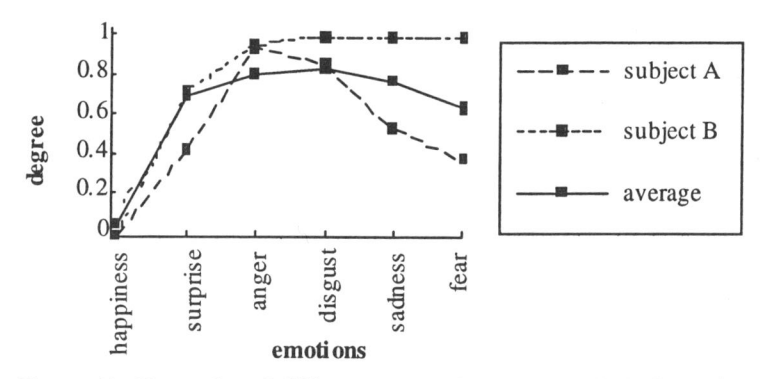

Figure 11. Examples of different scales of degree per kind of emotion in a facial expression.

3 Combination of Recognition Model of Emotions and Route Decision System

In this section the combination of the recognition model of emotions and the route decision system [19] is considered as an application of the recognition model of emotions, and an example of human-computer interaction using facial expressions.

3.1 Route Decision System

The route decision system searches a route to an instructed destination and expresses a system state with line drawing facial expressions. The route decision system consists of five sections, i.e., a receiving section of initial instructions, a route decision section, a route memorization section, a communication section and a face expressions section.

3.1.1 Receiving Section of Initial Instructions

The route decision system receives instructions on a route from a start point to a destination by natural language through the receiving section of initial instructions. Linguistic instructions used here are, for example, *make its way only a little, a little, some little, a few Km, about x Km, a good distance, turn to the right, find something*. Fuzziness of the meaning of linguistic instructions is dealt with a fuzzy set as shown in Figure 12 and Figure 13.

This section divides linguistic instructions into each part of speech of words in order to understand given instructions. The receiving section of initial instructions has a dictionary in which words used in instructions are entered beforehand. Words in the dictionary have somewhat different information according to the part of speech of words as shown in Table 5, where the action in the verb row means one of three actions that are *make its way*, *turn to the right* (or *the left*, or *the back*), or *find something*. For example, let us consider instructions such as *make its way a little and turn to the right at the intersection with a mailbox*. Actions are divided into *make its way a little, find a mailbox, find an intersection*, and *turn to the right at the intersection* as shown in Table 6.

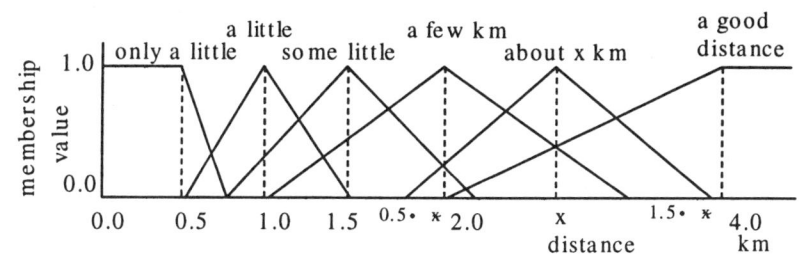

Figure 12. Fuzzy sets representing distance.

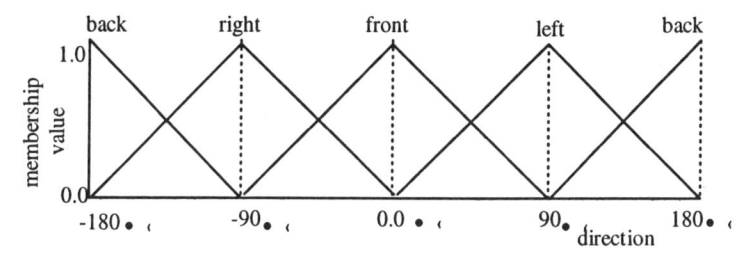

Figure 13. Fuzzy sets representing direction.

Table 5. Structure of dictionary.

Part of speech	Information
Auxiliary verb and postpositional particle	Entry of the word
Verb	Entry of the word, reading of the word, inflection of the word, action which word means
Noun	Entry of the word, actions associated with the word, definition of a direction by a fuzzy set when the word means a direction
Adverb	Entry of the word, actions associated with the word, definition of a distance by a fuzzy set when the word means a distance

3.1.2 Route Decision Section

The route decision section regards a distinct feature, e.g., a crossing, as a sub-goal, and searches a route to a given destination comparing given instructions and the route decision system's visual scene. And this section specifies the route decision system's own location by the use of given instructions and perceived information on route surroundings. Even humans, however, cannot measure the distance and the direction accurately without the measure. Therefore, the route decision section deals with fuzziness that is inherent in not only linguistic instructions but also perceived information on the distance and the direction of a street. The route decision section represents the meaning of fuzziness of these pieces of information by the use of the same fuzzy sets as shown in Figure 12 and Figure 13.

Table 6. An example of division of actions.

No. of Instructions	1
Action	Make its way
Instruction	Make its way a little
Fuzzy set associated with the instruction	A fuzzy set *a little* shown in Figure 12
No. of Instructions	**2**
Action	Find something
Instruction	Find a mailbox
Fuzzy set associated with the instruction	None
No. of Instructions	**3**
Action	Find something
Instruction	Find an intersection
Fuzzy set associated with the instruction	None
No. of Instructions	**4**
Action	Turn to the right(or the left or the back)
Instruction	Turn to the right at the intersection
Fuzzy set associated with the instruction	A fuzzy set *right* shown in Figure 13

The comparison is performed by the use of the degree of matching (*DM*) defined by

$$DM = \frac{1}{2}\{Sup(A \cap B) + Inf(A \cup B^C)\} \quad (2)$$

where A is a fuzzy set representing the meaning of given instructions, B is a fuzzy set representing perceived information on the distance or the direction of a street, and B^C is a complement of B. The symbol $\cap$ is an intersection of fuzzy sets, and $\cup$ is a union of fuzzy sets. The *DM* is used in order to give the priority order to alternative routes. The route whose *DM* is the largest is selected as the route which the system should take. The selected route, however, is not necessarily the correct route. When the route decision system takes a wrong way by selecting the largest *DM*, then the route decision section selects the second largest one.

The *DM* is also used in order to compare the *DM* with a given threshold. Let us assume that the route decision system receives the instructions *make its way to the intersection about 1Km ahead from the point B_0* and starts at the point B_0 as shown in Figure 14. Values of *DM*s of intersections B_1 and B_2 are calculated by the use of fuzzy sets shown in Figure 15. The *DM* value of the intersection B_1 is 0.17. If the given threshold is assumed to be 0.25, the *DM* of the intersection B_1 is under the threshold. The route decision system passes the intersection B_1. The system goes to the section B_2. The *DM* of the intersection B_2 is 0.66 over the threshold. The system takes the next action at the intersection B_2. If the intersection B_2 is found to be wrong, the route decision system uses information in the route memorization section as mentioned later.

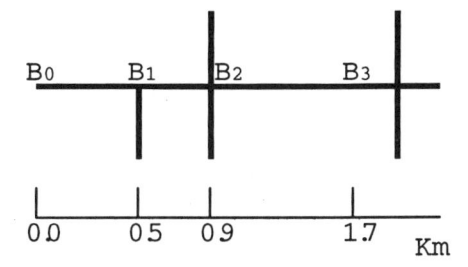

Figure 14. Example of intersection about 1km ahead.

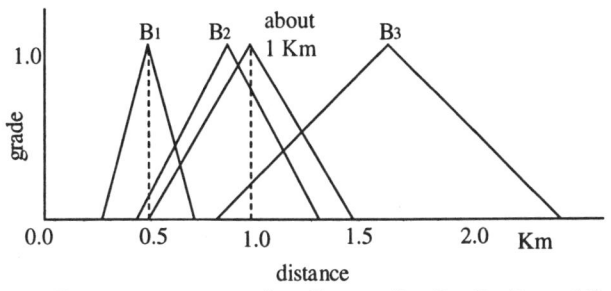

Figure 15. Fuzzy sets representing distance B1-B0, B2-B0 and B3-B0.

Selection of Shortcut

The route decision section has an ability to take a shortcut by perceiving surrounding situations. Based on given instructions the route decision section estimates a destination direction from the intersection that the route decision system stands. If the *DM* of the estimated direction is over the given threshold, the section takes a shortcut at the intersection. After taking a shortcut, the route decision section takes its route as it approaches the destination.

Selection of Roundabout Route

When the route decision system finds an instructed route impassable, the system tries to find another route, i.e., the roundabout route, based on memorized pieces of information on intersections that the model passes. The system selects the route of which estimated direction approaches the destination. The *DM* of the route is calculated in the same way as the selection of a shortcut. If the *DM* is over the given threshold, the route is selected.

3.1.3 Route Memorization Section

The route memorization section memorizes given instructions and information on objects along a route which the route decision system passes. When the route decision system finds that it takes a wrong way, the route decision system can search another route by the use of information in the route memorization section.

3.1.4 Communication Section

When the route decision system loses its way, the communication section asks an instructor or a third party about another route from the lost spot to a sub-goal or a destination. And when a third party addresses the route decision system, this section also responds to it. As mentioned later, the route decision system also has communication with the recognition model of emotions through the communication section about emotions under route decision system's situations.

3.1.5 Face Expressions Section

The face expressions section expresses emotions through a face according to the situation of the route decision system. The facial expressions section uses the following information (X); the *DM* of the distance at the standpoint(x_1), the *DM* of the distance that the route decision system has actually made its way(x_2), the *DM* of the guessed distance from a start point to a destination(x_3), the *DM* of the guessed distance from a standing point to a destination(x_4), reaching a destination(x_5), losing its way(x_6), finding an instructed route impassable(x_7), success in taking a roundabout route(x_8), finding a shortcut(x_9), success in taking a shortcut(x_{10}), many times addresses by a third party(x_{11}), and sudden address by a third party(x_{12}), where x_4 is estimated by x_2 and x_3. These pieces of information from the route decision system are transformed into parameter values of happiness(y_1), sadness(y_2), fear(y_3), surprise(y_4), disgust(y_5), and anger(y_6).

The relationships between information from the route decision system and parameters values in [0, 1] of emotions, happiness, sadness and fear, are represented by linguistic rules in order to estimate the degree of emotion. Linguistic rules about the relationships among x_1, x'_4 and happiness are shown in Table 7, where x'_4 is defined by $x'_4 = x_4 / x_3$ and $x'_4 = 1$ if $x'_4 > 1$. Table 8 shows linguistic rules about relationships among x_1, x'_4 and fear. Table 9 depicts linguistic rules about relationships among x_1, x'_4 and sadness. Meanings of linguistic expressions in these tables are represented by fuzzy sets shown in Figure 16. The min-max method is used as the fuzzy reasoning method and parameters values of emotions are obtained by the defuzzification of fuzzy reasoning results.

Table 7. Linguistic rules for happiness where the DM increases.

X_1	X_4'				
DM	VS	S	M	B	VB
VS	M	M	S	VS	VS
S	M	M	S	S	VS
M	B	B	M	S	S
B	VB	B	B	M	M
VB	VB	VB	B	M	M

VS:Very Small, S:Small, M:Medium, B:Big, VB:Very Big

Table 8. Linguistic rules for fear where the DM decreases.

X₁ DM	X₄'				
	VS	S	M	B	VB
VS	M	M	B	VB	VB
S	M	M	B	B	B
M	S	S	M	M	M
B	VS	VS	S	S	S
VB	VS	VS	S	S	S

VS:Very Small, S:Small, M:Medium, B:Big, VB:Very Big

Table 9. Linguistic rules for sadness where the DM decreases.

X₁ DM	X₄'				
	VS	S	M	B	VB
VS	M	M	B	VB	VB
S	M	M	B	B	B
M	S	S	M	M	M
B	VS	VS	S	S	S
VB	VS	VS	S	S	S

VS:Very Small, S:Small, M:Medium, B:Big, VB:Very Big

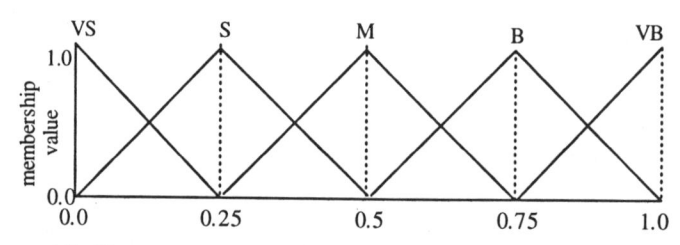

Figure 16. Fuzzy sets representing linguistic terms in Tables 7,8 and 9.

Although fatigue is not emotion, the drop size of sweat is expressed according to the degree of fatigue, which is defined by:

$$y_7 = \begin{cases} 0.0, & x_2 \leq x_3 \\ \dfrac{x_2 - x_3}{2 \times x_3}, & x_3 < x_2 \leq 3 \times x_3 \\ 1.0, & 3 \times x_3 < x_2 \end{cases} \quad (3)$$

Each degree of emotion is used for expression of emotion through a face. A threshold value is considered for the degree of emotions so that the following can be realized. If the threshold value for the degree of emotions, e.g., anger, is assigned low, anger appears in a facial expression easily. On the other hand, if the threshold value for the degree of emotions, e.g., surprise, is assigned high, surprise is difficult to appear in a facial expression. The assignment of the threshold value imagines, for example, a person who is touchy or does not show surprise on a face easily. The practical degree of emotion y_i, ($i = 1,2,3$), which is used for expressions of feelings through a face, is obtained by:

$$y_i = (degree\ of\ emotions\ obtained\ by\ the\ fuzzy\ reasoning)\quad (i = 1,2,3) \qquad (4)$$
$$+ (0.5 - the\ threshold\ value\ assigned\ to\ emotions)$$

Some special situations are considered as shown in Table 10. Numerical value on [0, 1] corresponding to special situations of the route decision system are added to y_i ($i = 1, 2, ..., 6$) as shown in Table 10. The threshold value is also used for the following case. For example, if the degree of happiness becomes large according to the situation of the route decision system, anger and disgust are usually difficult to be felt and to be expressed through a face. Therefore, it is appropriate that threshold values for anger and disgust are increased; for example, from 0.5 to 0.7, in order to implement this situation. On the other hand, if the degree of fatigue becomes large, anger and disgust are easy to be felt and to be expressed. Therefore, it is appropriate that the threshold values for anger and disgust are decreased; for example, from 0.5 to 0.3 in this situation.

Table 10. Relationships between special situations and practical degree of emotions.

Situations	y_i
Losing its way (x_6)	$y_2 = 0.3 \times (1.5 -$ the threshold assigned to sadness$) + y_2$, $y_3 = 0.3 \times (1.5 -$ the threshold assigned to fear$) + y_3$
Finding out no passing (x_7)	$y_4 = 1.0$
Success in taking a roundabout way (x_8)	$y_1 = 0.3 \times (1.5 -$ the threshold assigned to happiness$) + y_1$
Finding out a shortcut (x_9)	$y_3 = 0.3 \times (1.5 -$ the threshold assigned to fear$) + y_3$
Success in taking a shortcut (x_{10})	$y_1 = 0.3 \times (1.5 -$ the threshold assigned to happiness$) + y_1$
Many times addresses by a third party (x_{11})	$y_5 = 0.5 + 0.1 \times ($addressing times$-5) + y_5$, $y_6 = 0.5 + 0.1 \times ($addressing times$-5) + y_6$
Sudden addresses by a third party (x_{12})	If $y_3 >$ the threshold assigned to fear, $y_4 = 0.5 \times (1.5 -$ the threshold assigned to surprise$) + y_4$

The transformation procedure from information on the route decision system to parameter value changes about face features is shown in Figure 17.

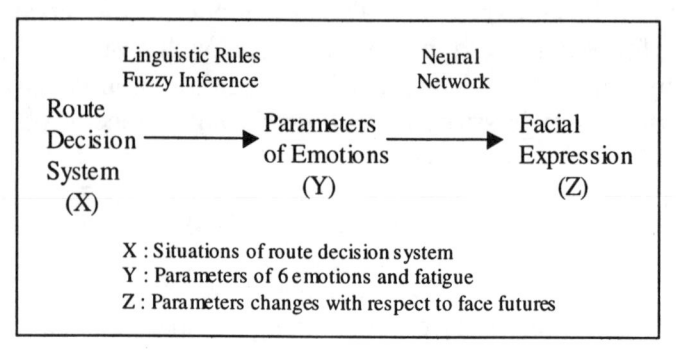

```
Linguistic Rules              Neural
Fuzzy Inference               Network
Route          Parameters              Facial
Decision  ───▶ of Emotions   ───▶     Expression
System            (Y)                    (Z)
(X)

X : Situations of route decision system
Y : Parameters of 6 emotions and fatigue
Z : Parameters changes with respect to face futures
```

Figure 17. Structure of face expressions section.

The neural network model is used in the face expression section. Inputs to the neural network model are parameters values of practical degree of emotions and outputs of the model are changes of eight parameters values with respect to face features, i.e., the size of eyes, the slant of eyes, the shape of eyes, the slant of eyebrows, the shape of eyebrows, the size of a mouth, the width of a mouth and the shape of a mouth. Eight kinds of neural network models are learned by the use of data, which are obtained by the questionnaire asking the correspondence between degrees of 6 basic emotions and 150 face expressions performed by 7 subjects. This questionnaire asks subjects only about the relationship between facial expressions and emotions without considering situations. This questionnaire is different from the one described in Section 2.3. Each network has an input layer, an output layer and a hidden layer that has 20 nodes. The output of each neural network model is one of 8 parameters values of features in a face expression.

3.2 Combination Model

The recognition model of emotions and the route decision system are combined into a communication model considering facial expressions and situations, where situations are limited into the ones with respect to the route decision. The recognition section of emotions in Figure 18 corresponds to the recognition model of emotions described in Section 2. The recognition sections of facial expressions and the situations are necessary in order to combine the route decision system and the recognition model of emotions. Eight parameters values, which are obtained from facial expressions through the recognition section of facial expressions, are inputs to the recognition section of emotions. The recognition section of situations understands the situations of the route decision system using information from the communication section in the route decision system. Since situations are transformed into emotions as described in Section 2.2, other 32 situations as shown in Table 11 are considered as situations in the combination model. Questionnaire about these 32 situations is performed in order to obtain the relation between situations and emotions. Two or more situations sometimes happen at once. Then, the total degree of emotions is obtained by adding the degree of each emotion under each situation.

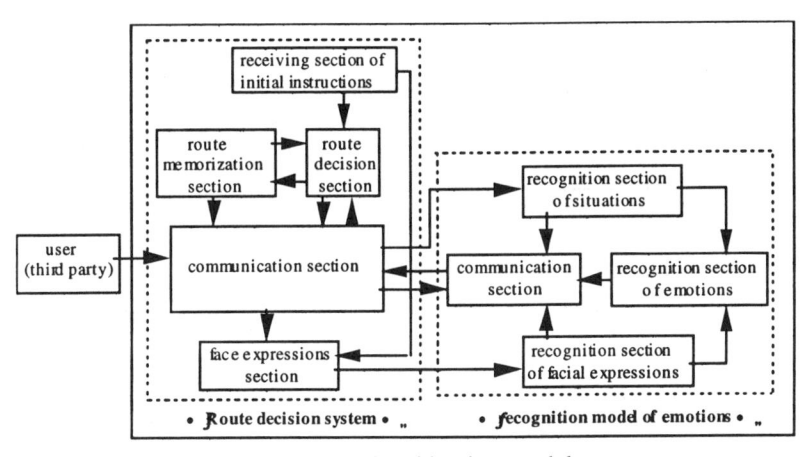

Figure 18. Combination model.

Interactions between the route decision system and the recognition model of emotions are performed by the following procedures.

1. The recognition model of emotions estimates the degree of emotions of the route decision system from only its facial expressions. If the estimated degree of emotions is over a constant value given beforehand, e.g., 0.5, the recognition model of emotions addresses the route decision system according to the degree of emotions through the communication section in the recognition model of emotions, for example, *"You appear happy. What happens to you?"*

Table 11. 32 Kinds of Situations.

No.	Situations
1	The RDS is addressed suddenly when it doesn't lose its way.
2	The RDS is addressed suddenly when it may lose its way.
3	The RDS is addressed many times when it doesn't lose its way.
4	The RDS is addressed many times when it may lose its way
5	The RDS finds out a building that is not an objective.
6	The RDS finds out an objective.
7	The RDS finds a route impossible.
8	The RDS is taking a roundabout route.
9	The RDS succeeds in taking a roundabout route.
10	The RDS finds out a short cut.
11	The RDS is taking a short cut.
12	The RDS succeeds in taking a short cut.
13	The RDS fails in taking a short cut.
14	The RDS fails in taking a roundabout way.
15	The RDS may lose its way.
16	The RDS continues making its way though it feels that it takes a wrong way.
17	The RDS goes back because it takes a wrong way.
18	The RDS finds that the model takes another way.
19	The RDS is given another route for a destination.
20	The RDS takes a wrong way and asks third party. But it is ignored.
21	The RDS may lose its way and asks third party. But it is ignored.
22	The RDS starts for a destination.
23	The RDS may lose its way but it can do next instruction.
24	The RDS takes a wrong way but it is given another route.
25	The RDS may lose its way but it is given another route.
26	The RDS takes a wrong way.
27	The RDS reaches a destination.
28	The RDS asks to third party because it may lose its way.
29	The RDS is given another way but it goes back because it may lose its way.
30	The RDS is given another way but the model may lose its way and can do next instruction.
31	The RDS is given another way but it may lose its way.
32	The RDS continues making its way though it feels that it takes a wrong way.

RDS: Route Decision System

2. After being addressed, the route decision system shows its situation and compares its degree of emotions with the import of the talk of the recognition model of emotions. The route decision system replies according to the comparison result, for example, *"I am not happy."*
3. The recognition model of emotions estimates the degree of emotions from facial expressions and situations of the route decision system. According to the estimated result, the recognition model of emotions addresses the route decision system again, for example, *"You appear surprised, don't you?"* Or *"Are you sad?"*
4. The route decision system replies with *"yes"* or *"no"*.
5. When the route decision system's reply is negative, the recognition model of emotions addresses the route decision system again, referring to the estimated degree of emotions only from facial expressions or only from situations.

Conversation is continued until the route decision system's reply is positive or the recognition model of emotions has nothing to talk any more. In this combination model, a user (a third party) can also address the route decision system at any time as shown in Figure. 18.

As mentioned in 2.4, when three or more unpleasant emotions out of the four (anger, disgust, sadness and fear) are obtained as the estimated result, *a feeling of unpleasant* is employed. The recognition model of emotions expresses the estimated result using adverb *"very"* or *"a little"* according to the degree as shown in Table 12.

Table 12. Degree of emotions and corresponding adverb.

Degree of emotions	Adverb
0.3<the degree of emotions≤0.5	a little
0.5<the degree of emotions≤0.7	(none)
0.7<the degree of emotions	very

4 Simulation Examples

4.1 Example I

In the map shown in Figure 19, the following instructions are given to the route decision system:

(1) Go to the intersection about 0.5 Km ahead from here, and turn to the right there.
(2) Go to the intersection about 1 Km ahead from there, and turn to the left there.
(3) Go to the intersection about 1.5 Km ahead from there, and turn to the right there. There is a bank at the intersection.
(4) Go straight a little, and find a building. And go straight a little still more, and find a hospital.

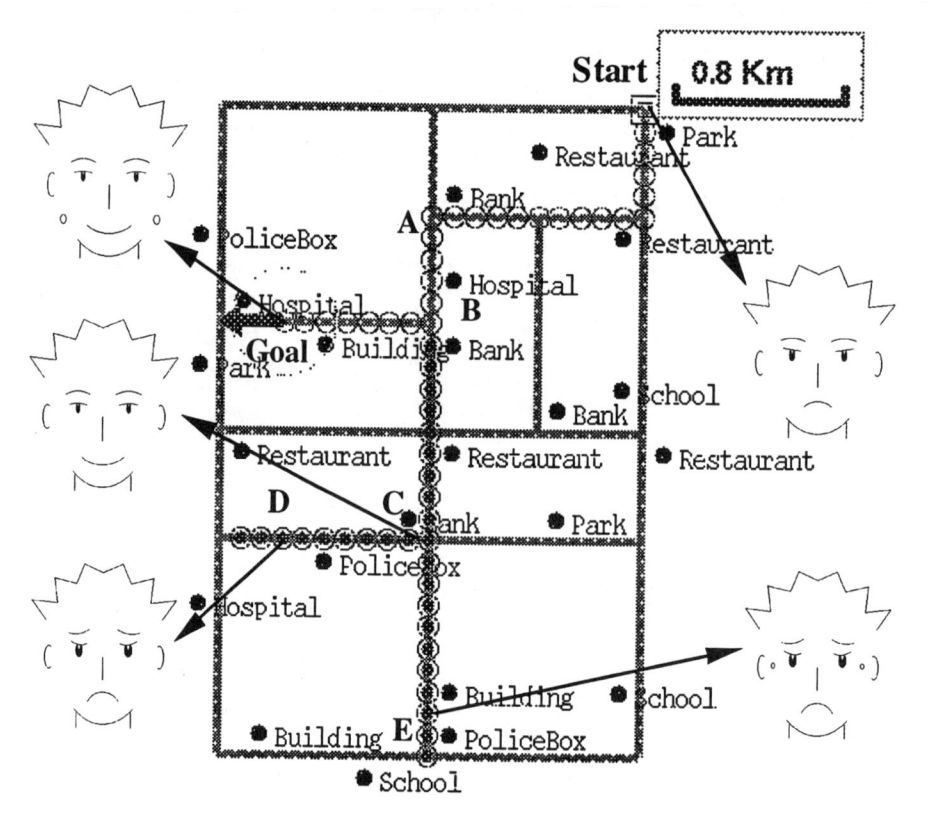

Figure 19. Example I.

The instructed route is Start-A-B-Goal in Figure. 19. In the instruction (3), however, the wrong distance "about 1.5 Km" is given to the route decision system instead of the actual distance about which is 0.5 Km. Therefore, the route decision system turns to the right at point C and makes its way to point D. But it cannot find the instructed building. Then the route decision system goes back to point C and makes its way to the point E in order to find the bank at the intersection given in the instruction (3). But it cannot find the intersection. The route decision system takes the route Start-A-B-C-D-C-E as symbols **O**s show. In this example, the threshold values for anger and disgust are assigned low, and the threshold values for other emotions are assigned normal; that is, anger and disgust are easy to appear in a face. Therefore, the route decision system's face already looks like angry at the start point. Conversation examples at points C, D and E are shown as follows, where the RME stands for the recognition model of emotions and the RDS stands for the route decision system.

<Point C>
RME: You appear happy. What happens to you?
RDS: I just find the bank at the intersection. I am searching a next objective.
RME: You are happy, aren't you?
RDS: Yes, I am very happy.
RME: I see. Good luck.
<Point D>
RME: You appear unpleasant. What happens to you?
RDS: No, I am not unpleasant. I will go back because I seem to take a wrong way. There is, I believe, a building near here. How long is it from here?
RME: You are disgust and fearful a little, aren't you? (The RME does not reply the RDS's question.)
RDS: Yes, I am very disgust and fear. There is, I believe, a building near here. How long is it from here?
RME: I am very sorry. (The RME does not reply the RDS's question again.)
<Point E>
RME: You appear unpleasant. What happens to you?
RDS: No, I am not unpleasant. I will go back because I seem to take a wrong way. There is, I believe, a building near here. How long is it from here? Please tell me the answer properly.
USER: Go straight a few Km. Find a bank at the intersection and turn to the left there. Go straight only a little.

At point C, the route decision system believes that there is the objective near and it appears happy. The recognition model of emotions recognizes happiness. At point D, the route decision system understands that it took a wrong way. Then it goes back. When the recognition model of emotions addresses the route decision system, *"You appear unpleasant, don't you?"*, the route decision system asks the route from the lost spot to the objective, the building. The recognition model of emotions, however, has not map information. Therefore, it cannot reply.

At point D, because the route decision system has not the answer from the recognition model of emotions when the recognition model of emotions addresses the route decision system again, the route decision system says *"Please tell me the answer properly."* A user (a third party) instead of the recognition model of emotions shows the route decision system the route from the point E to the sub-goal, the building. Then the route decision system can reach the destination according to the second instructions.

4.2 Example II (Shortcut)

The route decision system is given the following instructions in the map shown in Figure 20.

(1) Turn to the right at the intersection about 1.5Km ahead from here. There is a restaurant at the intersection.
(2) Turn to the left at the intersection about 2.0Km ahead from there. There is another restaurant.
(3) Go straight a little and find the school on the left side of the street.

The instructed route is Start-A-B-C-D-Goal in Figure. 20. At point A the route decision system finds the possibility of a shortcut A-C instead of the route A-B-C. The route decision system takes the route A-C-D-Goal as symbols **O**s show. In this example, the threshold values for six kinds of emotions are assigned normal. Conversation examples at points A and C are shown as follows:

<Point A>
RDS: I find a shortcut.
RME: You don't feel any emotion, do you?
RDS: Yes, I feel some emotions.
RME: You appear a little happy and a little fearful, don't you?
RDS: Yes, I feel happy and a little fear.
RME: I see. Good luck.

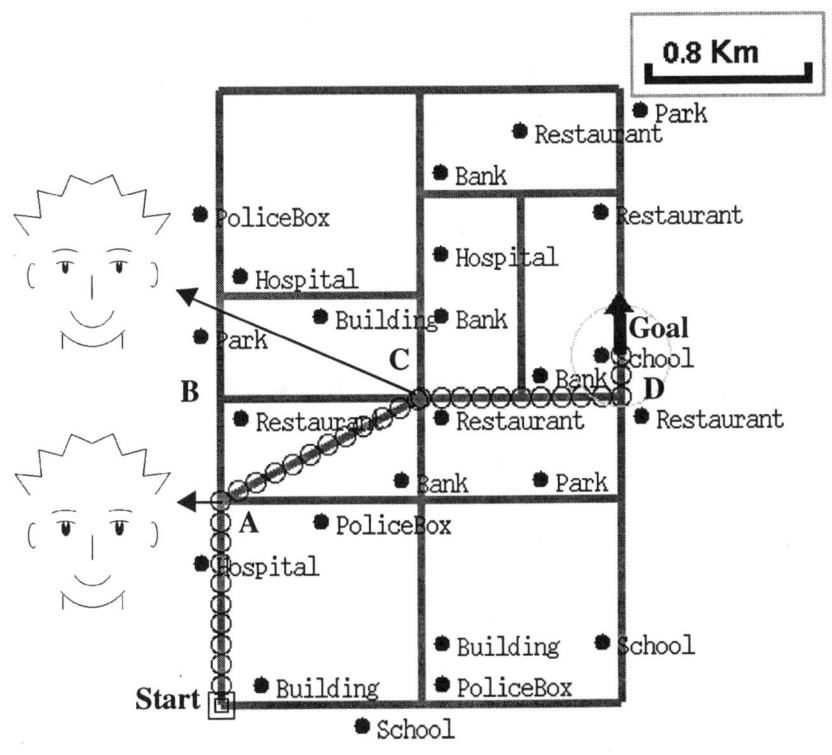

Figure 20. Example of shortcut.

<Point C>
RDS: I seem to succeed in taking a shortcut.
RME: You appear much happy and a little fearful, don't you?
RDS: Yes, I am very happy, but I am not fear.
RME: You appear very happy, don't you?
RDS: Yes, I am very happy.
RME: I see. Good luck.

At point A, the route decision system understands that it finds a shortcut. The recognition model of emotions addresses the route decision system, *"You don't feel any emotion, do you?"*. The model does not recognize emotions well. The model tends to recognize the degree of emotions low depending on the questionnaire results. The reply of the route decision system leads the recognition model of emotions to address the route decision system again, *"You appear a little happy and a little fearful, don't you?"*, referring to estimated emotions degrees from only facial expressions or from only situations.

The route decision system feels happiness for finding a shortcut and a little fear whether the route A-C is a shortcut or not.

At point C, the route decision system understands its success in taking a shortcut. The recognition model of emotions addresses the route decision system, *"You appear much happy and a little fearful, don't you?"*, as the recognition result of emotions comes not only from facial expressions but also from situations. In this situation the recognition model of emotions recognizes fear as well as happiness. The route decision system, however, denies fear. Then the recognition model of emotions addresses the route decision system again, *"You appear very happy, don't you?"*, as the recognition results of emotions come only from facial expressions or only from situations.

4.3 Example III (Roundabout Way)

The route decision system is given the same instructions as in example II shown in Figure 21. But the route B-D is not passable. Therefore, the route decision system goes by a roundabout route at point A since it memorizes the intersection A. The route decision system turns to the left at the intersection C so as to go to the direction of the goal. The route decision system takes the route Start-A-B-A-C-D-E-Goal. In this example, the threshold values for six kinds of emotions are also assigned normal. Conversation examples at points B, A and D are shown as:

<Point B>
RDS: This route is not passable.
RME: You appear much surprised, don't you?
RDS: Yes, I am much surprised.
RME: I see.
<Point A>
RDS: I will go by a roundabout route here since point B is not passable.
RME: You don't feel any emotion, do you?
RDS: Yes, I feel some emotions.
RME: You appear disgusted a little, don't you?
RDS: I am not disgusted.
RME: I don't recognize your emotions.
RDS: I have a fear.
<Point D>

RDS: I seem to succeed in taking a roundabout way.
RME: You don't feel any emotion, don't you?
RDS: Yes, I feel some emotions.
RME: You appear happy, don't you?
RDS: Yes, I am happy.
RME: I see.

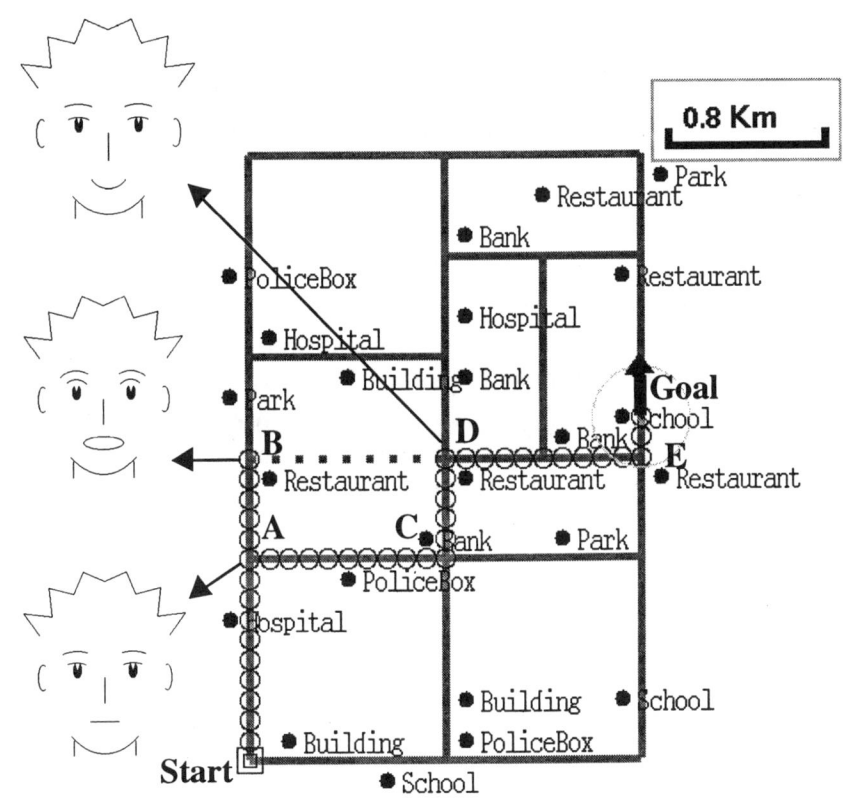

Figure 21. Example of roundabout way.

At point B, the route decision system finds the instructed route impassable. Therefore, it is surprised. The recognition model of emotions recognizes surprise as the recognition result of emotions from not only facial expressions but also situations.

At point A, the route decision system goes by a roundabout way since it memorizes the intersection A. The recognition model of emotions does not recognize emotions of the route decision system well. As mentioned before, the model tends to recognize the degree of emotions low depending on the questionnaire results. Since the recognition model of emotions receives the reply from the route decision system, *"Yes, I feel some emotions"*, the model addresses the route decision system again, *"You appear disgusted a little, don't you?"*, referring to estimated emotions degrees from only facial expressions or from only situations. But the route decision system denies disgust. Then the recognition model of emotions addresses the route decision system again, *"I don't recognize your emotions"*. The difference between the model of the face expressions section in the route decision system and the recognition model of emotions, i.e., the questionnaire results, leads to this conversation. This is a conversation example among people who don't understand each other well. The route decision system feels fear since it is not certain whether the route goes to the destination or not.

At point D, the route decision system understands that it succeeds in taking a roundabout way. Therefore, the route decision system feels happy. But the recognition model of emotions does not recognize emotions of the route decision system. Towards the response of the route decision system, *"Yes, I feel some emotions"*, the recognition model of emotions addresses the route decision system again, *"You appear happy, don't you?"*, recognizing emotions only from facial expressions or only from situations.

4.4 Remarks

Using examples it is found that the recognition model of emotions recognizes route decision system emotions to some extent, but does not sometimes recognize them since the model tends to recognize the degree of emotions low depending on the questionnaire results. This is also due to a difference between the model of the face expressions section in the route decision system and the recognition model of emotions. The latter means the conversation example among people who don't understand each other well. In the above examples, although the conversations between the route decision system and the recognition model of emotions are restricted to the contents of emotions since the combination model has not enough ability to have smooth conversation yet, a user (a third party) can also participate in the conversation.

5 Conclusions

The recognition model of emotions, which recognizes emotions from not only facial expressions but also from situations, is presented. Information on situations is transformed into six basic emotions, happiness, surprise, anger, disgust, sadness and fear by the use of questionnaire data about situations. The recognition model of emotions is implemented using an artificial neural network model by the use of questionnaire data about facial expressions and situations. Inputs to the model are emotions transformed from situations and parameter values of face features. The output of the model is the degree of emotions. The performance of our models is verified by other questionnaire data.

The recognition model of emotions and the route decision system are combined for the application of the recognition model of emotions, and as an example of human computer interaction considering facial expressions and situations. It is found that the recognition model of emotions recognizes emotions to some extent in the combination model which is based on recognized emotions. The interaction between the recognition model of emotions and the route decision system is performed. A user (a third party) can also participate in the conversation. As the human computer interaction, the combination model has not enough ability to have smooth conversation yet.

In the future the construction of a bi-direction human computer interaction system as well as the increase of interactive ability are considered.

References

[1] Kurokawa, T. (1994), *Non-Verbal Interface*, Ohmusha, Tokyo.

[2] Special Issue on Face. (1997), *The Trans. of the Institute of Electronics, Information and Communication Engineers*, A, Vol. J80-A, No.8, D-II, Vol.J80-D-II, No.8.

[3] Uwechue, O.A. and Pandya, S.A. (1997), *Human Face Recognition Using Third-Order Synthetic Neural Networks*, Kluwer Academic Publishers, Boston.

[4] Kobayashi, H. and Hara, F. (1993), "The Recognition of Basic Facial Expressions by Neural Network," *Trans. on the Society of Instrument and Control Engineers*, Vol.29, No.1, pp.112-118.

[5] Kobayashi, H. and Hara, F. (1993), "Measurement of the Strength of Six Basic Facial Expressions by Neural Network," *Trans. of the Japan Society of Mechanical Engineers (C)*, Vol.59, No.567, pp.177-183.

[6] Kobayashi, H. and Hara, F. (1993), "Recognition of Mixed Facial Expressions by Neural Network," ibid., pp.184-189.

[7] Kawakami, F., Morishima, S., Yamada, H., and Harashima, H. (1994), "Construction of 3-D Emotion Space Using Neural Network," *Proc. of the 3rd International Conference on Fuzzy Logic, Neural Nets and Soft Computing*, Iizuka, pp.309-310.

[8] Ueki, N., Morishima, S., and Harashima, H. (1994), "Expression Analysis/Synthesis System Based on Emotion Space Constructed by Multilayered Neural Network," *Systems and Computers in Japan*, Vol.25, No.13.

[9] Vanger, P., Honlinger, R., and Haken, H. (1995), "Applications of Synergetics in Decoding Facial Expressions of Emotions," *Proc. of International Workshop on Automatic Face- and Gesture-Recognition*, Zurich, pp.24-29.

[10] Izumitani, K., Mikami, T., and Inoue, K. (1984), "A Model of Expression Grade for Face Graphs Using Fuzzy Integral," *System and Control*, Vol.28, No.10, pp.590-596.

[11] Yamada, H. (1993), "Visual Information for Categorizing Facial Expression of Emotion," *Applied Cognitive Psychology*, Vol.7, pp.257-270.

[12] Yamada, H. (1994), "Psychological Models for Recognition of Facial Expressions of Emotion," *Journal of the Society of Instrument and Control Engineers*, Vol.33, No.12, pp.1063-1069.

[13] Ushida, H., Takagi, T., and Yamaguchi, T. (1993), "Facial Expression Model Construction Based Associative Memories and Linguistic Instructions Learning," *T. IEE. Japan*, Vol.113-C, No.12.

[14] Black, M.J. and Yacoob, Y. (1995), "Recognizing Facial Expressions under Rigid and Non-Rigid Facial Motions," *Proc. of International Workshop on Automatic Face- and Gesture-Recognition*, Zurich, pp.12-17.

[15] Essa, I.A. and Pentland, A. (1995), "Facial Expression Recognition Using Visual Extracted Facial Action Parameters," ibid., pp.35-40.

[16] Yacoob, Y., Lam, H.H., and Davis, L.S. (1995), "Recognizing Faces Showing Expressions," ibid., pp.278-283.

[17] Fernandez-Dols, J.M., Wallbott, H., and Sanchez, F. (1991), "Emotion Category Accessibility and the Decoding of Emotion from Facial Expression and Context," *Journal of Nonverbal Behavior*, Vol.15, No.2, pp.107-123.

[18] Carroll, J.M. and Russel, J.A. (1996), "Do Facial Expression Signal Specific Emotion? Judging Emotion from the Face in Context," *Journal of Personality and Social Psychology*, Vol.70, No.2, pp.205-218.

[19] Onisawa, T., Masuda, Y., and Iwata, M. (1996), "A Route Decision System with Human Interaction and Facial Expression Model," *Proc. of the 1996 International Fuzzy Systems and Intelligent Control Conference*, pp.137-146.

[20] Gill, K.S. (ed.) (1996), *Human Machine Symbiosis, The Foundations of Human-Centered Systems Design*, Springer, London.

[21] Ito, M. Ushida, H., Yamamori, A., Ono, T., Tokusumi, A., and Ikeda, K. (1994), *Recognitive Science 6, Jodo (Emotions)*, Iwanami, Tokyo.

INDEX

- A -

accuracy of matching, 378
acquisition,
    of faces, 233
    of images, 11
active shape models, 240
adjudicators, 326
alignment-based methods, 228
appearance-based methods, 229
associative networks, 250
authentication systems, 11

- B -

binarization, 22, 159
binary image enhancement, 22
binary to skeleton processing, 41
biometric test plan, 95
Bochum database
branch point location, 46
bunch graphs, 342
    elastic bunch graphs, 357-391
    face bunch graphs, 366

- C -

classification,
    of fingerprints, 13
    FBI classification formula, 116
    of minutiae, 48
classifiers,
    continuous n-tuple, 262
    nearest neighbor, 262
cleaning,
    pore detection, 45
coarse coding, 305
coarse to fine matching, 343
committees of networks, 255
compensating methods, 50
component level testing, 91, 97
condition of skin, 51

confidence measures, 326
confidence rating thresholds, 326
configurations,
    of minutiae, 68
    of pores, 72, 79
    probabilities, 82
    uniqueness, 68
connectionist approaches, 249
continuous n-tuple classifiers, 262
conversion,
    gray scale to binary, 39
convolutional networks, 261
correlation matching, 49, 59
correspondence maps, 337-349
correspondence-based recognition, 338

- D -

databases, 373
    Bochum database, 376
    design of face databases, 233
    FERET database, 375, 384
    testing, 91, 97, 103
deformable templates, 240
degradation,
    of images, 51
detection,
    of directional field, 123
    of faces, 236
    of minutiae, 23, 169
    of pores, 45, 46
    of singular points, 119, 124
    of singularity, 21
difference of Gaussian, 243
dimensionality reduction, 181
direction calculation, 20
directional field detection, 123
displacement estimation, 363
distribution,
    binomial, 79
    of distances between pores, 73
    of intra-ridge pores, 76

of points, 344
of pores, 72, 79
dynamic link graphs, 245

- E -

edge extraction, 302
eigenfaces, 241, 261
elastic bunch graph matching, 357-391
elastic graphs, 341
emotion recognition model, 427
    structure, 427
end point location, 46
enhancement, 121
    of binary image, 22
    of gray-level, 22
enrollment, 57
ensemble-based networks, 255
error rates, 63, 64
extraction,
    of edges, 302
    of facial characteristics, 409
    of fingerprint features, 17, 45
    of graphs,
        schedule, 371
    of gray-scale minutiae, 164
    of minutiae, 45, 155-187
    minutiae-based, 17
    of pores, 45
    of ridges, 195-213
    of segments, 49

- F -

face,
    acquisition, 233
    bunch graphs, 366
    databases, 233
    detection, 236
    expressionless face, 404
    individual faces, 365
    reasoning, 248
    recognition, 219-261, 287-308
        comparison, 259, 380
        from correspondence maps, 337-349
        by elastic bunch graph matching, 357-391
        rotated in depth, 385
        same view, 381
        transformation-invariant, 301
        typical system, 292
        various systems, 296
        with NNs, 287-308
    representation, 238, 365
        generating, 367
    segmentation, 236
face unit RBF networks, 317-333
    adding, 332
    as adjudicators, 326
    network model, 318
    network types, 319
    removing, 333
    terminology, 321
    updating, 332
facial characteristic points, 409
facial expression, 440
    questionnaire, 428
    recognition (FER), 400, 402, 404, 425-455
    synthesis (FES), 399-421
        2-D expressions, 401
facial information,
    multimodal, 235
FBI classification formula, 116
feature-based approach, 239
feature selective filtering, 195-213
features, 339
    area, 64
    extraction, 45
    of fingerprints, 12, 17
    Gabor features, 138
    preprocessing, 402
    statistics, 67
    uniqueness, 68
FERET database, 375
    performance comparison, 384
field testing, 94
filter design, 198
filtering,
    feature selective, 195-213

Index

of minutiae, 155-187
    NN based, 180
fingerprints, 4
    classification, 13
    features, 12
        extraction, 17, 45
        processing, 37-105
    identification, 137-147
    identification systems, 11
    matching, 16, 57
    poor quality, 50
    position on finger, 53
    quality determination, 55
    recognition, 3-28
    sub-classification, 109-131
FRR analysis, 84

- G -

Gabor,
    features, 138
    filter, 137-147
    wavelets, 244, 360
Gaussian receptive fields, 243
graph extraction, 371
graph similarity function, 369
gray-level enhancement, 22
gray-scale images, 155-187
gray-scale minutiae extraction, 164
gray-scale to binary conversion, 39

- H -

healing,
    broken skeleton segments, 44
    of the skeleton, 43
    of wrinkles, 44
hidden Markov models, 260
hierarchical neural networks, 251
human – computer interaction, 425-455

- I -

identification, 4, 140
    of fingerprints, 137-147
identification systems, 11

image,
    acquisition, 11
    degradation, 51
    normalization, 237
    vectorization, 237
    warping, 411
inherent reliability, 84
interaction,
    human – computer, 425-455
intra-ridge pores, 73
    distribution form models, 76
isodensity regions, 304

- J -

jets, 360
    comparing, 362

- L -

Laplacian operator, 243

- M -

matching,
    accuracy, 378
    coarse to fine, 343
    correlation matching, 49, 59
    elastic bunch graph matching, 357-391
    of fingerprints, 16, 57
    of minutiae, 61
    of pores, 61
    procedure, 369
    routine, 64
    techniques, 248
maximum determination, 167
minutiae, 17
    classification, 48
    configurations, 68
    detection, 23, 169
    extraction, 45, 155-187
    filtering, 155-187
        NN based, 180
    matching, 61
    reduction, 24

multilayer perceptrons, 250
multilevel verification, 60
multimodal facial information, 235

- N -

natural basis functions, 243
nearest neighbor classifier, 262
negative evidence selection, 319
negative examples, 256
neighborhood normalization, 181
neural models, 345
neural networks (NNs), 109-131
    back-propagation, 404
    for face recognition, 287-308
    for facial expression
        recognition, 404
    hierarchical NNs, 251
    for minutiae filtering, 180
    model, 430
        performance, 432
    NN classifier, 184
    for pattern recognition, 290
    probabilistic decision-based
        NNs, 262
normalization of images, 237

- O -

object recognition, 226

- P -

pattern recognition, 288
    with neural networks, 290
performance, 67, 86, 432
    comparison, 384
    criteria, 25
    evaluation, 171
personal identification, 4
plasticity, 65
point distribution models, 344
pores,
    configurations, 72, 76
    detection, 45, 46
    distribution, 72, 76

    binomial, 79
    extraction, 45
    intra-ridge, 73, 76
    matching, 61
    pruning, 48
    ridge-independent, 79
poroscopy, 37-105
postprocessing,
    of skeleton, 42
preprocessing, 39, 121, 360
    of features, 402
    of questionnaire data, 430
principal component analysis, 241
probabilistic decision-based NNs, 262
probabilities of configurations, 82
pruning,
    of pores, 48
psychological evidence, 230

- Q -

quality of fingerprints, 50, 53
    determination, 55
questionnaire,
    about facial expressions, 428
    about situations, 428
    data preprocessing, 430

- R -

radial basis function (RBF) networks,
    252, 263, 317-333, 399-421
receptive field, 242
    Gaussian, 243
recognition, 372
    correspondence-based, 338
    of emotions (model), 427, 436
    of faces, 219-265, 287-308,
        337-349, 357-391
    of facial expressions, 400, 402,
        408, 425-455
    of fingerprints, 3-28
    of objects, 226
    of patterns, 288
reduction,
    of dimensionality, 181

Index

of minutiae, 24
reliability, 66
    of algorithm, 85
    FRR analysis, 84
    inherent reliability, 84
representation of faces, 238, 365
    generating, 367
resolution, 64
ridges,
    counting, 115, 128
    extraction, 195-213
ridge-line following algorithm, 165
rotation, 65, 385
route decision system, 436, 437

- S -

sampling,
    space-variant, 235
scale, 65
scanning resolution, 64
schedule of graph extraction, 371
search,
    area, 64
    procedure, 119
sectioning, 167
segment extraction, 49
segmentation, 21, 121
    of faces, 236
shift and scale-varying data, 324, 330
singular points, 113
    detection, 124
        methods, 119
singularity detection, 21
skeleton, 41
    healing, 43
    healing broken segments, 44
    postprocessing, 42
    tracking, 46
skin condition, 51
space-variant sampling, 235
stop criteria, 168
sub-classification,
    of fingerprints, 109-131
    techniques, 113

synthesis,
    of 2-D expressions, 401
    of facial expressions, 399-421
system level testing, 94, 96, 100, 102

- T -

tangent direction computation, 168
task requirements, 224
template-based approach, 239
temporal networks, 257
testing, 91
    algorithm testing, 97
    biometric test plan, 95
    component level, 91, 97
    database testing, 91, 97, 103
    field testing, 94, 100, 102
    flow chart, 95
    preliminary testing, 96, 102
    system level, 94, 96, 100, 102
thinning, 159
topology, 340
tracing of whorls, 116
tracking,
    of skeleton, 46
transformation-invariance, 301
translation, 65

- U -

uniqueness of features, 68

- V -

vectorization of images, 237
verification, 4
    multilevel verification, 60

- W -

warping of images, 411
wavelets, 244, 360
whorl tracing, 116
wrinkles,
    healing, 44

Library of Congress Cataloging-in-Publication Data

Intelligent biometric techniques in fingerprint and face recognition /
   edited by L. C. Jain ... [et al.].
         p.   cm. — (International series on computational intelligence)
      Includes bibliographical references.
      ISBN 0-8493-2055-0 (alk. paper)
      1. Human face recognition (Computer science). 2. Fingerprints—
   Identification—Data processing. 3. Biometry—Data processing.
   I. Jain, L. C. II. Series: CRC Press International series on
   computational intelligence.
   TA1650.I53 1999
   006.4'2—dc21   99-23136
                                                                              CIP

   This book contains information obtained from authentic and highly regarded sources. Reprinted material is quoted with permission, and sources are indicated. A wide variety of references are listed. Reasonable efforts have been made to publish reliable data and information, but the author and the publisher cannot assume responsibility for the validity of all materials or for the consequences of their use.

   Neither this book nor any part may be reproduced or transmitted in any form or by any means, electronic or mechanical, including photocopying, microfilming, and recording, or by any information storage or retrieval system, without prior permission in writing from the publisher.

   All rights reserved. Authorization to photocopy items for internal or personal use, or the personal or internal use of specific clients, may be granted by CRC Press LLC, provided that $.50 per page photocopied is paid directly to Copyright Clearance Center, 222 Rosewood Drive, Danvers, MA 01923 USA. The fee code for users of the Transactional Reporting Service is ISBN 0-8493-2055-0/99/$0.00+$.50. The fee is subject to change without notice. For organizations that have been granted a photocopy license by the CCC, a separate system of payment has been arranged.

   The consent of CRC Press LLC does not extend to copying for general distribution, for promotion, for creating new works, or for resale. Specific permission must be obtained in writing from CRC Press LLC for such copying.

   Direct all inquiries to CRC Press LLC, 2000 N.W. Corporate Blvd., Boca Raton, Florida 33431.

   Trademark Notice: Product or corporate names may be trademarks or registered trademarks, and are used only for identification and explanation, without intent to infringe.

© 1999 by CRC Press LLC

No claim to original U.S. Government works
International Standard Book Number 0-8493-2055-0
Library of Congress Card Number 99-23136
Printed in the United States of America   1 2 3 4 5 6 7 8 9 0
Printed on acid-free paper

INTELLIGENT BIOMETRIC TECHNIQUES in FINGERPRINT and FACE RECOGNITION

Edited by
L.C. Jain • U. Halici
I. Hayashi • S.B. Lee
S. Tsutsui

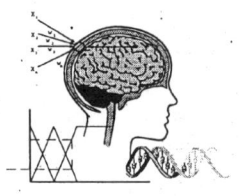

CRC Press
Boca Raton   London   New York   Washington, D.C.